Springer Series in

MATERIALS SCIENCE 136

Springer Series in

# MATERIALS SCIENCE



The Springer Series in Materials Science covers the complete spectrum of materials physics, including fundamental principles, physical properties, materials theory and design. Recognizing the increasing importance of materials science in future device technologies, the book titles in this series reflect the state-of-the-art in understanding and controlling the structure and properties of all important classes of materials.

Please view available titles in *Springer Series in Materials Science* on series homepage http://www.springer.com/series/856

Pavel Lejček

# Grain Boundary Segregation in Metals

With 80 Figures

Springer

Professor Pavel Lejček
Academy of Sciences of the Czech Republic, Institute of Physics
Na Slovance 2, 18221 Prague 8, Czech Republic
E-mail: lejcekp@fzu.cz

*Series Editors:*



Springer Series in Materials Science ISSN 0933-033X
ISBN 978-3-642-12504-1 e-ISBN 978-3-642-12505-8
DOI 10.1007/978-3-642-12505-8
Springer Heidelberg Dordrecht London New York

Library of Congress Control Number: 2010930167



*Cover design*: eStudio Calamar Steinen

Printed on acid-free paper

Springer is part of Springer Science+Business Media (www.springer.com)

*To Nad'a, Eva and Pavla*

# Preface

*Paulus servus Christi Iesu, vo catus apostolus, segregatus in evangelium Dei. . .*
*(Nova Vulgata, Ad Romanos Epistula Sancti Pauli 1–1)*

The idea of writing a book on various aspects of grain boundary segregation occurred in summer 2004 after my 20 years of experience in this field. In fact, somebody else suggested it to me because I am sure I am not capable enough myself to take it up seriously. However, once this idea was conceived, it started to gnaw my mind as a worm in an apple. Soon, I became deeply involved in thinking about the form and content of such a book, searched the literature sources extensively, and started to write at the same time. Now – after several years – I decided to complete the writing and you can see the result.

Although the consequences of the phenomenon of grain boundary segregation were observed more than 100 years ago and have been described theoretically and observed experimentally for decades, there still exists a lot of confusion and misconception, leading to false conclusions and wrong recommendations. I have mainly attempted to summarize the fundamentals of *equilibrium grain boundary segregation in metals*, which would enhance the understanding of the related thermodynamic functions, and to describe the *influence of individual variables on this phenomenon*. I am convinced that this approach can help the reader to understand all aspects of this phenomenon and see the obtained results – both theoretical and experimental – in a general light, despite the fact that the book provides an experimentalist's perspective of the phenomenon.

The book is composed of several chapters. After a brief introduction (Chap. 1), information on the structure, and geometric and thermodynamic description of the grain boundaries is given in Chap. 2, which is necessary for further reading. Chapter 3 is devoted to an overview of experimental methods used for measurement of interfacial chemistry and of computer simulations of grain boundary segregation. The current thermodynamic description of equilibrium grain boundary segregation is outlined in Chap. 4. A discussion of the various effects on this process is described in Chap. 5. Thorough attention is paid here to the anisotropic behavior of grain boundary segregation and its prediction. Kinetics of grain boundary segregation and the fundamental features of non-equilibrium segregation, including the interaction of grain boundary segregation and migration is provided in Chap. 6. Chapter 7 is devoted to the discussion of the consequences of grain boundary segregation for selected physicochemical properties and metallurgical phenomena.

I deeply believe that this book will be accepted by the readers as a serious attempt to elucidate many – although not all – aspects of grain boundary segregation.

## Acknowledgement

Writing a scientific book is a complex task that cannot occur as result of *a kiss of a muse*: Besides experimental work, one needs to study and understand the subject. It is necessary to spend time reading papers of other colleagues, to discuss the subject with them, and to enhance the knowledge which results in ideas and formulations. In this respect, I would like to acknowledge the help of those colleagues who contributed most extensively to my understanding of the field. First of all, I wish to thank my colleague Václav Paidar of the Institute of Physics, Academy of Sciences of the Czech Republic, Praha, Czech Republic, for his co-operation in the study of the field of grain boundary properties. Special thanks to my teacher, co-worker, and friend, Siegfried Hofmann of the Max-Planck-Institute for Metals Research, Stuttgart, Germany. I have been working with him since my first stay in Stuttgart during the Alexander-von-Humboldt Fellowship in 1989. I received invaluable suggestions for my professional as well as personal life from him. I would also like to acknowledge other colleagues who contributed to my knowledge (although they may not consider it in the same way as I do), Li-Shin Chang, Johan DuPlessis, Anna Fraczkiewicz, Hans-Jürgen Grabke, Wolfgang Gust, Olivier Harduin Duparc, Yuri Ivashchenko, Jozef Janovec, Radomila Konečná, Aleksander Krajnikov, Charles McMahon, Jr., Matthias Militzer, Carol Nichols, Tatiana Orlova, Antony Paxton, Micha Polak, Louisette Priester, Eugen Rabkin, David Seidman, Lasar Shvindlerman, Boris Straumal, Adrian Sutton, Shigeru Suzuki, Mojmír Šob, Sadahiro Tsurekawa, Vašek Vitek, Tadao Watanabe and Paul Wynblatt. I would also like to remember with gratitude my co-workers at the Institute of Physics, Slavica Kadečková, Jan Adámek, Gabriela Urbanová, Vlastimil Zachař and Karel Brambora as well as Bernhard Siegle of the MPI Stuttgart. Without the technical help of these colleagues in growing bicrystals, preparing samples, and measuring grain boundary segregation, I would have never been able to conduct the experiments and obtain the results, which are also mentioned in this book.

Many results were obtained under numerous granted projects; their financial support is gratefully acknowledged: the Alexander-von-Humboldt and Max-Planck Fellowships, grants of the Commission of the EC (CIPA CT93 0112, Copernicus ERB IC15 CT98 0812), COST (OC517.40), US-Czech Science and Technology Program (95 008), Czech Science Foundation (202/94/1177, 202/99/1665, 106/99/1178, 106/02/0253, 106/02/0687, 106/05/0134, 202/06/0004, 106/08/0369), Grant Agency of the AS CR (19096, 110404, A1010708, IAA1010414), Ministry of Industry and Trade of the Czech Republic ("Progress" FF–P2/053) and, last but not least, of the Academy of Sciences of the Czech Republic (AV0Z-1-010-914, AV0Z 10100520 and KAN300100801).

Prague
May 2010

*Pavel Lejček*

# Contents


**1 Introduction** ........ 1

**2 Grain Boundaries: Description, Structure and Thermodynamics** ........ 5
2.1 Crystallographic Description of Grain Boundaries ........ 5
2.2 Atomic Structure of Grain Boundaries ........ 8
2.2.1 Low-Angle Grain Boundaries ........ 9
2.2.2 High-Angle Grain Boundaries ........ 10
2.3 Classification of High-Angle Grain Boundaries ........ 13
2.3.1 Coincidence-Site Lattice Model ........ 13
2.3.2 Interplanar Spacing ........ 16
2.3.3 Hierarchy of Grain Boundary Planes ........ 17
2.4 Basic Thermodynamics of Grain Boundaries ........ 22

**3 Approaches to Study Grain Boundary Segregation** ........ 25
3.1 Basic Definitions and Terminology ........ 25
3.2 Experimental Methods for Study of Grain Boundary Segregation ........ 26
3.2.1 Surface Analysis Techniques ........ 27
3.2.2 Microscopic Techniques ........ 35
3.3 Theoretical Approaches to Study Grain Boundary Segregation ........ 38
3.3.1 Quantum Mechanical Methods ........ 39
3.3.2 Interatomic Potentials ........ 41
3.3.3 Simulation Techniques ........ 43

**4 Models of Equilibrium Grain Boundary Segregation** ........ 51
4.1 Thermodynamics of Grain Boundary Segregation ........ 51
4.2 Gibbs Adsorption Isotherm ........ 52
4.3 Langmuir–McLean Types of Segregation Isotherm ........ 55
4.3.1 Physical Meaning of Thermodynamic State Functions Appearing in Segregation Isotherms ........ 61
4.3.2 Langmuir–McLean Model ........ 74
4.3.3 Guttmann Model of Grain Boundary Segregation in Interacting Multi-component Systems ........ 81
4.3.4 Fowler Model ........ 86

4.3.5 Other Models for Grain Boundary Segregation ................ 89
4.3.6 Models for Different Segregation Sites ........................ 92
4.4 Models for Thermodynamic Functions of Interfacial Segregation ..... 94
4.4.1 Model of Wynblatt and Ku ..................................... 95
4.4.2 Model of Seah ................................................ 97
4.4.3 Model of Miedema ............................................. 98
4.4.4 Model of Kumar ............................................... 98
4.4.5 Model of Mukherjee and Morán-López ........................... 99
4.4.6 BFS Model ....................................................100
4.4.7 SMA–TBIM Approach ............................................101

**5 Effect of Variables on Equilibrium Grain Boundary Segregation**........103
5.1 Temperature and Bulk Composition ....................................103
5.2 Pressure ............................................................107
5.3 Magnetic Effects ....................................................108
5.4 Grain Boundary Energy ..............................................111
5.5 Nature of Segregating Element .......................................129
5.5.1 Truncated BET Isotherm .......................................129
5.5.2 Grain Boundary Segregation Diagram ...........................131
5.5.3 Enthalpy–Entropy Compensation Effect ..........................135
5.6 Grain Size ..........................................................141
5.7 Prediction of Grain Boundary Segregation ............................146
5.7.1 Model of Lejček and Hofmann ..................................146
5.7.2 Model of Wynblatt and Shi ....................................150

**6 Principles of Non-equilibrium Segregation** .............................153
6.1 Kinetics of Grain Boundary Segregation ..............................153
6.1.1 Semi-infinite Solution of Fick Equation .......................153
6.1.2 Layer-by-Layer Model (Model of Hofmann and Erlewein)....156
6.1.3 Model of Limited Reaction Rates ...............................158
6.2 Non-equilibrium Segregation .........................................158
6.2.1 Thermally Induced Segregation ................................159
6.2.2 Radiation-Induced Segregation .................................162
6.2.3 Stress-Induced Segregation ....................................165
6.2.4 Grain Boundary Segregation and Migration ......................167

**7 Grain Boundary Segregation and Related Phenomena**......................173
7.1 Grain Boundary Cohesion .............................................173
7.1.1 Grain Boundary Cohesion and Temper Embrittlement...............173
7.1.2 Hydrogen Embrittlement .......................................181
7.1.3 Other Types of Non-reactive Environmental Decohesion ......183
7.2 Grain Boundary Corrosion ............................................185
7.2.1 Corrosion ....................................................186
7.2.2 Stress Corrosion Cracking ....................................188

7.3 Grain Boundary Diffusion ....................190
7.3.1 Fundamentals of Grain Boundary Diffusion ....................190
7.3.2 Relationship Between Grain Boundary Diffusion and Segregation ....................195
7.4 Grain Boundary Engineering ....................196

**Appendix A Parameters of Grain Boundary Segregation** ....................203
A.1 Enthalpy, Entropy and Gibbs Energy of Segregation ....................203
A.2 Interaction Coefficients ....................210

**Appendix B Predicted Values of Enthalpy and Entropy of Segregation** .....213

**References** ....................215

**Index** ....................235

# Chapter 1
# Introduction

Metallic materials have been used by human being for eight millennia. Starting by simple mining of pure metals from the earth and their forming, man has soon learnt melting and alloying as well as treating them thermomechanically to obtain materials with required properties. We can say, therefore, that metals represent the first "artificial" construction materials used in history by human being [1]. During ages, huge practical experience has been obtained in preparing and treating extended spectrum of metallic materials such as various steels, nickel-based superalloys, aluminium and its alloys. Metals remain the most efficient materials for technological purposes in many areas even at beginning of the twenty-first century [2].

Scientific effort in understanding the mechanisms of the processes, which occur at every stage of production and application of metals, started in the "post-medieval" age in sixteenth century. First, the attempts to describe systematically these processes were done as for example, by Agricola [3]. Among others he also reported on intergranular cracking ("hot shortness") of iron during forging above a "cherry-red" temperature and proposed to avoid this effect by using "bitumen" for refining iron instead wood charcoal [3,4]. In fact, this example may represent the first notice of the phenomena induced by grain boundary segregation! In nineteenth century, various effects of external conditions and additives on materials behaviour were already detected and formulated. Although individual features of materials microstructure such as dislocations and grain boundaries were far not revealed, the effects of their existence were observed. Let us mention that at that time, the harmful effect of trace elements on mechanical properties was also detected as is demonstrated for examples of bismuth[1] [5] and antimony[2] [6] in copper. In 1914, the first report

[1] Was ... das metallische Wismut betrifft, so übt dieses schon in äußerst kleinen Mengen den nachteiligsten Einfluss auf die Dehnbarkeit des Kupfers aus. (As regards metallic bismuth, it exhibits the worst effect on the ductility of copper even in negligible amounts.)

[2] One thousandth part of antimony converts the first rate best selected copper into the worst conceivable.

of liquid metal embrittlement has appeared[3] [7]. In the second half of twentieth century, the scientific effort reached the high level which was adjusted by practice and was able to alternate it in proposing novel materials and procedures of thermal treatment for practical applications. Let us remind, for example, intermetallic compounds [8] and shape memory alloys [9]. We can agree with Cottrell that in twentieth century the materials were understood scientifically for the first time: what they are and why they perform [10]. Further progress in development of various metallic materials is expected not only in wide application of new structural materials such as composites, intermetallics and shape memory alloys but also in improving the properties of traditional materials by detailed controlling of their chemical composition and structural state. One of the visions for future is materials science as a rigorous and quantitative field in which materials synthesis, processing and performance are guided by predictive models capturing critical fundamental aspects of materials structure. Great majority of energy-related materials in future will be polycrystalline as today. Bi-material interfaces will be formed to join together materials with different structures and compositions. Most future energy applications, as at present, will entail the use of materials for extended periods in high temperatures, high stresses, corrosive environments, radiation fields, electromagnetic fields, high currents, or some combinations of these conditions. Hence, the goal is set high for a quantitative science and the associated large-scale computer simulations that will be needed to guide the manufacture and predict the performance of materials for future energy applications. Interfaces are a key, and perhaps least understood aspect of the structure of materials [11].

There are many structural features that strongly affect the properties of materials such as dislocations in heavily plastically deformed sheets or fine precipitates in dispersion strengthened materials. Among the most important components that create the micro-structure of polycrystalline materials, the internal interfaces and, especially, the grain boundaries separating adjacent crystals (grains) belong to the most important defects in solids. Due to their nature as conjugation interfaces between crystal lattices of two misoriented grains, the bonds between individual atoms in the grain boundary are changed as compared to the bulk crystal [12]. Due to the different energetic states, grain boundaries differ in properties as compared to the crystal interior. For example, a system tends to reduce its total energy by interaction of grain boundaries with other lattice defects such as solute or impurity atoms. As a consequence, these atoms accumulate (*segregate*) at grain boundaries to such an extent that the boundary may become qualitatively different in chemical nature as compared to the bulk. It can result in further changes of grain boundary properties such as loss of plasticity (temper embrittlement) or reduction of chemical resistance (corrosion cracking) [13] but also in dramatic increase of critical current density in high-temperature superconductors [14]. Since the grain boundaries in polycrystals create a three-dimensional net spreading throughout the material, they represent

[3] When ... beta-brass doped by 2%Al was treated with mercury, the crystals separate at grain boundaries without the crystal being attacked in any way.

a self-standing link of the material structure with different mechanic, electric and magnetic properties, and therefore, they can substantially contribute to the behaviour of the material. It is worth noting that the properties of the grain boundaries with respect to materials applications are in majority of cases worse than those of the bulk crystal, and therefore, the presence of grain boundaries ultimately limits the application of a material. This is the reason for usage of single crystalline blades in jet turbines although their production is rather expensive and time-consuming [15].

Let us mention that the term *segregation* is used in metallurgy and materials science to describe various phenomena. The reader can often meet this term in relation to solidification processes. During solidification of a molten alloy, concentration differences occur in the cast on the "metre" scale. Similarly, the concentration differences can be found on the "micro-metre" scale between dendrites and interdendritic space. These effects are called *macro-segregation* and *micro-segregation*, respectively [16], and arise from the differences between the chemical compositions of solidus and liquidus at the melting temperature. Comparing to the length scale, grain boundary segregation, which is driven by reduction of the grain boundary energy, is limited to the "nanometre" scale but can reach much larger concentration differences than those in the case of macro- or micro-segregation. From the points of view of the length range and of the above terminology, the grain boundary segregation can be called *nanosegregation* (Fig. 1.1) [17].

*Grain boundary segregation* is thus a very important phenomenon that affects the behaviour of polycrystals to a large extent. In fact, it covers only part of the

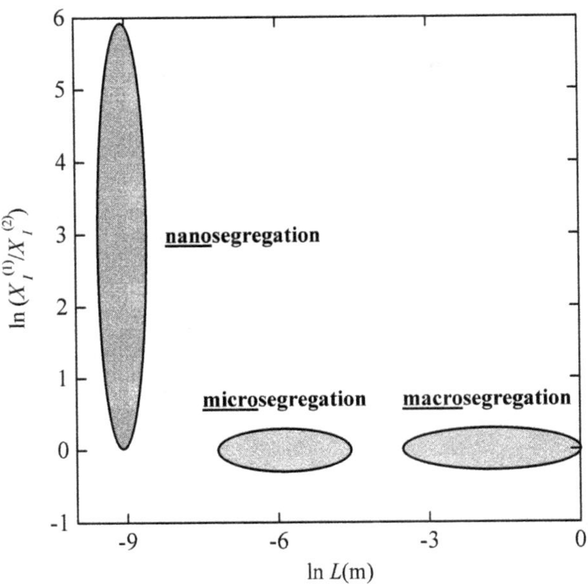


**Fig. 1.1** Schematic depiction of individual types of segregation. $X_I^{(1)}$ and $X_I^{(2)}$ represent the concentrations of compared regions, $L$ represents the length scale of the effect [17]

subject called *chemical composition of solid interfaces*. Grain boundary segregation has, however, an important impact also for other material properties, for example, it may promote successful sintering of ceramic powders while it substantially reduces the electric conductivity of semiconductors and high-temperature superconductors. Specific field represents *surface segregation*, which plays a dominant role, for example in catalytic ability and following reactions in a broad spectrum of systems. A very important aspect is *non-equilibrium segregation* mainly in connection to problems of irradiated materials in nuclear power reactors [18].

# Chapter 2
# Grain Boundaries: Description, Structure and Thermodynamics

*Grain boundary* in a solid crystalline material is a region separating two crystals (grains) of the same phase. These two grains differ in mutual orientations and the grain boundary thus represents a transition region, where the atoms are shifted from their regular positions as compared to the crystal interior [12, 19, 20]. Grain boundaries represent the simplest interface: If the adjoining grains differ in chemical composition and/or in parameters of the crystal lattice, the interface between them is called *phase boundary* (*interphase boundary, heterophase boundary*). The grain boundary is also called *homophase boundary* in this classification. In general, interfaces represent a crystallographic and/or chemical discontinuity with an average width less than two atomic diameters [21, 22] although they may be sometimes more diffuse spreading over appreciable number of interplanar spacings [12, 23]. In this context, *free surface* is the interface between solid and vacuum [20].

Only those aspects of grain boundaries that are necessary for further reading will be mentioned in this chapter. For more thorough information other sources are recommended to the reader, especially the comprehensive book of Sutton and Balluffi [12] and selected parts of the book edited by Wolf and Yip [24].

## 2.1 Crystallographic Description of Grain Boundaries

To describe grain boundary crystallographically, a number of variables must be specified. Generally, the grain boundary can fully be characterised by five independent parameters (macroscopic degrees of freedom, DOFs), which provide us with information how to prepare the *bicrystal* (i.e. a sample containing two grains with the required orientation of the planar separating interface) from given single crystals (e.g. [12, 22, 25–27]). Three of them specify mutual misorientation of the adjoining grains $A$ and $B$ (Fig. 2.1). This misorientation is represented by a rotation, which brings both grains in perfect matching. It is defined by the rotation axis $\boldsymbol{o}$ (2 DOFs) and angle $\theta$ (1 DOF). Let us mention that there always exists at least one way how to describe such relationship in the case of non-enantiomorphic crystals. The orientation of the grain boundary between these misoriented grains is defined by the normal $\boldsymbol{n}$ to the grain boundary plane (2 DOFs).

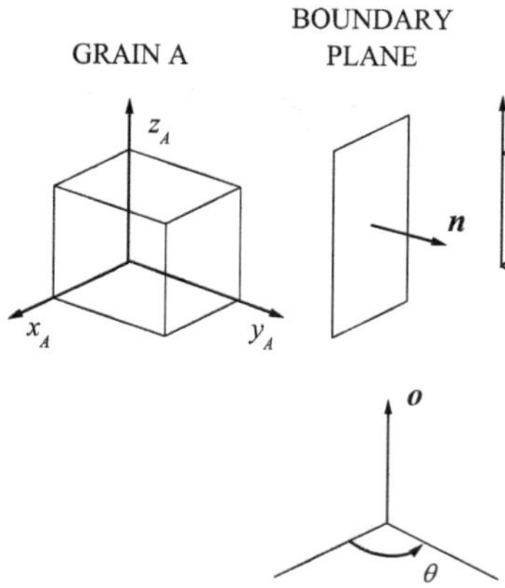


**Fig. 2.1** Variables that define a grain boundary. $x_A, y_A, z_A$ and $x_B, y_B, z_B$ are the axes of the co-ordinates parallel to crystallographic directions in grains $A$ and $B$, respectively. $\boldsymbol{o}$ is the rotation axis and $\theta$ is the rotation (misorientation) angle necessary to transfer both grains to an identical position. $\boldsymbol{n}$ determines the orientation of the grain boundary plane [20]

Following this characterisation, we can completely and unambiguously describe any grain boundary by the notation $\theta°[h_o k_o l_o], (h_{nA} k_{nA} l_{nA})$[1]. The grain misorientation is defined by the common axis $\boldsymbol{o} = [h_o k_o l_o]$, which is identical in both grains and expressed in a chosen co-ordinate system. The information about the grain boundary plane is related to one of these two grains only. From practical point of view, it is sometimes useful to identify the other part of the grain boundary, that is the joining plane related to the other grain, so that the boundary is described by $\theta°[h_o k_o l_o], (h_{nA} k_{nA} l_{nA})/(h_{nB} k_{nB} l_{nB})$. This notation represents an over-determination, indeed, but can give an easier and quicker view on the grain boundary crystallography.

It is necessary to add that besides the above mentioned five independent macroscopic DOFs, three other microscopic parameters exist that are represented by a vector $\boldsymbol{T}$ characterising a rigid body translation of both grains relatively one to the other, parallel and perpendicular to the grain boundary plane. The latter one represents, in fact, a volume expansion [28]. These translations are independent of macroscopic DOFs but controlled by the energetic reasons and cannot be chosen arbitrarily: for each grain boundary, few mutual translations may only exist that generate equilibrium atomic structures of the grain boundary under actual external conditions such as temperature, pressure and chemical composition. Thus, they cannot be considered as independent DOFs [12, 27, 29, 30].

Five DOFs that are necessary to describe completely the crystallography of a grain boundary imply existence of a huge number of different grain boundaries. Therefore, it is sometimes reasonable to categorise the grain boundaries into groups according to the relationships among individual DOFs. For example,

[1] In case of hexagonal structures, the notation should be modified according to the corresponding description of crystal planes and directions by four Miller indices.

the relationship between the rotation axis, $\boldsymbol{o}$, and the grain boundary normal, $\boldsymbol{n}$, leads to definition of the *tilt* grain boundaries ($\boldsymbol{o}\perp\boldsymbol{n}$) and the *twist* grain boundaries ($\boldsymbol{o}||\boldsymbol{n}$); the interfaces that do not fit with any of these relationships belong to the group of *mixed* grain boundaries [12, 19]. In the latter case, it is sometimes useful to find the tilt and twist components of the mixed relationship by imagining two successive rotations about perpendicular axes, one being located in the boundary plane, the other perpendicular to it [12]. When the boundary plane represents the plane of the mirror symmetry of the crystal lattices of two grains, it is described by the same Miller indices from the point of view of both adjoining grains. This boundary is called *symmetrical.* Its notation can be then simplified as $\theta°[h_o k_o l_o]$, $\{h_n k_n l_n\}$. The other grain boundaries are *asymmetrical.* Systematically, the categorisation of the grain boundaries can be represented by so-called *interface-plane scheme* based on relationship of the Miller indices of individual contacting planes 1 and 2 in a bicrystal and the twist angle $\varphi$ of both planes as proposed by Wolf and Lutsko [31]. They distinguish four categories of grain boundaries

| | |
|---|---|
| Symmetrical tilt grain boundary | $\{h_1k_1l_1\} = \{h_2k_2l_2\}$ and $\varphi = 0$ |
| Asymmetrical tilt grain boundary | $\{h_1k_1l_1\} \neq \{h_2k_2l_2\}$ and $\varphi = 0$ |
| Twist grain boundary | $\{h_1k_1l_1\} = \{h_2k_2l_2\}$ and $\varphi \neq 0$ |
| Random (i.e. mixed) grain boundary | $\{h_1k_1l_1\} \neq \{h_2k_2l_2\}$ and $\varphi \neq 0$ |

Note that the mirror symmetry can be reached only at pure tilt grain boundaries. Let us mention that the last-mentioned grain boundary is called "random" or often "general": This brings terminological ambiguity, because the same terms are used to specify the character of the grain boundaries from the point of view of the properties (e.g. [12, 13, 20, 21, 32]). Therefore, we propose to call this class of the grain boundaries "mixed".

The above categorisation is, however, somewhat simplified and not unambiguous. From mathematical point of view, the total misorientation $\theta$ is given – in analogy to orientation matrix of a single crystal – by a $3 \times 3$ orthonormal matrix $M$. The matrix columns are the directional cosines of the crystal 2 in relation to the respective co-ordinates of the reference grain 1. Then the pair of the angle and axis is obtained from the matrix according to the relation [26]

$$\cos\theta = \frac{a_{11} + a_{22} + a_{33}}{2} \tag{2.1}$$

and the elements of the matrix $M$, $a_{ij}$ $(i, j = 1, 2, 3)$, are related to the rotation axis $\boldsymbol{o}$ and its components $o_i$ by

$$\boldsymbol{o} = o_1 : o_2 : o_3 = (a_{32} - a_{23}) : (a_{13} - a_{31}) : (a_{21} - a_{12}). \tag{2.2}$$

If $\theta = 180°$,

$$o_1 : o_2 : o_3 = (a_{11} + 1)^{1/2} : (a_{22} + 1)^{1/2} : (a_{33} + 1)^{1/2}. \tag{2.3}$$

Due to the symmetry of cubic structures, it is possible to describe the same boundary by 24 various but equivalent notations and characters [12]. These notations result from the product of the symmetry matrices $T_i$ with $M$ [26]

$$M' = T_i M, \tag{2.4}$$

where $i = 1, 2, \ldots, 24$ in cubic structures. The 24 matrices determine 24 pairs of the rotation axis and misorientation. For example, the same interface characterised as 36.87°[100]{013} *symmetrical tilt grain boundary* can also be described as 53.13°[100]{013} *symmetrical tilt grain boundary* or 180°(013) *twist grain boundary* [20, 33, 34], etc. Although all descriptions are entirely equivalent, the lowest-angle solution or lowest-index rotation axis is conventionally used for description of the grain boundary (i.e. 36.87°[100]{013} in the above example). This notation will also be used throughout this work. For detail crystallographic description of the grain boundaries, the reader is referred to more thorough papers dealing with the systematic of grain boundaries, for example to Sutton and Balluffi [12], Randle [22, 26, 35], Wolf [29] and Wolf and Merkle [36].

## 2.2 Atomic Structure of Grain Boundaries

As was mentioned above, individual atoms in the grain boundary core are shifted from their regular crystal positions as compared to the crystal interior. Then, a question arises: How are the atoms arranged there? The early models assumed that the grain boundaries are amorphous (e.g. [37], cf. also [19]). Amorphous layers as an equilibrium arrangement of silicon atoms at the grain boundaries also resulted from recent computer simulations [38–40], although they could exist obviously only in a metastable state [41]. Later, the grain boundary was considered as composed of the regions of the "good" and "bad" material [42]: this idea was developed in the dislocation models of the grain boundary structure [25]. At present, however, it is undoubtedly established that the structure of grain boundaries is crystal like [28–31]. This model is strongly supported by theoretical as well as experimental evidences of anisotropic behaviour of grain boundaries (e.g. [12, 20, 21, 43–46]). From the point of view of actual atomic structure, two groups can be distinguished, *low-angle grain boundaries* and *high-angle grain boundaries*. Let us notice that we can meet dual terminology in the literature corresponding to these groups (a) *low-angle* and *high-angle* grain boundaries (e.g. Gleiter and Chalmers [21], Wolf and Merkle [27], McLean [19], Smith [46], Flewitt and Wild [47]) and (b) *small-angle* and *large-angle* grain boundaries (e.g. Sutton and Balluffi [12], Cahn [39], Gottstein and Shvindlerman [48], Finnis and Rühle [49]). In the present work, the terminology of type (a) is used.

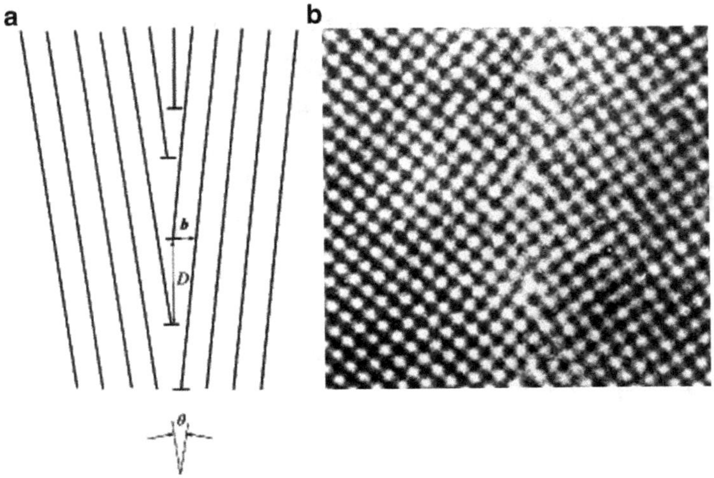


**Fig. 2.2** Structure of a low-angle grain boundary (**a**) schematic illustration, (**b**) image of a [100] low-angle grain boundary in molybdenum revealed by the high-resolution electron microscopy. The distance between individual atoms is about 0.2 nm. (J.M. Pénisson, with permission)

### *2.2.1 Low-Angle Grain Boundaries*

When the angle $\theta$ between two adjoining grains is low enough, this misorientation can be accommodated by an array of dislocations [12, 50, 51]. It is well known that equilibrium arrangement of discrete dislocations (so called *primary grain boundary dislocations*) with Burgers vector $\boldsymbol{b}$ having the same sign can be represented by a periodic row (wall) as it is shown in Fig. 2.2. Tilt grain boundaries are formed by edge dislocations, while twist grain boundaries consist of an array of screw dislocations.

The angle $\theta$ is related to the size of the Burgers vector $|\boldsymbol{b}|$ and the dislocation spacing $D$ by the expression

$$\sin\frac{\theta}{2} = \frac{|\boldsymbol{b}|}{2D}. \tag{2.5}$$

For low $\theta$, $\sin(\theta/2)$ can be approximated by $\theta/2$ and thus, $\theta \approx |\boldsymbol{b}|/D$.

The energy of a low-angle grain boundary can simply be derived on basis of the theory of elastic continuum. The energy of an edge dislocation can be expressed as

$$E_{\mathrm{ed}} = \frac{\mu \boldsymbol{b}^2}{4\pi(1-\nu)} \ln\frac{D}{r_0} + E_{\mathrm{c}}, \tag{2.6}$$

where $\mu$ is the shear modulus, $\nu$ is the Poisson ratio, $r_0$ and $E_{\mathrm{c}}$ are the radius and the energy of the dislocation core. Supposing that $r_0 \approx |\boldsymbol{b}|$ and that the number of dislocations per unit length in the low-angle grain boundary is $n = 1/D \approx \theta/|\boldsymbol{b}|$, we can write the energy of a low-angle tilt grain boundary per unit area as

$$\sigma_{\mathrm{LATGB}} = \frac{\theta}{|\boldsymbol{b}|}\left[\frac{\mu \boldsymbol{b}^2}{4\pi(1-\nu)}\ln\frac{1}{\theta} + E_{\mathrm{c}}\right] = \theta(A - B\ln\theta), \tag{2.7}$$

where $A = E_{\mathrm{c}}/|\boldsymbol{b}|$ and $B = \mu|\boldsymbol{b}|/4\pi(1-\nu)$ [25].

With increasing $\theta$, $D$ decreases to such an extent that dislocations loose their character and the dislocation theory cannot be further used to describe the grain boundary structure. Generally, it is assumed that the limit for successful application of the dislocation model lies between 13° and 15° [52] that corresponds approximately to the value $D \approx 4|b|$. This limit has been supported experimentally: For example, measurements of the contact angle at the grain boundary trace at free surface in bismuth showed the transition between low-angle and high-angle grain boundaries at 15° [53]. Recently, the measurements of migration of planar grain boundaries in aluminium showed a sharp limit between low-angle and high-angle $\langle 112\rangle$ and $\langle 111\rangle$ tilt grain boundaries at 13.6° [54].

### 2.2.2 *High-Angle Grain Boundaries*

Overcoming the above-mentioned limit of the misorientation angle between two adjoining grains, the dislocation model of the grain boundary structure fails because individual dislocations are no more distinguishable and overlap one with the other. As a result, the angular dependence of grain boundary energy does not fit with the course proposed by (2.7). Systematic computer modelling of numerous grain boundaries in face-centred cubic (fcc) crystals [55–60] resulted in development of the *structural unit model* to describe atomic arrangement of the high-angle grain boundaries [55–66]. According to this model, a high-angle grain boundary is formed by repeated structural units that represent particular arrangements of limited number of atoms. In fact, only a few types of basic structural units exist [43, 46, 66–68] and therefore, there is a limited number of grain boundaries formed exclusively by single structural units. Ashby et al. [66] describe seven different convex polyhedra – tetrahedron, regular octahedron, pentagonal bipyramid, tetragonal dodecahedron, capped trigonal prism, capped Archimedian prism and icosahedron – as basic objects forming the structural units. The structures of the majority of grain boundaries consist of combinations of the simple structural units. Between two *delimiting* grain boundaries formed by single structural units of the types $|A|$ and $|B|$, the structural units of the other grain boundaries will be described as $|A_x B_y|$ to accommodate corresponding misorientation angle $\theta$ [12]. Examples of the structures of high-angle grain boundaries obtained by computer modelling are depicted in Fig. 2.3. Here, three [100] symmetrical tilt grain boundaries in the body-centred cubic (bcc) structure are shown, the {013} delimiting grain boundary (structural unit of type $|A|$), the {024} delimiting grain boundary (structural unit of type $|B.B|$), and the {037} grain boundary. As it is apparent, the structure of the latter grain

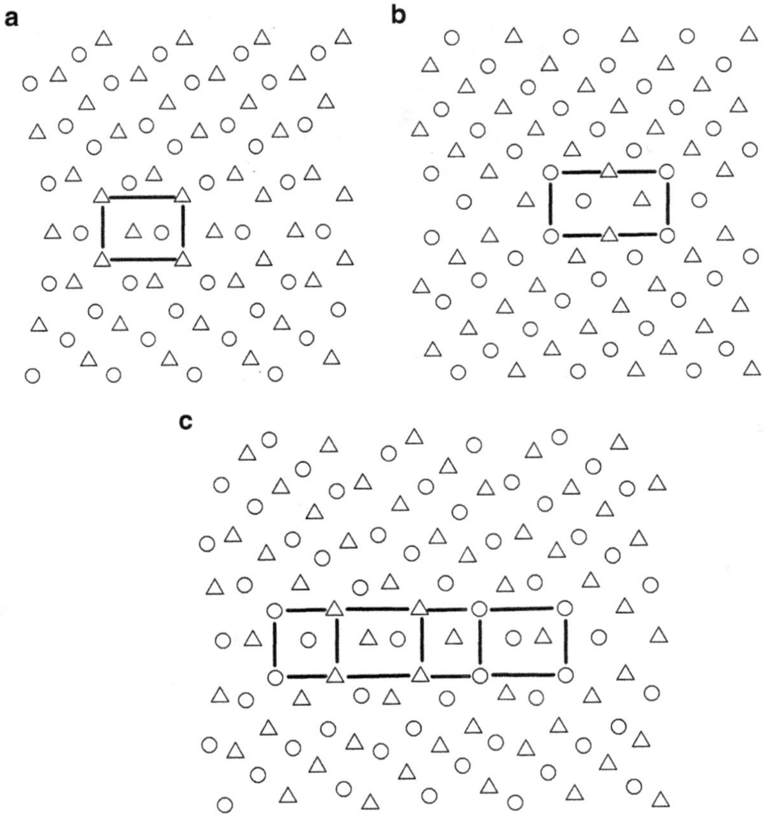


**Fig. 2.3** Structure of three [100] symmetrical tilt grain boundaries in a bcc bicrystal (**a**) {013}, (**b**) {024}, (**c**) {037}. *Circles* and *triangles* represent two parallel (100) planes [20]. (According to J. Erhart)

boundary is formed by combination of structural units of neighbour delimiting grain boundaries, {013} and {024}, according to the scheme $|AA| + |B.B| \rightarrow |AB.AB|$. This scheme copies the combination of Miller indices of the grain boundary planes, $\{013\} + \{024\} \rightarrow \{037\}$ [20]. Since the atomic arrangement of the grain boundary differs from that of the bulk crystal, the grain boundary possesses higher energy as compared to the crystal interior. A low energy of the grain boundary suggests that this arrangement is close to that of crystal interior. The (delimiting) grain boundaries consisting of single structural units and exhibiting sharp minima at energy–orientation dependence are called *singular* [12]. An example of the atomic structure of a singular grain boundary is shown in Fig. 2.4. The high-energy grain boundaries are then called *general*. Their structure is composed of combination of two or more structural units of singular grain boundaries (cf. the {037} grain boundary in Fig. 2.3). There is another group of grain boundaries representing transition in energetic behaviour of singular and general grain boundaries, so called *vicinal*

**Fig. 2.4** Atomic structure of the 70.5° [110], {112} symmetrical tilt grain boundary in molybdenum. The spacing between parallel rows of atoms is 0.25 nm. The *bright belt* in the *centre* of the image is computer-simulated structure [69]. (T. Vystavěl, with permission)

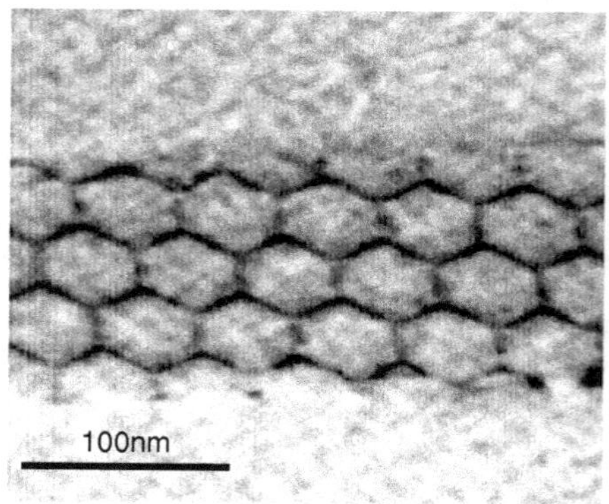


**Fig. 2.5** Array of screw dislocations in the tilt grain boundary in molybdenum vicinal to the 70.5° [110], {112}. Transmission electron microscopy [69]. (T. Vystavěl, with permission)

grain boundaries [12, 70]. The structure of a vicinal grain boundary is composed of the structural unit of the closest singular grain boundary modified by an array of *secondary grain boundary dislocations* (low-angle grain boundary) [12]. An example of such dislocation array is shown in Fig. 2.5. The combination of the single structural unit and the dislocation array transforms to the structural unit of a general grain boundary at much lower misorientations from the singular grain boundary than it corresponds to the limit of existence of low-angle grain boundaries 10–15°. Because the Burgers vectors of the secondary grain boundary dislocations for positive and negative deviations from the singular orientation may be different, the cusps on the structural dependence of grain boundary energy (and other interface properties) can be asymmetric [71].

In addition to the categories of grain boundaries described above, we can often find in literature the term *special* grain boundaries (e.g. [21, 27, 32, 47, 72]). This term denotes those grain boundaries that exhibit sharp extremes at any property–orientation dependence, for example, fracture toughness, diffusivity, propensity for segregation, migration rate, sliding rate as well as corrosion rate. It is expected that each singular grain boundary is special albeit not vice versa. Sometimes, we can also find – mainly in earlier sources – the term *random* grain boundaries (e.g. [21, 32]) as a synonym for *general* grain boundaries. The term "general grain boundaries" is more straightforward when related to the behaviour of the interface. Speaking about the occurrence of the grain boundaries in a material, the term "random" can be used to document randomness of nucleation of individual grains in cast material.

## 2.3 Classification of High-Angle Grain Boundaries

In Sect. 2.2.2, the high-angle grain boundaries were classified into singular, vicinal and general. Although numerous computer simulations and experimental studies revealed some of the singular grain boundaries, their complete set has not been unambiguously established until now. This set can differ according to the lattice structure, type of material, etc. Therefore, a large effort has been paid to develop a simple geometrical classification of grain boundaries to specify the grain boundaries of singular (special) character. Let us summarise here the most important attempts.

### 2.3.1 *Coincidence-Site Lattice Model*

Historically, the first model used for identification of special grain boundaries, the *coincidence-site lattice* model, was proposed in 1949 by Kronberg and Wilson [73]. It is based on simple assumption that the grain boundary energy is low when the coincidence of atomic positions in both adjoining grains is high because the number of bonds that are broken across the boundary is small [74]. This is understandable if we accept that the minimum Gibbs energy of the system corresponds to the state of perfect arrangement of the atoms in the lattice positions. Therefore, a grain boundary will possess lower energy when more atoms will coincide with the positions of the perfect crystal than in a non-coincident state. Let us assume that two grains are misoriented by a chosen angle $\theta$ around a chosen axis $\boldsymbol{o}$. At superposition of these crystals some atomic sites coincide: such sites are called *coincidence sites*. They are spread regularly throughout the whole superimposition and create a superlattice called *coincidence-site lattice* (CSL). A two-dimensional (2-D) example of the CSL is shown in Fig. 2.6. In 3-D space, the CSL in cubic structures is frequently tetragonal.

It is apparent from Fig. 2.6 that in 36.87° [100] misorientation relationship, each fifth position is the coincidence site. The density of coincidence sites, or better, its reciprocal value $\Sigma$ is an important parameter characterising the CSL. Choosing an elementary cell, we can determine the value of $\Sigma$ as

$$\Sigma = \frac{\text{number of coincidence sites in an elementary cell}}{\text{total number of all lattice sites in an elementary cell}}. \qquad (2.8)$$

In cubic lattices, it can simply be evaluated from the Miller indices of the symmetrical tilt grain boundary corresponding to a given misorientation [75],

$$\Sigma = \delta(h^2 + k^2 + l^2), \qquad (2.9)$$

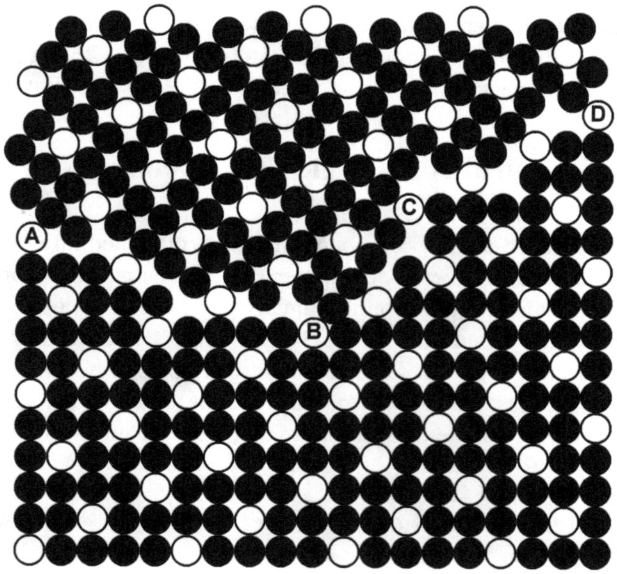


**Fig. 2.6** Geometric model of the 36.87° [100] tilt bicrystal with simple cubic lattice. The *circles* represent the positions of individual atoms in misoriented crystals, the *empty circles* denote the coincidence sites. The orientation of the grain boundary varies from {013} symmetrical (A–B), through (001)/(034) asymmetrical (B–C) to {012} symmetrical (C–D) [20]

where $\delta = 1$ if $(h^2 + k^2 + l^2)$ is odd and $\delta = 1/2$ if $(h^2 + k^2 + l^2)$ is even because in cubic systems, all $\Sigma$ values are odd [76]. For example, the symmetrical grain boundaries in the 36.87° [100] misorientation relationship of the two grains in Fig. 2.5 are {013} and {012}. Both combinations of Miller indices result in value $\Sigma = 5$. Sometimes, a parameter $\Gamma$ is used defined as surface density of the density of coincidence sites in the grain boundary plane [77].

It is apparent that orientation dependence of $\Sigma$ is not monotonous. There alter misorientations characterised by low values of $\Sigma$ with those possessing high values. Since the ideal crystal can be displayed as a bicrystal with 0° misorientation, each atomic position in its overlapping superlattice is the coincidence site and thus, $\Sigma = 1$. The interfacial energy of an imaginary grain boundary in such misorientation relationship (i.e. a crystal plane) is naturally equal to zero and this "interface" can be considered as singular grain boundary. Various experiments proved that the twin grain boundaries in both bcc and fcc structures characterised by the value $\Sigma = 3$ are singular. Similarly, special behaviour was often detected for the grain boundaries characterised by low values of $\Sigma$ (e.g. [32, 48, 72, 78]). Therefore, it has widely been accepted that *low value of $\Sigma$ indicates special grain boundary*.

Due to the purely geometrical character of CSL it is clear that any small change of grain misorientation from the singular grain boundary results in drastic change in coincidence and consequently, in corresponding change of the value of $\Sigma$. On the other hand, it usually does not bring substantial change of the properties of the grain boundary because its structure is changed by introducing individual dislocations. It is thus reasonable to relate the character of such (vicinal) boundaries to the basic singular one and to define the range of existence of low-$\Sigma$ CSL relationship despite small deviations from the true coincidence. This range of existence of low-$\Sigma$ CSL relationship is characterised by the maximum angular deviation, $v_{\mathrm{m}}$, which is supposed to conserve its character by addition of an array of secondary grain boundary dislocations. The density of dislocations is then related to $\Sigma$ [79]. The relationship between $v_{\mathrm{m}}$ and $\Sigma$ is usually introduced rather empirically,

$$v_{\mathrm{m}} = \frac{v_0}{\Sigma^{\xi}}. \quad (2.10)$$

Generally, it is mostly accepted that $v_0 \approx 15°$ represents the angular limit for low-angle grain boundary so that all low-angle grain boundaries are described as $\Sigma = 1$. The mostly adopted Brandon criterion [80] uses the value $\xi = 1/2$, while other authors proposed to use other values, for example Ishida and McLean [81] the value of $\xi = 1$, Deschamps et al. [82] the value of $\xi = 2/3$ and Palumbo et al. [83] the value of $\xi = 5/6$.

As mentioned above, $\Sigma$ is strictly geometric criterion [49,78] and its value characterises exclusively mutual misorientation of two adjoining crystals. We can see in Fig. 2.6 that CSL gives no information on actual grain boundary orientation or on its atomic structure: the grain boundary plane changes its orientation while the value of $\Sigma$ remains constant. This is the reason why $\Sigma$ often fails as the characteristic parameter for classification of grain boundaries [12,71,84,85]. For example, the spectrum of the $\Sigma = 5$, 36.9°[100] tilt grain boundaries includes not only the {012}, {013} and (011)/(017) special grain boundaries but also the (0 3 11)/(097) and (018)/(047) general ones [86]. In turn, a typically non-coincidence relationship 45°[100]($\Sigma \to \infty$) includes also the special (001)/(011) grain boundary [87]. Despite this basic failure, however, the CSL approach is still used to characterise individual grain boundaries in the concept of Grain Boundary Engineering (e.g. [32,88], see Chap. 7) and in many structural studies [89,90] although other schemes were proposed for classification of grain boundaries (e.g. [12,84]).

Let us notice that the 45°[100] grain boundaries belong among the irrational or incommensurate interfaces [91]. The symmetrical tilt grain boundary corresponding to this misorientation relationship can only be described as {0 *kl*}, since the ratio of the Miller indices $k$ and $l$ is irrational, $k/l = \sqrt{2}$.

The CSL concept is sometimes generalised, and considered as the *O-lattice* [75] or the DSC-lattice[2] (DSCL). O-lattice, which has been used mainly to analyse the dislocation structure of grain boundaries, is defined as the space array of coincidence points of two interpenetrating misoriented lattices of grains $A$ and $B$ albeit not only atomic sites. A grain boundary can be geometrically constructed by discarding atoms of lattices $A$ and $B$ and the points of registry in the boundary will be given by the intersections of the grain boundary plane with the O-lattice. Similarly to CSL model, O-lattice is independent of the position of the grain boundary. For

[2] DSC: Displacement Shift Complete.

a chosen misorientation of the two grains, there exists a large number of O-lattices due to a variety of transformations of the space lattice of one grain to that of the other grain. Each point of the O-lattice can serve as an origin of this transformation [88]. Therefore, the O-lattice is a "lattice of origins" (O for Origin) [49]. From this point of view, the CSL is a sublattice of the O-lattice. The periodicity of the structure of the boundary does not coincide with the period of the O-lattice but with that of the CSL. The advantage of the O-lattice is that it is a continuous function of the transformation relating the nearest neighbour atoms of space lattices of grains $A$ and $B$. In contrast to CSL, the spacing of the O-lattice thus varies continuously with misorientation of the two grains.

DSCL was proposed to describe isolated dislocations and steps on grain boundaries [75]. In fact, small misorientation from ideal CSL relationship does not bring a dramatic change of the coincidence but this deviation is compensated by periodic arrangement of dislocations. These dislocations need to have Burgers vectors that conserve the original CSL relationship when forming a low-angle grain boundary (in fact, this is concept of vicinal grain boundaries). In this way, the energy of the grain boundary increases. There are rather small vectors that conserve the CSL relationship – the vectors of the DSCL. Supposing two misoriented grains $A$ and $B$, the DSCL is defined as the grid including all points of grains $A$ and $B$, that is by the minimum displacement vectors preserving the CSL relationship. DSCL, thus, defines all relative displacements of these two grains supposing the overall pattern of atoms produced by the interpenetrating space lattices to remain unchanged. These patterns conserve the displacements: any displacement of the space lattice of the one grain relatively to the other one by a DSCL vector defined by the DSCL represents a complete pattern shift. Additionally, each vector joining the points of misoriented space lattices of grains $A$ and $B$ is a vector of the DSCL [49, 92]. It is worth noting that the DSCL vectors define the possible Burgers vectors of the grain boundary dislocations, which may appear in its structure [49]. Another important property of the DSCL is that the interplanar spacing in direction perpendicular to the misorientation axis varies with decreasing CSL spacing. It means that the degree of coincidence of two misoriented grains decreases with reducing DSCL vectors. The subject of the O-lattice and the DSCL approach is treated in detail elsewhere [12, 93].

### 2.3.2 *Interplanar Spacing*

Another geometrical parameter that was applied to characterise individual grain boundaries is *interplanar spacing d(hkl)*. Interplanar spacing is defined as the shortest distance between two parallel crystal planes (*hkl*) and can be simply evaluated. For example, for cubic structures we can express $d(hkl)$ as

$$d(hkl) = a\frac{\varepsilon}{\sqrt{h^2 + k^2 + l^2}}, \tag{2.11}$$

where $a$ is the lattice parameter and $\varepsilon = {}^{1}/_{2}$ or 1 depending on the particular combination of odd and even Miller indices of the grain boundary plane [27]. When considering the asymmetrical interfaces, the effective interplanar spacing is used averaging $d(hkl)$ of both component boundary planes

$$d_{\mathrm{eff}} = \frac{d(h_1k_1l_1) + d(h_2k_2l_2)}{2}. \tag{2.12}$$

High values of the ratio $d/a$ correspond to densely packed boundary planes and are accepted to indicate special interfaces [20].

A comparison of the dependence of the standard enthalpy, $\Delta H_{\mathrm{P}}^{0}$, (see Chap. 4) of phosphorus grain boundary segregation in α-iron [94] on $d_{\mathrm{eff}}/a$ and $\Sigma$ is shown in Fig. 2.7. We can see that high values of $d_{\mathrm{eff}}/a$ characterise special tilt grain boundaries much better than low values of $\Sigma$ [20,95].

To be rigorous, anisotropic grain boundary properties should be correlated to the *relative grain boundary volume* $\delta V/A$ [96]. Unfortunately, there is only little information about this quantity from computer simulations as well as from high-resolution electron microscopy observations so that this kind of correlation has not been done yet. Therefore, the first rough approximation for the correlation of grain boundary properties such as solute segregation, to $d(hkl)/a$ and/or to $d_{\mathrm{eff}}/a$ seems to be quite promising.

Let us mention that $d_{\mathrm{eff}}/a$ similarly to $\Sigma$, is exclusively a geometrical parameter and does not reflect completely the specific properties of grain boundaries. Although it seems to classify the tilt grain boundaries in a better way than $\Sigma$, it completely fails in case of twist grain boundaries when any misorientation around a chosen axis provides identical value of $d$ while both the grain boundary properties and the value of $\Sigma$ change [70,96].

### 2.3.3 *Hierarchy of Grain Boundary Planes*

Based on analysis of numerous experimental data and results of computer simulations, Paidar [84,97] proposed a geometrical classification of tilt grain boundaries according to hierarchy of their planes. In principle, it is based on the scheme of formation of structural units of individual grain boundaries as was suggested in Sect. 2.2.2. In case of symmetrical grain boundaries, the hierarchy starts at the "ideal" singular grain boundaries such as $\{001\}$ and $\{011\}$. Supposing the 2-D space (i.e., [100] misorientation), these interfaces are the starting interfaces. Due to their exclusivity, they may be ascribed to the 0th *classification level* (CL). Combining them (in relation to bcc structure) according to the "reaction"

$$\{002\} + \{011\} \rightarrow \{013\}, \tag{2.13}$$

the grain boundary corresponding to the first CL is specified. As it was already mentioned in Sect. 2.2.2, this "reaction" is accompanied by the combination of structural

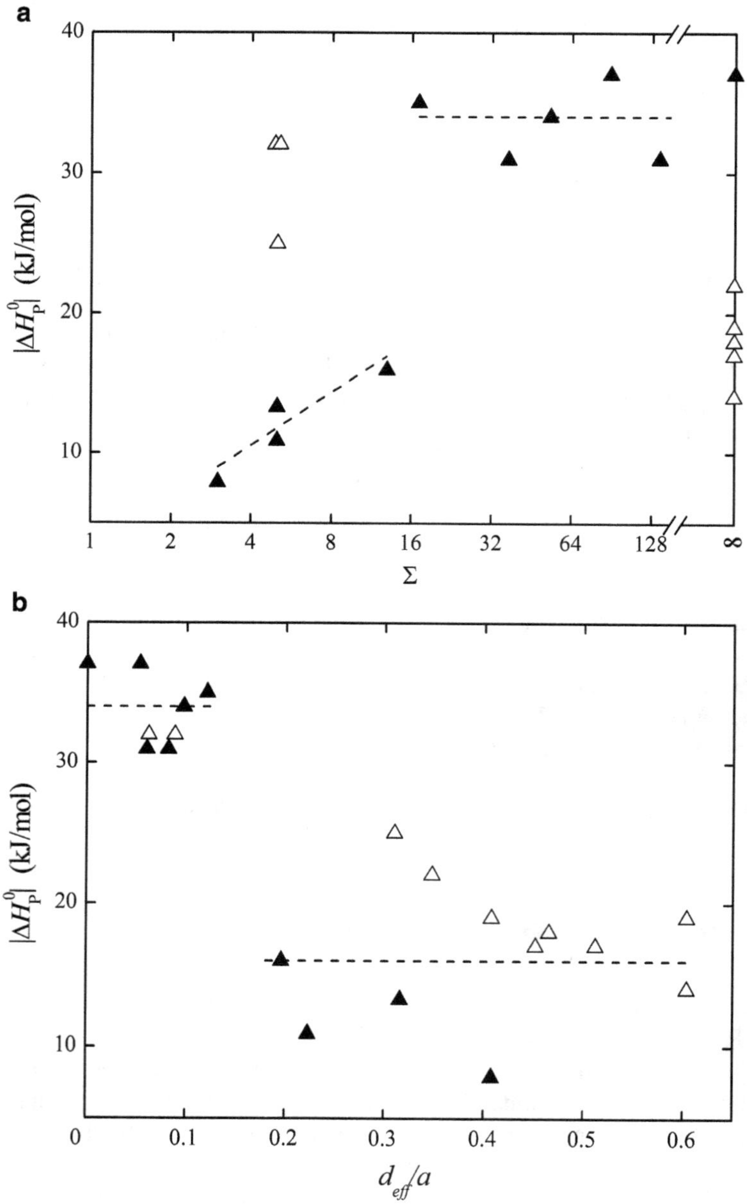


**Fig. 2.7** Dependence of absolute value of enthalpy of phosphorus segregation at grain boundaries of α-iron, $|\Delta H_P{}^0|$, (**a**) on reciprocal density of coincidence sites $\Sigma$, and (**b**) on effective interplanar spacing, $d_{\mathrm{eff}}/a$, [20,95]. *Solid symbols* depict symmetrical grain boundaries, *empty symbols* denote asymmetrical grain boundaries. Low values of $\Sigma$ and high values of $d_{\mathrm{eff}}/a$ should indicate special grain boundaries

units of the grain boundaries of the 0th CL. In case of the {013} grain boundary, however, a new simple structural unit is formed and therefore, this grain boundary is considered as singular.

Combination of the grain boundary plane belonging to the first CL with the neighbour grain boundary planes of the lower CL gives

$$\{002\} + \{013\} \rightarrow \{015\} \tag{2.14}$$

and

$$\{013\} + \{011\} \rightarrow \{024\}. \tag{2.15}$$

{015} and {012} (which are considered as identical with the {024}) are the grain boundary planes corresponding to the second CL. As was shown above, the {012} grain boundary also possesses its own structural unit and is singular. Further application of the proposed scheme leads to identification of the grain boundary planes of the third CL,

$$\{002\} + \{015\} \rightarrow \{017\}, \tag{2.16}$$

$$\{015\} + \{013\} \rightarrow \{028\}, \tag{2.17}$$

$$\{013\} + \{024\} \rightarrow \{037\} \tag{2.18}$$

and

$$\{024\} + \{011\} \rightarrow \{035\}. \tag{2.19}$$

This scheme can continue by specifying the grain boundary planes of higher CLs. In this way, each [100] tilt grain boundary can be classified. Similarly, applying this scheme onto the whole 3–D space of orientations, all grain boundaries can be specified. All grain boundaries belonging to the CLs 1–4 in cubic structures are listed in Table 2.1.

It is demonstrated above, for example for the {037} grain boundary plane, that the structural unit of this grain boundary is composed of the structural units of the grain boundaries from the lower CL, that is this grain boundary is considered as general. The main message that we can draw from this scheme is that *singular grain boundaries belong to the lowest CLs*: *Once a plane of a general grain boundary appears in the hierarchy, the resulting grain boundaries on the higher CLs cannot be singular* [84]. It was also shown that there is a close relationship between this hierarchy and the value of $d/a$ [84]. On the other hand, the relationship between this hierarchy and the value of $\Sigma$ is not straightforward (cf. Table 2.1). Although this hierarchy well reflects formation of structural units of individual grain boundaries, it is not unambiguously determined which CL represents the border between the singular and general grain boundaries. Nevertheless, the grain boundary plane hierarchy offers a clear instruction where the singular grain boundaries can be found.

As regards asymmetrical tilt grain boundaries, singular interfaces should be exclusively formed by the planes of singular symmetrical grain boundaries [97]. Let us represent the orientation of the asymmetrical grain boundaries by the map

**Table 2.1** Classification of symmetrical tilt grain boundary planes [84, 97]

| CL | fcc structure | | bcc structure | |
|---|---|---|---|---|
| | Plane | $\Sigma$ | Plane | $\Sigma$ |
| 1 | {111} | 3 | {112} | 3 |
| 2 | {113} | 11 | {123} | 7 |
| | {133} | 19 | {013} | 5 |
| | {012} | 5 | {111} | 3 |
| | | | {114} | 9 |
| 3 | {112} | 3 | {134} | 13 |
| | {115} | 27 | {235} | 19 |
| | {122} | 9 | {012} | 5 |
| | {155} | 51 | {015} | 13 |
| | {013} | 5 | {113} | 11 |
| | {023} | 13 | {116} | 19 |
| | {135} | 35 | {233} | 11 |
| | | | {334} | 17 |
| | | | {124} | 15 |
| 4 | {117} | 51 | {145} | 21 |
| | {114} | 9 | {257} | 39 |
| | {337} | 67 | {358} | 49 |
| | {335} | 43 | {347} | 37 |
| | {355} | 59 | {035} | 17 |
| | {377} | 107 | {037} | 29 |
| | {144} | 33 | {014} | 17 |
| | {177} | 99 | {017} | 25 |
| | {034} | 25 | {338} | 41 |
| | {035} | 17 | {3 3 10} | 59 |
| | {025} | 29 | {115} | 27 |
| | {014} | 17 | {118} | 33 |
| | {179} | 131 | {122} | 9 |
| | {157} | 75 | {455} | 33 |
| | {124} | 21 | {556} | 43 |
| | {357} | 83 | {223} | 17 |
| | {123} | 7 | {129} | 43 |
| | {159} | 107 | {127} | 27 |
| | {139} | 91 | {138} | 37 |
| | {137} | 59 | {147} | 33 |
| | {134} | 13 | {136} | 23 |
| | | | {124} | 21 |
| | | | {349} | 53 |
| | | | {237} | 31 |
| | | | {239} | 47 |
| | | | {457} | 45 |
| | | | {345} | 25 |
| | | | {356} | 35 |

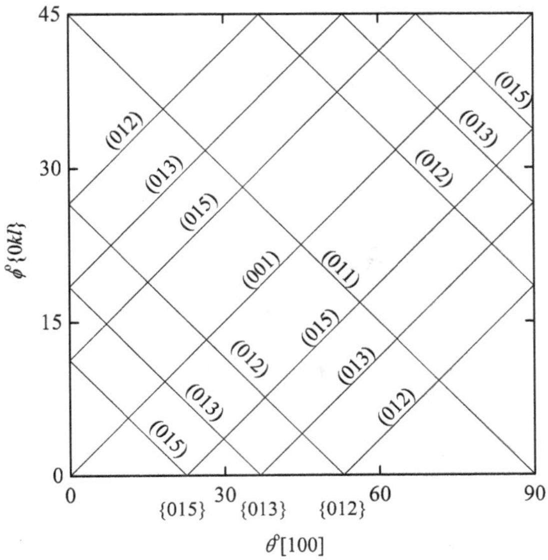


**Fig. 2.8** Orientation map of [100] tilt grain boundaries in cubic structures. $\theta$ is the misorientation angle between two adjoining grains, $\phi$ is the deviation of the grain boundary plane from the chosen symmetrical orientation. The *lines* parallel to the *diagonals* denote the grain boundaries having one of the planes identical [94,95]

shown in Fig. 2.8. Here, the angle $\theta$ means the misorientation angle between two adjoining grains and the angle $\phi$ is the deviation of the grain boundary plane from the chosen symmetrical position (for [100] tilt grain boundaries in cubic structures $0 \leq \theta \leq 90°, 0 \leq \phi \leq 45°$). The lines parallel to the diagonals in this plot represent the grain boundaries in which one of the planes is kept identical. According to this scheme and accepting that the {012}, {013} and {015} grain boundaries are singular [94], singular asymmetrical grain boundaries are expected to be formed by mutual combinations of the (001), (011), (012), (013) and (015) planes.

Recently, this classification of asymmetrical tilt grain boundaries was proved experimentally by measurements of anisotropy of grain boundary segregation of phosphorus, silicon, and carbon in bcc iron [94, 95, 98–100]. These experiments showed, that besides the {013}, {012} and {015} symmetrical tilt grain boundaries that belong to the first, second and second CLs (Table 2.1), respectively, the (001)/(013) grain boundary and all boundaries containing the (011) boundary plane exhibit special behaviour (Fig. 2.9). It means that all symmetrical grain boundaries denoted as singular were found to exhibit special segregation behaviour. In the case of asymmetrical tilt grain boundaries, the combinations of the lowest CL planes, (011)/(001), (011)/(013) and (001)/(013) were found to form special grain boundaries, too. Therefore, these grain boundaries can be considered as singular. In addition, all other asymmetrical grain boundaries formed by the (011) grain boundary plane were found to be special albeit they may not be all singular. Similarly, all asymmetrical grain boundaries formed by the (001) grain boundary plane exhibit segregation behaviour, which can be characterised as "transition", that is they can be considered as kind of "vicinal" grain boundaries. The segregation experiments leading to this conclusion will be discussed in Chap. 5 in more detail.

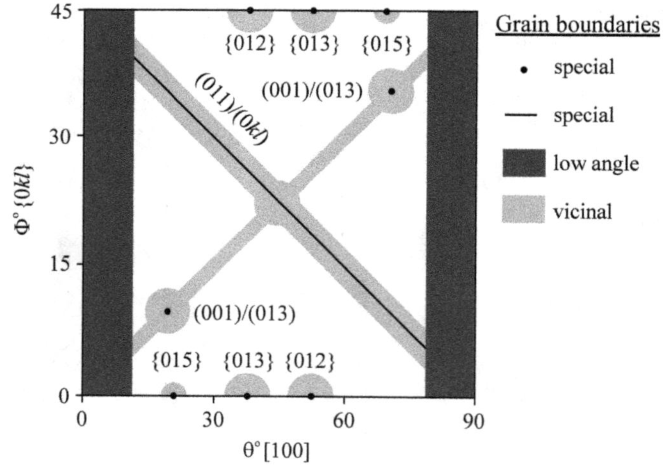


**Fig. 2.9** Schematic map of character of [100] tilt grain boundaries in bcc iron, indicated by different shadings and symbols. *Black line* and *points* refer to the special grain boundaries. *Light grey* area depicts vicinal grain boundaries and *dark grey area* does low-angle interfaces. All boundaries in the white area are general interfaces. The orientations in the map are defined in Fig. 2.8 [94, 95]

## 2.4 Basic Thermodynamics of Grain Boundaries

Since the grain boundaries represent one type of general interfaces, their thermodynamic treatment should be consistent with the general thermodynamics of interfaces, for example with thermodynamics of free surfaces that is thoroughly developed [12]. On the other hand, each interface has its own specifics that should also be taken into account. In contrast to free surfaces, there is much larger variety of different grain boundaries due to presence of the other crystal in contact with the grain "surface" in the grain boundary in comparison to vacuum in case of the free surfaces.

However, the grain boundaries represent a non-equilibrium crystal defect and thus, it may seem questionable whether the equilibrium thermodynamics can be principally applied to describe their states and processes. In contrast to equilibrium point defects (vacancies and interstitials) that can simply be equilibrated due to local fluctuations, the grain boundaries represent extended defect, which can be only removed by application of external forces. For example, grain boundaries form a 3-D network throughout a polycrystal that is additionally pinned at free surfaces: this network represents a stable object characterised by a local minimum of an appropriate potential and therefore, its components such as grain boundaries can be treated thermodynamically [48, 72].

The first thermodynamic description of interfaces was proposed by Gibbs [101]. His treatment is based on construction of a "dividing surface" between the interface and bulk crystal, which has to be later subtracted from the values of the thermodynamic parameters [12]. This rather laborious treatment was replaced by an

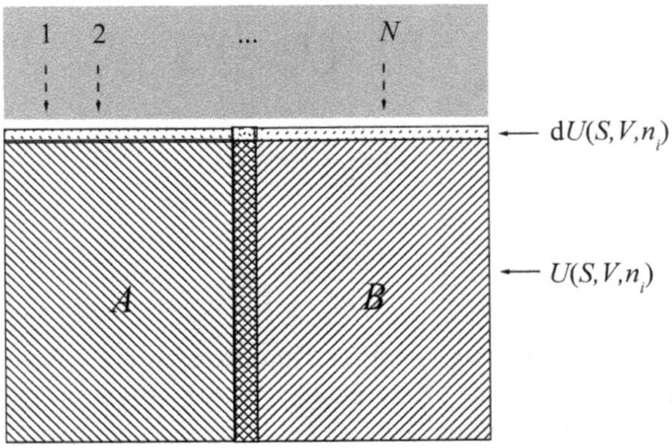


**Fig. 2.10** A growing bicrystal for introduction of the grain boundary energy

equivalent but more straightforward one developed by John Cahn [102, 103]. For introduction of thermodynamic functions of grain boundaries, we will adopt his modern formalism [12, 102].

To introduce thermodynamic state functions of grain boundaries, let us consider an open system composed of two $N$-component grains $A$ and $B$ of an identical phase differing exclusively in orientations that are separated by a planar grain boundary (bicrystal, Fig. 2.10). If this bicrystal grows as a result of equilibrium transport of components $1, 2, \ldots, N$ from reservoirs under constant temperature $T$, hydrostatic pressure $P$ and chemical potentials $\mu_i$ of each component $i = 1, 2, \ldots, N$, the increase of the internal energy of the system can be expressed as [12],

$$\mathrm{d}U = T\mathrm{d}S - P\mathrm{d}V + \sum_{i=1}^{N} \mu_i \mathrm{d}n_i + \sigma \mathrm{d}A, \tag{2.20}$$

where $S$ is the entropy of the system, $V$ is its volume, $n_i$ is the amount of the component $i$ ($i = 1, 2, \ldots, N$) and $A$ is the grain boundary area. $\sigma$ is the *grain boundary internal energy per unit area* (of the grain boundary) and represents the change of the internal energy of the closed system with the change of grain boundary area at constant entropy and volume [102, 103],

$$\sigma = \left(\frac{\partial U}{\partial A}\right)_{S,V,n_i}. \tag{2.21}$$

Adopting the fundamental relationship among the thermodynamic state functions $U$, $H$ (enthalpy), $F$ (Helmholtz energy) and $G$ (Gibbs energ) [104], $H = U + PV$, $F = U - TS$, $G = U + PV - TS$, we can additionally express the grain boundary energy as

$$\sigma = \left(\frac{\partial F}{\partial A}\right)_{T,V,n_i}, \tag{2.22}$$

or

$$\sigma = \left(\frac{\partial H}{\partial A}\right)_{S,P,n_i}, \tag{2.23}$$

or

$$\sigma = \left(\frac{\partial G}{\partial A}\right)_{T,P,n_i}. \tag{2.24}$$

Equation (2.24) gives probably the most important representation of the grain boundary energy per unit area as the change of the Gibbs energy of the system with the change of the grain boundary area under constant temperature and pressure in a closed system.

# Chapter 3
# Approaches to Study Grain Boundary Segregation

## 3.1 Basic Definitions and Terminology

As was shown in Chap. 2, grain boundaries are characterised by increased Gibbs energy (grain boundary energy, $\sigma$) comparing to the bulk crystal. The system containing grain boundaries will tend to reduce this energy in many ways. One group of them is interaction of the grain boundaries with other crystal defects. If the interaction of the grain boundaries with the point defects – foreign atoms – results in reduction of $\sigma$, an accumulation of foreign atoms at the grain boundary can occur. This thermodynamically preferred process then leads to segregation of alloying or impurity elements at the grain boundary. Under the term grain boundary *segregation*, we will understand here *all changes of concentration of any element at a grain boundary in the form of a solid solution* [13]. Let us mention that the limit of solute solubility in the basic material may be different for grain boundaries and for the bulk [105]: It can allow formation of few solute atom layers at a grain boundary that is still considered as a single phase. If secondary phase particles (precipitates) are situated at grain boundaries, we will speak about grain boundary *precipitation*. In fact, grain boundary precipitation is a continuation of the process of interfacial enrichment started by the grain boundary segregation when *new particles appear at the interface due to supersaturation of the above-mentioned grain boundary solid solution* [106].

According to the nature of the forces causing grain boundary segregation, we can distinguish two basic types of this phenomenon (a) equilibrium and (b) non-equilibrium grain boundary segregation. The term *equilibrium* grain boundary segregation will be used here to denote the very local redistribution of solutes at grain boundaries caused by minimisation of the total Gibbs energy of the system that is quite analogous to the adsorption phenomena well-known in surface science. It is supposed that the chemical potentials of all species involved in solid solution are constant throughout the system. At equilibrium, there is a partitioning that results in enrichment of grain boundaries by the surface-active species. The levels of the enrichment are defined only by the system parameters at equilibrium and not by the history of the material. The chemical enrichment at any interface can be reproduced simply by re-establishing the identical physicochemical conditions.

The nature of the *non-equilibrium* grain boundary segregation consists in interaction of solute atoms with the excess of vacancies in the system. A solute concentration is thus built up around the boundary because of dragging the vacancy–solute pairs when the vacancies flow to the grain boundaries in order to preserve their thermal equilibrium value. This phenomenon may occur in consequence of certain metallurgical operations such as quenching and irradiation. In contrast to the equilibrium segregation, non-equilibrium segregation may extend up to the distances of several μm across grain boundaries. This effect is sensitive to the rate of cooling from a high temperature [13, 18] and is briefly presented in Chap. 6.

Grain boundary segregation and precipitation represent only particular examples of different chemical composition of interfaces. At heterophase boundaries such as oxide/metal interfaces, the chemical differences are driven by chemical reactions at the contact plane between metal and oxide components (i.e. between oxygen or metal atomic layers) and can also alter according to their crystallography [20, 107]. Similar differences occur at the interfaces between metals on the one hand, and superconductors or semiconductors on the other hand, or at the grain boundaries of various chemical compounds, for example polymers [20, 108]. Such changes of chemical composition will not be treated here.

There are two basic approaches to study chemical composition of grain boundaries: experimental measurement and theoretical simulation. In the following, we will briefly address the main methods of both these approaches.

## 3.2 Experimental Methods for Study of Grain Boundary Segregation

Grain boundary segregation has been detected in the last century by various methods starting from indirect and mostly qualitative approaches. At first, the effort has been made to detect the suspected contaminants at the grain boundaries, which gave rise to various metallurgical effects. These techniques covered such measurements as electrode potential on grain boundary fracture surfaces, variations in X-ray lattice parameter with varied grain size, internal friction, micro-hardness, spectrographic analysis of material extracted from the grain boundary region, autoradiography and deductions made from anomalous grain growth behaviour and from subtle metallographic features observed at the grain boundaries [13, 21]. Although these measurements provided us with numerous interesting results, the main drawback of them is principal impossibility to determine the nature and concentration of the segregated element. However, this knowledge is fundamental for understanding the role of grain boundary chemistry in changes of a property. It is also hardly applicable to ternary and more complex systems, where the effects of individual elements may overlap. The importance of indirect methods has been superseded in the last decades with the development of a range of sophisticated direct techniques of surface analysis based upon various electron and ion emission processes. In these methods, the segregated elements are qualitatively and quantitatively detected from an appropriate spectrum.

There are some specific features characterising the grain boundary (a) the grain boundary is buried in the sample and (b) the changes of the chemical composition joined with the segregation phenomena are confined in a very thin layer of one or few atomic layers along the interface. Therefore, that technique must be suitable for the study of interfacial chemistry that enables the direct approach to the interface and the localisation of the analysis with atomic resolution in one dimension at least.

The necessary requirements to experimental techniques to study grain boundary segregation following from the above characteristics of grain boundary segregation were identified by Hondros [109] more than 30 years ago as

(a) High spatial resolution
(b) Elemental identification without a priori knowledge of segregating element(s)
(c) Quantitative measurement
(d) Non-exposure of a grain boundary by fracture in order to study weakly embrittling as well as non-embrittling species
(e) Determination of chemical bonding

Unfortunately, not all techniques used to study grain boundary segregation fulfil all these requirements. Principally, we can distinguish two groups of methods used for direct study of grain boundary segregation (a) surface analysis techniques and (b) microscopic techniques [13].

### *3.2.1 Surface Analysis Techniques*

As the name of this class of techniques suggests, the methods under consideration have been mainly developed to study chemistry of free surfaces. Similarly to grain boundaries, many features of surface chemistry are confined in a monatomic surface layer. In contrast to buried grain boundaries, however, free surfaces are better accessible to analytical tools. The surface analysis techniques can only be applied to the grain boundaries revealed by brittle intergranular fracture. Indeed, the above requirement (d) of Hondros [109] is not fulfilled and such methods can be applied exclusively on segregation-embrittled systems. Since the fresh fracture surface may be very often – and mainly in case of metals – quickly covered by a layer of environmental oxygen and/or carbon (for example in fractions of seconds under normal pressure in case of iron-based materials), the measurements of untouched grain boundary chemistry have to be performed in high vacuum.

There are two main techniques, Auger electron spectroscopy (AES) and secondary ion mass spectroscopy (SIMS), that have frequently been used to study grain boundary segregation and each of which provides little bit different information. Both these techniques operate in ultra-high vacuum (UHV) of the order of $10^{-8}$ Pa or better. To study grain boundary chemistry, these facilities have to have a possibility to expose the interfaces by fracture in the UHV of the instrument using a suitable fracture stage. If the segregated element embrittles the grain boundary, this fracture may be achieved fairly readily either at room temperature or by cooling the sample to liquid nitrogen temperatures. In the absence of intrinsic embrittlement, the grain boundary in some materials may be open by impact after cathodic charging the solid with hydrogen to promote hydrogen embrittlement [110]. Unlike the studies of free surfaces, new sample must be prepared for each measurement of the grain boundary state characterised by the heat-treatment time and the temperature. The relative ease of studying free surfaces explains the extent of effort devoted to free surfaces compared with that to grain boundaries [13].

The fact that the grain boundary has to be open by intergranular brittle fracture thus limits the experimental studies to the systems of embrittled grain boundaries. From experimental point of view, it evokes an additional requirement to the experimental techniques of surface analysis – an ultimate lateral resolution of imaging the area to be analysed. The fracture process can introduce various subtle defects to the structure that also appear at the fracture surface to be measured. Such defects affect substantially the measurements of grain boundary chemistry, and thus, suitable area should be chosen for the analysis that is free of any defect [111, 112]. To be absolutely fair and get well-interpretable results, extended experiments should be performed on bicrystals ensuring the identical grain boundary is measured under different conditions. Using polycrystalline samples, different grain boundaries are measured in different samples without taking into account the anisotropy of grain boundary segregation [113].

#### 3.2.1.1 Auger Electron Spectroscopy

The analysis of the intergranular fracture surface in UHV is usually undertaken with the most powerful and popular technique, AES [114]. A focused electron beam of energy of 1–30 keV is used in this technique to excite the atoms of an appropriate area of the target, creating holes in their inner core energy levels. The atoms subsequently decay by emission of either characteristic X-ray (which is the basis of bulk chemical analysis) or by emission of characteristic Auger electron from a thin surface layer: If the initial core hole is created in the orbital $w$ (Fig. 3.1), that is then filled by a higher energy electron from the orbital $x$, the energy balance is taken by the electron of the orbital $y$. This electron is ejected from the atom with the energy $E_{wxy}$ and is known as Auger electron. $E_{wxy}$ is characteristic of the particular atom and in this case is given by [115, 116]

$$E_{wxy} = E_w(Z) - E_x(Z') - E_y(Z') + \Phi_S, \qquad (3.1)$$

where $E_w$ is the binding energy of ionised level $w$, $E_x$ and $E_y$ are the energies of the interacting outer levels $x$ and $y$, $Z' \approx Z \pm 1/2$ corrects for the doubly ionised state and $\Phi_S$ is the analyser work function. As it is apparent from (3.1), the Auger transition covers three different electrons; therefore, the elements having less electrons (hydrogen and helium) do not exhibit this transition and, thus, cannot principally be analysed by this technique [117].

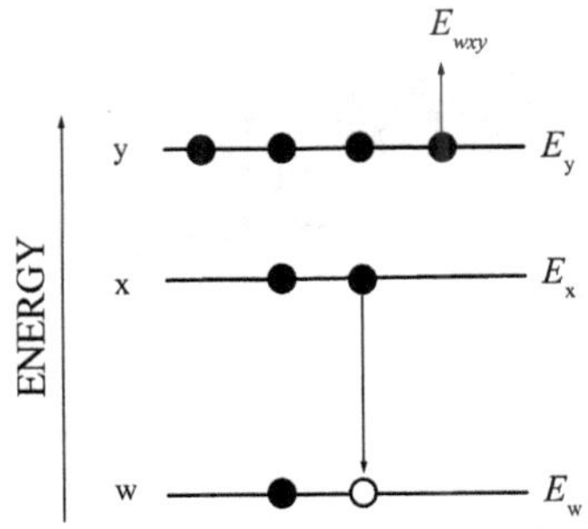


**Fig. 3.1** Schematic transition level diagram for electrons in AES

Although many transitions may occur in the volume irradiated by electrons, only few of them are strong enough for each element and these enable the surface atoms to be clearly identified from the emitted electron energy spectrum. Auger electrons ejected by atoms from the depths greater than few atomic layers (1 nm and more) are swallowed up in material and, thus, do not contribute to the emitted line spectrum. In this way, AES is characteristic for the outermost atomic layers of a solid, and by focusing the electron beam, for example via field emission gun (FEG), we may now localise the composition of the grain boundary fracture surfaces with a resolution that currently approaches less than 10 nm and the energy resolution 0.05% [118, 119]. Elements except for H and He as explained above, are generally detected with sensitivities ranging between 0.1 and 1 at.% depending on the nature of the element and the matrix [116, 120]. These concentrations are well reached in case of the grain boundary segregation.

Since its first metallurgical application to explain the origin of the fatal disaster of the Hinkley Point Power Station in the 1970s by Kalderon [121], a large progress has been achieved in both the instrumentation and the quantification of AES in upcoming decades [122]. Based on the inspiration evoked by the above example, AES was mainly used to study grain boundary segregation of impurity elements in iron and ferritic steels in connection to their temper embrittlement [123, 124]. Besides steels, AES studies also provided important results on equilibrium segregation of impurities at grain boundaries in nickel-based superalloys (e.g. [125, 126]), intermetallics (e.g. [127, 128]) and other materials (e.g. copper [105, 129] and molybdenum [130]). Besides the polycrystalline materials providing us with typical "faceted" fracture surfaces (Fig. 3.2), fine measurements on bicrystalline samples have been performed (Fig. 3.3) directed to study the anisotropic segregation phenomena (e.g. [94, 135]). During the time, AES became the most frequently applied technique for studies of grain boundary segregation. Precise quantification of obtained data is made using the standards as well as by means of standardless methods [109–111]. An interesting attempt to determine the grain boundary chemistry without breaking the samples was done using scanning Auger electron spectroscopy (SAES) in case of nickel interfaces decorated by chlorine, sulphur and oxygen [136]. More details on application of AES in grain boundary segregation are given in specialised works [114, 117, 122].

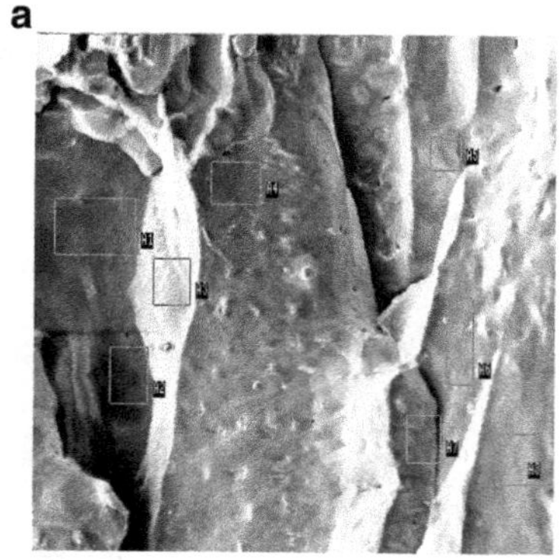

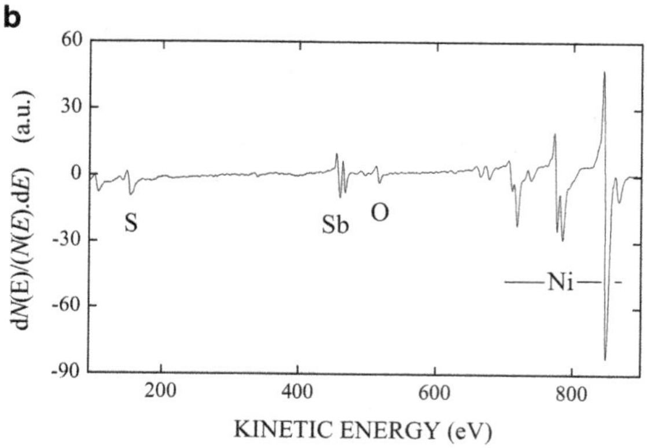


**Fig. 3.2** Example of the AES analysis. (**a**) intergranular fracture surface, (0.4 × 0.4 mm) (**b**) AES spectrum of one of the facet. Polycrystalline Ni–0.5at.%Sb alloy showing an extended grain boundary segregation of sulphur and antimony [131]

The thickness of the segregated layer can be determined by AES monitoring of the fracture surface during its slow peeling by in situ ion sputtering. In this way, a composition depth profile may be obtained for all elements present [137]. The results of such ion sputtering made by Seah showed that the segregated atoms are concentrated in the outermost plane of the fracture surface and hence, in the grain boundary plane [138]. The present state of the method also enables to detect fine variations in the content of segregated elements at different parts of both matching fracture surfaces of a bicrystal, that are connected with an asymmetry of the fracture process on the atomic scale [139].

#### 3.2.1.2 Secondary Ion Mass Spectroscopy

SIMS is a well-known technique for highly sensitive elemental analysis. It consists of the layer-by-layer peeling of the surface during its bombardment with primary

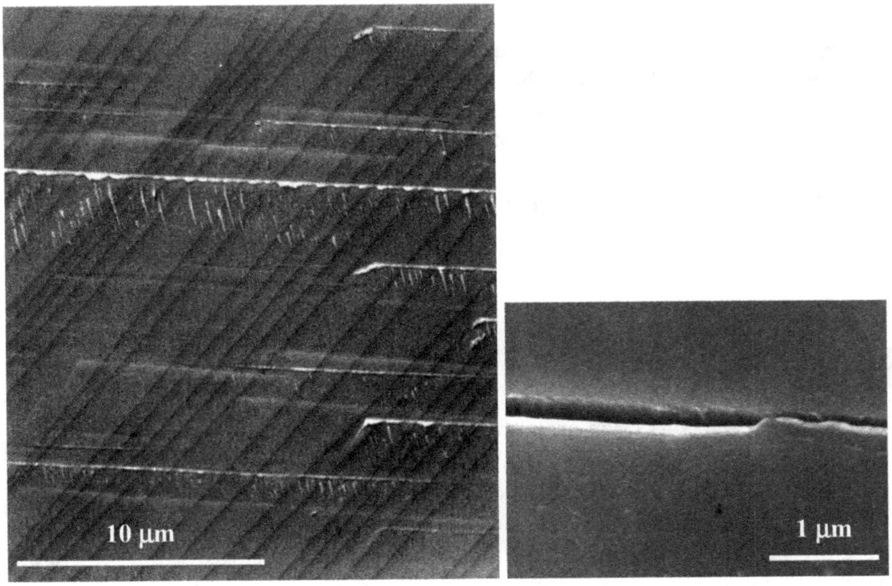


**Fig. 3.3** Grain boundary fracture surface of a bicrystal of an Fe–3.5at.%Si alloy. Individual defects introduced by the fracture process are apparent (cleavage tongues (see side detail), mechanical twins, deformation bands) [132–134]

ion beam, usually argon, with energy in the range 1–3 keV and current of units of nA/cm$^2$. During interaction of primary ions with the atoms of the target, a fraction of the ion energy is transferred to the lattice atoms. Atomic collisions thereby result in emission of secondary ions from the lattice. The secondary ions characterising the composition of the studied surface are then analysed in a mass spectroscope [140–143], usually of the time-of-flight type. Complex atom groupings of 20 or more atoms represent bonding arrangements at a surface [144] and so, in principle, the interactions of co-segregated elements can be studied. Low primary beam current yields very low sputtering rates that enable to analyse monatomic layers by this technique that can be so considered as *static* [142]. As a technique of surface analysis, it needs an intergranular fracturing of the sample before analysis. SIMS has some advantages over AES in (a) detecting hydrogen [145], (b) being more elemental sensitive than AES (up to ppb range) and (c) providing information about the chemical structure of the surface [143]. Due to many atoms grouping in complex ions, SIMS spectra may be complicated and quantification needs standards. On the other hand, information about the molecules present in the samples can be obtained that is important mainly in case of organic materials. In addition, ion focussing is not so sharp as that of electrons in AES and the measured area is also much larger (about 100 nm [47,146] although recent development of the instruments approaches to a resolution of about 20–50 nm [147]) so that the finding of an area for analysis that was not damaged during the fracture process may be complicated.

Therefore, SIMS has not yet become popular as a technique to study grain boundary segregation. Despite it, SIMS was used to detect grain boundary segregation of hydrogen in niobium [148] or of boron in 316L austenitic stainless steel [149], 718 cast superalloy [150] and nickel [151]. For another example, yttrium segregation in monatomic layer at alumina grain boundaries was found with relatively uniform lateral homogeneity [152] and silicon, aluminium, sodium, calcium, iron and potassium segregation was detected in yttria-stabilised zirconia [153]. SIMS high-resolution mapping enabled to detect segregation of magnesium at grain boundaries of $Al_2O_3$ [154].

#### 3.2.1.3 Other Spectroscopic Techniques

Besides AES and SIMS, the group of AEM covers other techniques such as X-ray photoelectron spectroscopy (XPS, also electron spectroscopy for chemical analysis, ESCA [114]), ion scattering spectroscopy (ISS or low-energy ion scattering, LEIS) [155], Rutherford backscattering spectroscopy (RBS) [156], Mössbauer spectroscopy [157], secondary neutron mass spectroscopy (SNMS) [158], fast atom bombardment – mass spectroscopy (FAB–MS), glow discharge mass spectroscopy (GDMS), glow discharge optical emission spectroscopy (GDOES) [159], proton-induced X-ray emission (PIXE), nuclear reaction analysis (NRA) and extended X-ray absorption fine structure (EXAFS) [160]. These techniques were only rarely used to study grain boundary segregation till now and therefore, no large details of them are given here. Although the techniques like ISS or RBS are able to analyse interfaces buried under the surface, many of the listed methods possess the common drawback of AEM techniques – the necessity to open the interface by intergranular brittle fracture. In comparison with AES and SIMS, in addition, they exhibit larger analysed area or lower depth sensitivity: These facts discriminate them for the present application to study grain boundary segregation. However, some of them may be improved in future to such an extent that they will be able to compete to the standard techniques of AES or SIMS [161].

ESCA is based on the photoelectric effect: A beam of X-rays with energy $E_X$ striking the sample surface under high vacuum induces emission of photoelectrons (core electrons) of kinetic energy $E_k$ and their escape from thin (2–10 monolayers) surface layer. The energy of these electrons can be used for analysis, and the characteristic binding energy

$$E_b = E_X - E_k - \Phi_S \tag{3.2}$$

is obtained. Similarly to AES, hydrogen and helium cannot be detected in this method. Present lateral resolution of ESCA is better than 3 $\mu$m and energy resolution is about 0.28 eV (with a monochromator). Currently, available computers can significantly improve the technological limits of the available instruments [119]. In connection with synchrotron radiation, the ultimate lateral resolution of 150 nm can be reached. The method enables to perform multi-element qualitative and quantitative analysis of surfaces and the depth distribution of segregated elements. The substantial advantage of this method is its ability to provide chemical bonding of the surface atoms and identification of their oxidation states [160]. ESCA was used to study grain boundary segregation of bismuth in copper [162, 163], of manganese, sulphur, antimony [164, 165] and phosphorus [166] in ferritic steels or of sulphur in nickel [167]. It was proved by this technique that phosphorus and tin segregate at grain boundaries of ferritic steels in their elemental states [168]. Recently, ESCA was applied to study grain boundary segregation of lead at grain boundaries of ancient silver coins, which causes a large embrittlement of these archaeological artefacts of the Roman age [169, 170].

A group of techniques is based on the ion scattering. In these techniques, a beam of ions of well-defined mass and energy is directed to a target (sample). The energy distribution of the particles scattered from the surface (or subsurface) layers in a specific direction with respect to that of the primary beam is measured to deduce information about the composition and atomic arrangement of solid surfaces [171]. The most used methods in this group are ISS and RBS.

In ISS, a well collimated mono-energetic (0.1–5 keV) primary ion beam of noble gases ($He^+$, $Ne^+$), alkali metals ($Li^+$, $Na^+$, $K^+$) or other light elements ($B^+$, $N^+$, $O^+$, $F^+$) of energy $E_0$ impinges on the surface. The ISS signal originates – similarly to AES – from the topmost layers only. The ions are scattered in collision from the target nuclei and their energy, $E$, defined by the mass of the surface target atom,

$$\frac{E}{E_0} = \frac{m^2}{(m+M)^2}\left[\cos\zeta + \left(\frac{M^2}{m^2} - \sin\zeta\right)^{1/2}\right]^{1/2}, \tag{3.3}$$

is detected in an analysing spectrometer at a scattering angle $\zeta$. The surface atoms of all elements can be identified from the energy spectra with high surface sensitivity (less than 1 nm) with detection limit ranging from $10^{-1}$ for light elements to $10^{-4}$ for heavy ones [171]. Although this method exhibits excellent surface sensitivity, provides simple quantification and is virtually non-destructive, it has only limited application in studies of interfacial composition [156]. The use of this technique can be demonstrated, for example on detection of strong segregation of copper at surfaces of Cu–Ni alloys [172].

In contrast to ISS, in RBS, high-energy light ions (e.g. $H^+$, $D^+$, $He^+$ or $C^+$ ions) accelerated usually by van de Graaff accelerator to energies 0.5–3 MeV, strike the target and penetrate into the sample interior. Some of the ions are back-scattered by the lattice atoms close to the sample surface and the scattering angle is related to the mass of the scattering atom via the ion energy. The detection limits are similar to those of ISS [171]. Unfortunately, the depth resolution of the method is of the order of 25 nm and the signal sensitivity mainly for the elements lighter than the primary ions is rather low although also here, all elements can be detected. The advantage of the low sensitivity to the light elements and the low-depth resolution is that the samples can be fractured on air without the danger of shadowing the information about the surface composition by oxygen or carbon adsorption on the fracture surface

[173]. Due to the low sensitivity and the basic requirements for experimental conditions, however, this method has only rarely been used to study grain boundary segregation. An example is the study of antimony segregation in manganese steel [174, 175].

EXAFS was used in studies of chemistry of Ni–Ti and Co–Ti interfaces [176] and for determination of atomic structural environment of yttrium and zirconium atoms segregated at the grain boundaries of $Al_2O_3$ [177]. EXAFS data indicate that yttrium and zirconium ions segregated at the grain boundary of an $Al_2O_3$ alloy, preferentially occupied the sites in the grain boundary core region, which are characterised by a well-defined dopant (Y,Zr)-O nearest neighbour co-ordinations, but the dopant (Y,Zr)-cation next nearest neighbour co-ordination is changing from site to site either within the same boundary or within different boundaries. The dopant (Y,Zr)-oxygen nearest neighbour bond length in the grain boundary, is on average fully relaxed to that in the parent oxides [178].

Combination of emission and transmission modes of Mössbauer Spectroscopy (MS) was used to study the surrounding of the $^{57}$Fe isotope that penetrated intergranularly into polycrystalline Fe–23.7at.%Si alloy during annealing at 700°C [157] and to deduce the grain boundary segregation [179]. Although this method is only indirect and rather time consuming, it brought the results that were very well comparable to the AES measurements (Fig. 3.4).

Recently, a very promising technique was proposed to identify individual atoms on sample surfaces on the basis of atomic force microscopy [180].

It was shown that surprisingly, the electron probe microanalysis (EPMA) method, which is basically focussed on characterisation of much thicker surface layers, is – in combination with the wavelength-dispersive X-ray spectroscopy (WDS) technique – also able to provide information about the monolayer segregation at the

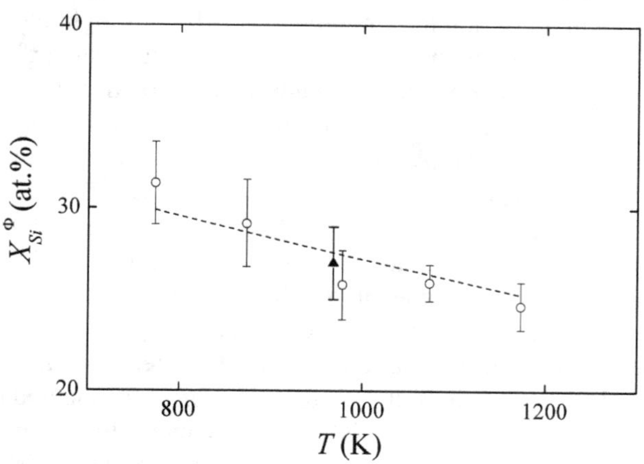


**Fig. 3.4** Temperature dependence of silicon grain boundary concentration in polycrystalline Fe–23.7at.%Si alloy. Experimental data were measured by AES (*circles*) and by MS (*solid triangle*). Theoretical curve was calculated according to the model of segregation in ordered alloys [157]

surfaces and interfaces [181, 182]. This technique consists of combination of a SEM microprobe with a number of WDS spectrometers, each of them containing several mobile crystals dispersing the X-rays according to their wavelength following the Bragg's law. The intensity of a particular line is obtained by measurement of diffracted X-ray intensity (measured using the gaseous proportional counters) at the spectral position of the top of the peak and subtracting the background intensity. Since the characteristic peak-to-background ratio of the WDS technique is about tens of ppm, the sensitivity of the method can be very good. The quantification of the data is usually done by means of the Stratagem™ [183] software. This technique was applied to study sulphur segregation at grain boundaries of an Fe–42%Ni alloy and documents high preciseness of the technique and its insensitivity to surface contamination [182].

More information on experimental details of individual techniques and their application to measurements of grain boundary chemistry can be found in specialised literature devoted to individual techniques.

## *3.2.2 Microscopic Techniques*

Two microscopic techniques have mainly been used in the study of grain boundary segregation (a) analytical electron microscopy (AEM) and (b) atom-probe field ion microscopy (AP FIM). The crucial advantage of these techniques over the techniques discussed above is that they can provide measurements for grain boundaries without fracture. They, thus, enable another family of systems to be studied.

### 3.2.2.1 Analytical Electron Microscopy

AEM represents a unique combination of high-resolution imaging of the high-resolution electron microscopy (HRTEM) (or scanning transmission electron microscopy (STEM)) dedicated to providing images with high spatial resolution, with highly localised analytical tools such as energy dispersive X-ray analysis (EDX) [184] or electron-energy loss spectroscopy (EELS) [185, 186]. Here a thin foil of a material (usually thinner than 100 nm) is prepared with the grain boundary perpendicular to the surface. The sample is then imaged in the electron microscope and analysed. In this technique, all requirements on the experimental methods to study grain boundary segregation [109] are fulfilled. The advantage of this technique does not only consist in its ability to analyse layers of segregation but also small precipitates and inclusions on the grain boundary by X-rays and electron diffraction and – in combination with modelling the atomic and electronic structures of grain boundaries – to localise the segregation sites of impurities and their bonding states [187]. On the other hand, however, it has some important disadvantages because it is not able to analyse light elements ($Z < 10$) and the preparation of the samples may be rather difficult and time consuming. In the last decade, the X-ray method was improved to identify quantitatively the crystallographic sites, distribution and types of impurities in many materials. This method using the incident electron beam orientation dependence of localised atomic characteristic X-ray emission is known as Atom Location CHannelling Enhanced MIcroanalysis(ALCHEMI) [188].

The present high-resolution microanalytical STEMs operate at acceleration voltages ranging between 100 and 400 keV. The electron beam can be focussed to the diameter less than 1 nm using FEG [189]. The spatial resolution and the minimum quantity of an analysed element are interrelated: The higher spatial resolution provides a smaller analysed volume, thus, reducing the signal intensity. In addition, the beam coming through the foil is broadened to some extent and the localised information is somehow dissolved, however, this broadening can later be removed by deconvolution methods [47, 190–192]. The segregation at the level of 0.01 monolayer can be identified with an electron beam of the diameter less than 2 nm by EDX [193], for example about 4% of molybdenum and 0.1% of phosphorus were detected at grain boundaries of a low alloy steel [194]. Let us mention that the recent progress in microshaping the samples by focussed ion beam (FIB) opened the possibility to prepare successfully the foils for AEM that enable to determine the chemistry of selected grain boundaries as well as of their triple junctions (Fig. 3.5) [195].

AEM offers an exclusive possibility to study the grain boundary segregation of solutes despite the fact whether they embrittle the matrix or not. It was applied to study the grain boundary segregation of manganese, tin, phosphorus and silicon in ferritic steels [196, 197], of boron in $Ni_3Al$ [198], bismuth in copper [199], titanium in Fe–Ni alloy [200], phosphorus, chromium, nickel and molybdenum in ferritic–martensitic steel (FV448) [201], sulphur in nickel [202], of silicon [200] and iron [203] in aluminium and the depletion of chromium in austenitic stainless steels [204, 205]. As for the non-metallic systems, grain boundary segregation was detected for arsenic in silicon and calcium in MgO [206], and titanium [191] and yttrium [178, 207–209] in $Al_2O_3$ in this way. In addition, varied copper concentration and depletion of oxygen along the grain boundaries were detected by AEM in $YBa_2Cu_3O_{7-\delta}$ high-temperature superconductor [210].

**Fig. 3.5** Ion-induced image of the progressing lift-out procedure for preparation of a TEM foil containing a triple junction. Let us notice that the triple junction is located within 5 μm under the sample surface [195]

If the dimensions of the segregated and matrix atoms differ substantially as, for example in case of segregation of bismuth in copper, the positions of large atoms can be identified at the grain boundary by HRTEM in combination with the theoretical study of a segregated boundary without any other analytical method. Using this combination, Luzzi et al. [211–213] described with a great accuracy the structure of the segregated grain boundaries containing multiple atomic sites. The cross-sectional analytical TEM was successfully applied to study the intergranular corrosion and cracking in 316 austenitic steel and nickel-based 600 alloy [214].

#### 3.2.2.2 Atom-Probe Field Ion Microscopy

Another microscopic technique applied to study the grain boundary segregation is AP FIM [215–217]. This method represents a combination of the field ion microscope (FIM) possessing atomic spatial resolution with a mass spectrometer. In FIM, the atoms of an inert gas (helium or neon) at pressure of about $10^{-3} - 10^{-2}$ Pa are ionised in a localised electric field ($4.5$–$5.0 \times 10^{10}\,\mathrm{Vm}^{-1}$) via tunnelling mechanism. Each ion created in this way is then ejected from positively charged surface of the sharp (but round) tip of the needle-like sample in the perpendicular trajectory to its local surface. In this way, the arrangement of individual atoms on the end of the sharp tip is imaged on a screen. The mass spectrometer provides chemical analysis of selected atoms removed from the tip by electric field evaporation, with the single atom sensitivity [216, 218]. In principle, individual atoms in vicinity of a grain boundary or a precipitate/matrix interface may, thus, be analysed and counted. In an alternative variant of the instrument, the atoms of the selected masses may be field-evaporated by ultra-short (10–25 ns) high-voltage ($3$–$6 \times 10^{10}\,\mathrm{Vm}^{-1}$ under high vacuum) pulses of electric field and analysed. In this way, the images of the trace of the grain boundary may be constructed with nearly atomic resolution [216, 218]: The lateral spatial resolution in FIM is of the order of $10^{-1}$ nm, but the AP FIM analysis is limited to the diameter of the probe hole, typically 1–10 nm with the depth resolution comparable with the interplanar spacing [219].

AP FIM technique meets many advantages – the possibility of observation of the interfaces with atomic resolution, crystallographic information about the interface and quick quantitative and standardless analysis of all elements. On the other hand, there are serious disadvantages that prevent the massive spreading of this method for application to segregation experiments: The procedure of the preparation of the sample in form of a thin needle with the chosen interface located just in the tip is rather complicated, and there is still a danger of mechanical failure of the samples by high-electric field required for pulse-field evaporation [218]. As regards the problems of the sample preparation, the FIB microshaping of the samples promises a large progress in preparing the needles for AP FIM [220].

A large progress of AP FIM was reached recently by its modification as the 3-D tomography (or tomographic atom probe, TAP [221, 222], or 3-D atom probe, 3DAP [223]). In this modification, the finest details of the grain boundary segregation are studied by consecutive field evaporation of individual layers from the tip of the material. All atoms removed from the needle tip are analysed in the positioned detectors. Using the computer treatment of the obtained data, a 3-D reconstruction of the distribution of segregated atoms in a small volume can be obtained with the resolution better than 0.5 nm [224, 225].

AP FIM was applied to detect the grain boundary segregation of boron in $Ni_3Al$ [216, 226], NiAl [227], 316L austenitic stainless steel [150, 228] and 2.25Cr–1Mo ferritic steel [229], carbon in NiAl [230], silicon in iron [231] and niobium and molybdenum in bcc iron [232]. Enrichment of molybdenum, nitrogen, silicon, boron and iron were measured at the grain boundaries of NiAl base material [233]. Grain boundary segregation of boron and carbon in IN 600 nickel-based superalloy was also shown [186] similarly to titanium, phosphorus and boron in 308 stainless steel [234]. In addition, palladium was detected in modified 4130 steel [235], oxygen in molybdenum [236] and osmium and nickel in tungsten [237]. This method was further applied to detect the Suzuki segregation of niobium and iron at stacking faults of cobalt [238–240] and to measure structural dependence of silicon segregation in bcc iron [241]. Phosphorus segregation in pressure vessel steels was found to be in good agreement with predictions based on McLean model of grain boundary segregation showing that the atom probe technique is a reliable method for characterising the solute segregation to interfaces [242].

Application of 3-D tomography to study the grain boundary segregation is still not yet extended. It was used to describe and quantify niobium and molybdenum segregation at high-angle grain boundaries in bcc iron [243]. These measurements enabled to determine the Gibbs energy of segregation of particular solutes at 800°C on basis of the Langmuir–McLean segregation isotherm (see Chap. 4 for more details). This method also unambiguously proved that the segregation effects are confined in approximately two monolayers along the grain boundary in equilibrium [244]. 3-D tomography was also used to study boron segregation to extended defects (e.g. dislocations) in FeAl base intermetallics [245, 246]. Atom probe tomography has also revealed segregation of zirconium, boron and carbon and depletion of oxygen at the grain boundaries of a Mo–0.15mass%Zr alloy in the base metal and in the heat-affected zone [247]. Recently, bismuth segregation was proved at the grain boundaries of nanocrystalline copper [248] and solute segregation was detected at various defects in irradiated Fe–Ni–P model alloy [249]. Laser-assisted 3-D tomography enabled to detect the segregation of the silicon enriched clusters at the grain boundaries of austenitic matrix of irradiated CW 316 stainless steel and their further enrichment by nickel [250].

## 3.3 Theoretical Approaches to Study Grain Boundary Segregation

Theoretical approaches to study the grain boundary segregation are principally identical to the simulations of the grain boundary structure and energetics in pure material. The only difference is that one or more foreign atoms are considered in the studied ensemble. These studies are fundamentally based on the research of the chemical and physical aspects of interatomic forces acting in vicinity of the interface [12]. The reliable methods of calculation of interfacial segregation should be based on two approaches [251]. The first approach – the *quantum mechanical calculations* from *first principles* (ab initio) – is aimed to solution of the Schrödinger equation. However, such calculations are complicated and time consuming. Therefore, suitable approximations and processes such as Hartree–Fock theory or density functional theory are applied. The second approach – the *molecular mechanics* – consists of two subsequent steps. At beginning, suitable interatomic potentials must be developed or applied describing reasonably the atomic bonds in an ensemble of atoms describing expected structure of a bicrystal with chosen grain boundary. The following step consists in calculations of the equilibrium structure. In this way, the energetics of the interface with regard to the location of the atoms of individual components is obtained. The reasonable potentials describing the structure and properties of the material can be developed on basis of ab initio quantum mechanical calculations or by using empirical or semi-empirical potentials describing the properties of the material. In the latter step, computer simulation methods like *molecular* (or *lattice*) *statics* (MS), *molecular dynamics* (MD) or *Monte Carlo* (MC) techniques are usually used. Let us state that the results on grain boundary structure and energetics obtained from atomistic simulations are never totally reliable, because there is generally too large number of degrees of freedom, so that all of them cannot be explored simultaneously [46]. Nevertheless, very good approximations of both the above features can be obtained in this way. Both above-mentioned approaches to the theoretical calculations of grain boundary segregation can also overlap, for example in ab initio molecular dynamics of quantum Monte Carlo method [251].

### 3.3.1 Quantum Mechanical Methods

Interfacial segregation can be calculated on basis of ab initio or first principles methods of quantum mechanics by accurate approximate solution of the many-electron Schrödinger equation. This is definitely the most precise way to calculate the segregation phenomena, however, time consuming in comparison to the methods based on the use of the potentials described below.

Quantum mechanical methods are based on the solution of time independent electronic Schrödinger equation

$$\hat{H}\Psi = E\Psi \tag{3.4}$$

where $\Psi$ is the wave function and $\hat{H}$ is the non-relativistic Hamiltonian in atomic units,

$$\hat{H} = \frac{1}{2}\sum_i \nabla_i^2 - \sum_i\sum_\alpha \frac{Z_\alpha}{|r_i - d_\alpha|} + \sum_i\sum_{j>1} \frac{1}{|r_i - r_j|} + \sum_\alpha\sum_{\beta>\alpha} \frac{Z_\alpha Z_\beta}{|d_\beta - d_\alpha|}. \tag{3.5}$$

In (3.5), $r_i$ are the electron positions, $d_\alpha$ are the nuclear positions and $Z_\alpha$ are the nuclear charges. The four terms in (3.5) represent the kinetic energy of the electrons, the electron–nuclear attractions, the electron–electron repulsions and the nuclear–nuclear repulsion, respectively. Due to complexity of the systems, the Schrödinger equation can only be solved approximately by using simple functions instead of the true wave functions.

A very suitable approximation is the *orbital approximation*. According to the *Hartree–Fock theory*, the many-electron wave function is expressed as the antisymmetrised product of one-electron wave function (Slater determinant)

$$\Psi(x_1, x_2, \ldots) = \frac{1}{\sqrt{N!}} \begin{vmatrix} \psi_1(x_1) & \psi_1(x_2) & \ldots & \psi_1(x_N) \\ \psi_2(x_1) & \psi_2(x_2) & \ldots & \psi_2(x_N) \\ \ldots & \ldots & \ldots & \ldots \\ \psi_N(x_1) & \psi_N(x_2) & \ldots & \psi_N(x_N) \end{vmatrix}. \tag{3.6}$$

Eventually, the Schrödinger equation (3.4) transforms to the self-consistent Hartree–Fock equation

$$\hat{f}\psi_i = \varepsilon_i \psi_i, \tag{3.7}$$

where $\varepsilon_i$ are the orbital energies and $\hat{f}$ is the Fock operator

$$\hat{f} = -\frac{1}{2}\nabla^2 + V_{\mathrm{nuc}} + J - K \tag{3.8}$$

containing the kinetic energy operator, the electron–nuclear potential energy, $V_{\mathrm{nuc}}$, the Coulomb term $J$ and the exchange term $K$. Hartree–Fock theory is relatively simple although the evaluation of a large number of 6-D integrals is required. This problem can be simplified by application of semi-empirical methods such as the Hückel theory or tight-binding technique, where many of these integrals are either simplified or even neglected or adjusted to reproduce some known thermodynamic properties of the system [251].

Presently, the most popular method to calculate the electronic properties of solids is the *density functional theory*. It is based on the theorem stating that the properties of the ground state of a many-electron system can be obtained by minimising the energy functional $E[\rho]$ of the electron density $\rho(r)$ [252]. Here, the electron density of the interacting system can be expressed in terms of one-electron wave function of the non-interacting system,

$$\rho(r) = \sum_{i=1}^{N} |\psi_i(r)|^2. \tag{3.9}$$

The energy functional is then

$$E[\rho] = T_{\mathrm{S}}[\rho] + V_{\mathrm{nuc}}[\rho] + J[\rho] - E_{\mathrm{xc}}[\rho], \tag{3.10}$$

where $T_S[\rho]$ is the kinetic energy of the non-interacting system, $V_{nuc}[\rho]$ is the electron–nuclear attraction, $J[\rho]$ is the Coulomb term and $E_{xc}$ is the exchange-correlation energy. The latter term is usually determined on basis of the *local density approximation*.

The first principles based methods were widely applied to compute the interfacial segregation and its consequences for various grain boundary affected phenomena. In this way, the effect of grain boundary segregation of boron and sulphur on grain boundary cohesion was explained on basis of the density functional theory [253] and the co-segregation of boron, titanium and oxygen at the grain boundaries of $\alpha$-iron was determined on basis of the density-of-states analysis [254]. Segregated gallium was found to draw charge from the surrounding aluminium atoms, thus, to reduce the cohesion of aluminium [255]. The first principles calculations of boron segregation at the $\{013\}$ tilt grain boundary in B2 FeAl intermetallics showed strong Fe–B interactions [256]. On the basis of the local density functional equations, several phenomena were also determined: the structure and the electronic properties of boron and sulphur at the coherent twin boundary in ferritic iron [257], the embrittlement of the same boundary induced by phosphorus segregation [258], hydrogen segregation [259] or the effect of boron on the cohesion of iron [260]. The first-principles quantum mechanical calculations showed that large bismuth atoms weaken the interatomic bonding by pushing apart the copper atoms at the interface [261].

The density-functional theory was further applied to study the geometric and magnetic structures of fully relaxed symmetrical tilt $\{013\}$ grain boundary in iron and $\{012\}$ grain boundary in nickel. In both cases, enhancements of the local magnetic moments of the atoms in the grain boundary plane were found. Calculated values of the segregation enthalpy of silicon and tin at these grain boundaries are in good agreement with experiment [262].

### 3.3.2 *Interatomic Potentials*

As the first step of the simulations of the grain boundary structure and chemistry, reliable models are required, which are able to evaluate sufficiently the energetics of various physically reasonable atomic configurations. The realisation of this approach is based on application of various interatomic potentials. The interatomic potentials can be obtained in various ways, for example by ab initio calculations representing the most accurate way to do this [263] or using empirical or semi-empirical approaches such as the *embedded atom method* (EAM) [264], *Finnis–Sinclair potentials* [265] and *glue model* [266]. Here, we will discuss the most applied type of the potentials based on the EAM technique. Usually, it is accepted in energy calculations that the electronic co-ordinates are mutually dependent variables (Born–Oppenheimer (adiabatic) variations). Then the total potential energy of the solid is the unique function of the atomic co-ordinates. The forces on individual atoms can be then calculated on basis of pure electrostatics (Hellmann–Feynman theorem), however, until now, it was performed only for systems consisting of few atoms or for pure metals. For more complicated systems, the use of empirical or semi-empirical methods seems to be more efficient [70].

#### 3.3.2.1 Pairwise Potentials

The simplest approach to describe the atomic structure of metals and some other materials and to search for the total internal energy of the system is application of *pairwise potentials* [70]. The internal energy has simple analytical form

$$U = \frac{1}{2}\sum_{ij.j\neq i} \phi_{ij}(R_{ij}), \tag{3.11}$$

where $\phi_{ij}(R_{ij})$ is the pairwise interatomic potential, which can be considered as the central-force repulsive energy between atoms $i$ and $j$ with $R_{ij} = |R_i - R_j|$ and which can be fitted to experimental data [70]. The potential $\phi_{ij}$ involves the parameters that should be fitted to suitable materials properties. This potential can rigorously be derived for metals with simple s–p bonding such as alkali metals, magnesium and aluminium from the electrostatics of the interacting spherical screening clouds of electrons. Here, it is accepted that the ion cores in these metals exhibit negligible effects on conduction electrons that screen them in a linear additive way. The electron density is then a superposition of the spherical charge densities and corresponding pairwise potentials can be obtained directly from ab initio calculations [70, 267–269]. Presence of non-spherical orbitals (d-electrons) deviates the screening from the linear way, and therefore, the ab initio potentials cannot be derived [70]. The tight-binding approach used by Masuda-Jindo to the calculation of segregation of various s–p metals to the 36.87° [100], {013} symmetrical tilt grain boundary in $\alpha$-iron [270] resulted in quite reasonable results that are comparable with the experimental data especially in the case of phosphorus and silicon [271].

Typical example of pairwise potential

$$\phi_{ij}(R_{ij}) = -\frac{A}{r^n} + \frac{B}{r^m} \tag{3.12}$$

was proposed by Lennard-Jones ($n = 6,\ m = 12$) [272]. Similar potential, although, in exponential form, was proposed by Morse [273]. Pairwise potential of the Morse type was used, for example for simulations of segregation of various solutes to a 38.21° [110], {122} symmetrical tilt grain boundary in copper showing that the solute atoms with larger atomic diameter than the matrix atoms decrease the grain boundary energy [274].

#### 3.3.2.2 Embedded Atom Method

EAM is a semi-empirical technique, which allows to calculate the total energy of an arbitrary arrangement of atoms [70, 264]. In this approach, the internal energy $U$ of a metal is composed of two terms,

$$U = \sum_{i} F_i \left( \sum_{j \neq i} \rho_j (R_{ij}) \right) + \frac{1}{2} \sum_{ij.j \neq i} \phi_{ij}(R_{ij}). \tag{3.13}$$

$F_i$ – the embedding (or glue) function – is the energy that is necessary to place the atom $i$ (embed) to the homogeneous environment (free electron gas) with the local electron density $\rho_j$ at the given atomic site. The function $\rho_j(R_{ij})$ is the radial density function of atom $j$ at distance $R_{ij}$ and it is considered as the electronic charge density of a free atom or as a measure of the squared interatomic matrix elements of a tight-binding Hamiltonian. It can be most simply interpreted as an empirical quantity, which defines the local density of atoms [70]. The three functions – $F$, $\rho$ and $\phi$ – are fitted to the experimental values or to the first-principles results, for example from simplified tight-binding description of the electronic densities of states.

EAM potentials were successfully applied to represent the metallic bonding, for example to determine the bismuth segregation in copper [253], to study the interfacial chemistry in $L1_2$-ordered intermetallics [252, 275] or to study the site competition effect of phosphorus, sulphur, silicon, oxygen, nitrogen and boron in $\alpha$-iron and tungsten and its consequences for the intergranular cohesion [276].

### 3.3.3 *Simulation Techniques*

To obtain an equilibrium grain boundary structure including the energetics of solute segregation at individual sites, an ensemble containing a number of atoms (simulation cell or box [70]) has to be defined. The simulation cell usually contains the periodically repeated structural unit of the chosen grain boundary inside several tens of atomic planes parallel to the interface on its both sides. This cell also repeats in three dimensions and the simulation represents the finite pattern of that interface. To define the border conditions, there are either two free surfaces introduced well distant from the interface to suppress their interaction with the grain boundary or two semi-infinite rigid crystals considered on both sides of the interface. In the latter scheme, the surface effects are eliminated. In some cases, Möbius or antiperiodic border conditions can be used [277].

Starting from a chosen structure, there are various ways how to calculate the structures of the segregated grain boundaries. The most advantageous are the techniques of molecular statics, molecular dynamics and Monte Carlo. They may provide us with the values of the enthalpy – and in some cases also the entropy – of the solute segregation to individual grain boundary sites.

#### 3.3.3.1 Molecular Statics

Molecular (or lattice) statics represents the simplest way of atomistic simulation of atomic structure of the grain boundaries possessing the lowest energy. It is based on determination of the atomic configuration providing minimum energy of the system, $\partial U/\partial r_i = 0$. Starting form a chosen initial configuration, the forces among atoms are calculated and atoms are slightly shifted in direction of these forces. A stable (equilibrium) configuration is reached when all forces vanish. During this minimisation entropy is neglected and therefore, obtained results are related to the temperature of 0 K (i.e. all vibrational effects are neglected). The minimisation, however, indicates only local energy minima albeit not the global minimum. Therefore, this technique is sensitive to the choice of the starting configuration. On the other hand, it is a very fast computational procedure and is widely used in atomistic simulations of grain boundary structure [251].

In the field of the grain boundaries and their properties, molecular statics is primarily used to simulate the atomic structure of the interfaces at 0 K. To apply this technique to studies of the grain boundary segregation, the grain boundary structure has also to be determined: first as relaxed without the solute, thus, providing the internal energy (enthalpy) of the system, $U_\alpha^\Phi$. Then one solvent atom at a chosen site in the structural unit is replaced by the solute atom and the simulation procedure is repeated. In this way, the energy $U_\beta^\Phi$ of such "segregated" interface is determined. The same procedure is performed for the bulk crystal with and without one solute atom thus providing the energies $U_\beta^\nu$ and $U_\alpha^\nu$, respectively. The segregation energy is then determined as

$$\Delta U_I = \left(U_\beta^\Phi - U_\alpha^\Phi\right) - \left(U_\beta^\nu - U_\alpha^\nu\right). \tag{3.14}$$

Molecular statics was used to describe hydrogen and helium trapping at the 28.07° [100], {035}, 38.94° [110], {114} and 50.48° [110], {113} tilt grain boundaries in nickel and showed that the grain boundaries are a major sink for interstitials [278]. Another study using this technique was devoted to solute segregation at the $\Sigma = 7$ symmetrical tilt grain boundary of copper [279].

#### 3.3.3.2 Molecular Dynamics

In molecular dynamics, the atom positions are determined as time dependent. For this purpose, numerical integration of the Newton equation of motion of each atom has to be performed. As a discrete time step used in integration, usually few percent of the atomic vibration frequency is applied. As a consequence, this technique is time consuming. The major advantage of this technique is the calculation of the thermodynamic averages of the equilibrium atomic co-ordinates at elevated temperatures [280].

The first step of this procedure is to describe all atoms in a chosen box of the starting configuration in a microcanonical ensemble by their positions $r_i$ and velocities $v_i$. For a given temperature, the velocities can be considered as randomly distributed according to a Maxwell distribution. The force $f_i$ on each atom (i.e. the acceleration) is calculated using the interatomic potentials. This force is assumed to remain constant during a chosen time step $\Delta t$ and corresponding $r_i$ and $v_i$ are found. The time step should be chosen smaller than typical periods of atomic vibrations ($10^{-12}$–$10^{-13}$ s) but too small time steps require very long computer times. On the other hand, large time steps may cause large truncation errors and result in instabilities when the atoms go near too closely. Typical time step is $10^{-15}$ s [251]. The process runs until the system properties converge to equilibrium.

The early attempt of applying the molecular dynamics to theoretical studies of grain boundary chemistry was done by Hashimoto et al. for phosphorus and boron segregation in $\alpha$-iron. It was shown that both phosphorus and boron segregate preferentially at interstitial sites of $\Sigma = 5$, {013} and $\Sigma = 9$, {114} symmetrical tilt grain boundaries [281]. This pioneering work started the effort in simulations of the segregation behaviour of the grain boundaries despite the obtained values of the segregation energies calculated by Hashimoto et al. are too high comparing to the experimental data. Quenched molecular dynamics providing the Helmholtz energy of segregation at 0 K was applied to simulate the segregation effects at the {113} and {233} symmetrical tilt grain boundaries in Ni–Ag and Ag–Ni alloys [274].

#### 3.3.3.3 Monte Carlo Simulations

Similarly to molecular dynamics, Monte Carlo (MC) simulations allow to obtain the grain boundary structures at elevated temperatures. Here, the dynamics of the system is not taken in account, and the time necessary for successful equilibration is much shorter compared to molecular dynamics. Therefore, this technique is optimal for determination of thermodynamic averages. According to statistical mechanics, the average value of a thermodynamic function, $\langle Q \rangle$, is given as

$$\langle Q \rangle = \int Q(Z)\Pi_i(Z). \tag{3.15}$$

The probability $\Pi_i$ of finding a particular configuration in a chosen statistical ensemble is given as

$$\Pi_i = \frac{\exp[-U(Z)/kT]}{\int \exp[-U(Z)/kT]\,\mathrm{d}Z} \approx \frac{1}{Z}\exp\left(-\frac{U(Z)}{kT}\right). \tag{3.16}$$

In (3.15) and (3.16), $Z$ is the partition function of the system and $U$ is the internal energy. In practice, series of configurations of the grain boundary structure are generated in these ensembles usually by means of the Metropolis algorithm [282]. Starting from an initial configuration, new configurations with changed atomic co-ordinates are randomly produced. The ratio of the probabilities for the initial and new configurations is

$$\frac{\Pi_n}{\Pi_i} = \exp\left(-\frac{U_n - U_i}{kT}\right). \tag{3.17}$$

In the calculations, only $\Pi_n > \Pi_i$ is accepted as a new initial configuration. If $\Pi_i > \Pi_n$, this configuration is accepted with the probability $\Pi_n/\Pi_i$. Each repetition of the procedure with new initial configuration is called MC step. In this procedure, the configuration used as initial ones (which were chosen arbitrarily) are not included in the ensemble. Due to some temporal correlation, the produced configurations are not completely independent, and therefore, it is necessary to consider only the configurations uncorrelated with those produced before and after them. In each MC step, atoms are chosen randomly and are displaced by random vectors of the lengths of fractions of the maximum permitted displacement.

Theoretically, the Monte Carlo technique simulates the equilibrium state of the system. Practically, the equilibrium may not be sometimes reached, for example if a highly metastable initial configuration is used. Since the grain boundary also represents high-energy metastable state comparing to the crystal interior, the equilibrium reached in MC simulations indicates generation of a stable grain boundary structure. Different configurations of the statistical ensemble implicitly cover thermal fluctuations in the system. Therefore, MC simulations can also be used to represent the "annealing" of the grain boundary structure – obtained, for example from molecular statics simulations – at chosen temperature and to testing its stability [219].

Similarly to molecular statics simulation, the ensembles with and without solute atoms are generated and their energetics are compared. The positions of the atoms, their chemical identities and the volume of the system can change in this procedure supposing constant temperature, pressure, total number of atoms and difference in chemical potentials, $\Delta\mu$, of atomic species. Using a transmutational ensemble with chosen bulk solute concentration, the grain boundary segregation can be simulated at chosen temperature with

$$\frac{\Pi_n}{\Pi_i} = \exp\left(-\frac{U_n - U_i + \Delta\mu}{kT}\right). \tag{3.18}$$

Here, $\Delta\mu$ represents the difference of chemical potentials of the solute and the solvent between the actual chemical potential and the chemical potential of an ideal gas with the same composition and volume [12].

Monte Carlo simulations with transmutational ensemble have been presently used to simulate solute segregation to twist grain boundaries [283–287]. Besides this procedure, the overlapping distribution MC (ODMC) method is used to determine the values of $\Delta F_I$ [219]. This is applied because the Helmholtz energy, $\Delta F_I$, cannot be expressed as ensemble average and, thus, cannot be directly calculated from MC simulations. According to statistical mechanics,

$$F = -kT\ln(Z), \tag{3.19}$$

where $Z$ is the partition function given as

$$Z = \int \exp\left(-\frac{\hat{H}}{kT}\right) \mathrm{d}p\mathrm{d}q \tag{3.20}$$

and $p$ and $q$ are the momentum and position vectors for $A$ atoms in the ensemble. $\hat{H}$ is the Hamiltonian, defined as

$$\hat{H} = \sum_{i=1}^{N} \frac{p_i^2}{2m_i} + U \tag{3.21}$$

with $m_i$ being the mass of the $i$th atom. If the atoms in the ensemble are identical, the partition function is

$$Z = \int \exp\left(-\frac{p^2}{2mkT}\right) \mathrm{d}p^N \int \exp\left(-\frac{U}{kT}\right) \mathrm{d}q = (2\pi mkT)^{3N/2} \int \exp\left(-\frac{U}{kT}\right) \mathrm{d}q. \tag{3.22}$$

The difference of the Helmholtz energy between the arbitrary ensembles $A$ and $B$ is

$$\Delta F = F_B - F_A = -kT \ln\left(\frac{Z_B}{Z_A}\right). \tag{3.23}$$

Supposing a single solute atom $B$ is present in ensemble $A$, the ratio $Z_B/Z_A$ is

$$\begin{aligned} \frac{Z_B}{Z_A} &= \left(\frac{m_B}{m_A}\right)^{3/2} \frac{\int \exp[-(U_B - U_A)/kT] \exp[-U_A/kT] \,\mathrm{d}q}{\int \exp[-U_A/kT] \,\mathrm{d}q} \\ &= \left\langle \exp\left(-\frac{U_B - U_A}{kT}\right)\right\rangle_A, \end{aligned} \tag{3.24}$$

that represents the thermodynamic average over the ensemble $A$. The difference in $\Delta F$ can be calculated for each atom in the simulation cell [219]. If there is a large overlap of the distributions, one can obtain reliable values of $\Delta F$.

Because the degree of the overlap is not known a priori, it is necessary to determine the thermodynamic averages for MC runs of various lengths to determine the number of steps per atom required for a given degree of accuracy.

The use of MC technique has widely been applied since 1990s starting by the first attempts of Foiles et al. [288]. Using this method, Foiles showed that copper segregation at $\Sigma = 5$, $\Sigma = 13$ and $\Sigma = 61$ (001) twist grain boundaries in nickel is confined in few atomic layers near the grain boundary and is stronger at high-angle grain boundaries [289]. MC calculations were also used to study solute segregation at 36.87°[100], {013} symmetrical tilt grain boundary in the Cu–Ag system [290] and in the NiAl intermetallics [291], or to determinations of solute segregation at low-angle [285] and high-angle [292] twist grain boundaries in the Ni–Pt system. Segregated magnesium at grain boundaries of $Ni_3Al$ intermetallics preferentially substitutes the aluminium atoms rather than nickel ones. The segregation of magnesium additionally reduces the grain boundary energy thus increases the grain boundary cohesion [293].

The ODMC method was applied to study platinum segregation at two sites of the $\Sigma = 5$, (001) twist grain boundary in gold at 850 and 1,650 K [294]. Ab initio calculated many-body potentials (linear muffin-tin-orbital method) and MC were successfully used to simulate the structure and energy of bismuth segregated twin grain boundary of copper [213].

#### 3.3.3.4 Minimisation of the Helmholtz Energy

Another approach to obtaining the values of the Helmholtz energy is so-called free energy minimisation (FEM) [295]. In this method, an approximation is used to calculate the entropy of the system. The atoms in a solid can be considered as harmonic oscillators. The vibrational contribution to the Helmholtz energy, $F_{vib}$, at the classical limit for high temperatures is given as

$$F_{vib} = kT \sum_{i=1}^{N} \sum_{j=1}^{3} \ln \left( \frac{h\omega_{ij}}{2\pi kT} \right), \tag{3.25}$$

where $\omega_{ij}$ are the vibrational eigenfrequencies of the atoms and $h$ is the Planck constant. The values of $\omega_{ij}$ are calculated on basis of the quasi-harmonic (QH) approximation by so-called local harmonic (LH) model. This model accepts the harmonic approximation keeping only the second-order terms but neglecting the coupling of vibrations between the atoms. Each atom is considered as an independent oscillator and the computation is reduced from the diagonalisation of the full $3N \times 3N$ dynamic matrix to the evaluation of $N$ local dynamic $3 \times 3$ matrices for $N$ atoms in the ensemble.

The configurational entropy $S_c$ is calculated as

$$S_c = -k \left\{ X_A(i) \ln [X_A(i)] + X_B(i) \ln [X_B(i)] \right\}, \tag{3.26}$$

where $X_A$ and $X_B$ represent the concentrations of "effective" atoms, which represent the time average probability that a site is occupied by an atom of type $A$ or $B$, respectively ($X_B = 1 - X_A$). Equation (3.26) is only valid for an ideal solid solution; real configurational entropy will possess different values.

In case of solute segregation the grand potential, $\Omega$, is given as

$$\Omega = U + F_{vib} - TS_c + \Delta\mu \sum_{i=1}^{N} X_A(i). \tag{3.27}$$

In very dilute solid solutions, $\Delta\mu = \text{const}$ and $S_c = 0$ so that [294]

$$\Omega = U + F_{vib}. \tag{3.28}$$

Computer simulation provides us with the values of the internal energy $U$ of the system as well as of the segregation at various types of the grain boundary sites. However, no information about the entropy of interfacial segregation, which represents an important component of the Helmholtz energy of segregation, can directly be obtained. In fact, the Helmholtz energy of segregation depends on temperature,

$$\Delta F_I(T) = \Delta U_I(T) - T\Delta S_I(T). \tag{3.29}$$

In general, both $\Delta U_I$ and $\Delta S_I$ depend on temperature although it may be weakly and therefore, they are often considered as constant. As the entropy of segregation cannot be calculated directly, $\Delta S_I$ can be determined as

$$\Delta S_I = \frac{1}{T}(\Delta U_I - \Delta F_I(T)). \tag{3.30}$$

The FEM method was applied to determination of copper segregation to (001) twist grain boundaries in nickel within the misorientation range 15°–45° at temperatures 600 K $< T <$ 1,100 K [296]: The obtained results are comparable to those obtained by MC simulations [289]. In fact, copper was found to segregate to 36.87°[100], {013} symmetrical tilt grain boundary even in copper-rich alloys [297]. Similar results were also obtained for segregation of gold in palladium while in the system Ag–Au, the majority element segregates at the grain boundary [298].

# Chapter 4
# Models of Equilibrium Grain Boundary Segregation

## 4.1 Thermodynamics of Grain Boundary Segregation

Interfaces, in general, represent structural defects of a crystal characterised by existence of unsaturated bonds. Therefore, the Gibbs energy of an interface will be higher than the Gibbs energy of the single crystal containing the same ensemble of atoms. The value of the Gibbs energy of an interface depends on various variables, particularly on its energy (i.e. type, orientation and atomic structure), composition, temperature and pressure. To minimise the total Gibbs energy of the system (i.e. Gibbs energy of the interface plus Gibbs energy of the volume), the interface interacts with other lattice defects such as dislocations, vacancies and foreign atoms. As mentioned previously, the latter interaction results in accumulation of the solute atoms in the interface region and is called *interfacial segregation* (e.g. [12, 19, 20]). As individual types of the interfaces possess the same nature (i.e. planar defects), the interaction of individual types with foreign atoms is expected to be qualitatively similar.

As was mentioned in Chap. 2, two basic forms of grain boundary segregation can be distinguished: *equilibrium* and *non-equilibrium* segregation. Equilibrium segregation occurs as a result of inhomogeneities in the solid giving rise to sites for which solute atoms have a lower Gibbs energy. These sites occur at interfaces such as free surface, grain boundaries and phase interfaces as well as at defect sites, dislocations and stacking faults. All of these regions then may exhibit concentrations of solute atoms that differ from each other and from that of the bulk materials. At any temperature, there is a unique value of the solute concentration for each of these sites that is asymptotically approached as time goes to infinity and at a rate governed by diffusion. On the other hand, non-equilibrium segregation depends on rate processes and kinetic events and, in general, disappears as time approaches infinity if diffusion processes are allowed to reach full equilibrium. There are number of discrete routes for producing this form of segregation, which include moderate rate quenching of samples from a high temperature, the growth of precipitates, the effect of stress at temperature, etc. In this chapter, we will deal with equilibrium segregation.

Phenomenological description of equilibrium segregation is qualitatively the same for all types of interfaces. The differences arise concerning the values of particular thermodynamic parameters resulting from different structural/bonding conditions at individual cases [299–301]. In the following, we show two main approaches to the thermodynamic description of the grain boundary segregation, the *Gibbs adsorption isotherm* and the *Langmuir–McLean types of segregation isotherm* (e.g. [13, 19, 20]).

## 4.2 Gibbs Adsorption Isotherm

Let us consider a macroscopic system large enough to keep constant pressure and plane interfaces [302]. Integration of (2.20) accounting Euler theorem results in

$$U = TS - PV + \sum_{i=1}^{N} \mu_i n_i + \sigma A. \tag{4.1}$$

Its differentiation and comparison with (2.20) provides us with the relationship

$$A\mathrm{d}\sigma = -S\mathrm{d}T + V\mathrm{d}P - \sum_{i=1}^{N} n_i \mathrm{d}\mu_i. \tag{4.2}$$

Let us note that (4.1) and (4.2) are related to the whole system. Since the Gibbs–Duhem equation

$$-S\,\mathrm{d}T + V\,\mathrm{d}P - \sum_{i=1}^{N} n_i \mathrm{d}\mu_i = 0 \tag{4.3}$$

is valid for each homogeneous phase [104], it should also be valid for both grains $A$ and $B$,

$$-S^A\mathrm{d}T + V^A\mathrm{d}P - \sum_{i=1}^{N} n_i^A \mathrm{d}\mu_i = 0 \tag{4.4}$$

and

$$-S^B\mathrm{d}T + V^B\mathrm{d}P - \sum_{i=1}^{N} n_i^B \mathrm{d}\mu_i = 0. \tag{4.5}$$

Comparison of (4.4) and (4.5) with (4.2) gives

$$-s^\Phi\mathrm{d}T + v^\Phi\mathrm{d}P - \sum_{i=1}^{N} \Gamma_i^\Phi \mathrm{d}\mu_i + \mathrm{d}\sigma = 0 \tag{4.6}$$

where $s^\Phi$, $v^\Phi$ and $\Gamma_i^\Phi = n_i^\Phi/A$ are the entropy, the volume and the amount of the component $i$ at the grain boundary $\Phi$, respectively, all normalised by the grain boundary area. The quantity $\Gamma_i^\Phi$ is, thus, the *surface density of the solute i at the interface* as introduced by Gibbs [101] or the *adsorption* as used by McLean [19]. Much later, a term *grain boundary excess* of component $i$ was introduced. This terminology is very unlucky as discussed in more detail in Sect. 4.3. Therefore, we will use the term *adsorption* for $\Gamma_i^\Phi$.

Let us define the interfacial quantities of the system as

$$U^\Phi = U - (U^A + U^B), \tag{4.7}$$

$$S^\Phi = S - (S^A + S^B), \tag{4.8}$$

$$n_i^\Phi = n_i - \left(n_i^A + n_i^B\right) \tag{4.9}$$

and

$$V^\Phi = V - (V^A + V^B) = 0. \tag{4.10}$$

The value of $V^\Phi$ is zero by definition ($V = V^A + V^B$). Combining (4.7)–(4.10) with (2.20) and subsequent expressions,

$$\mathrm{d}U^A = T\,\mathrm{d}S^A - P\,\mathrm{d}V^A + \sum_{i=1}^{N} \mu_i \mathrm{d}n_i^A \tag{4.11}$$

and

$$\mathrm{d}U^B = T\,\mathrm{d}S^B - P\,\mathrm{d}V^B + \sum_{i=1}^{N} \mu_i \mathrm{d}n_i^B \tag{4.12}$$

produces

$$\mathrm{d}U^\Phi = T\,\mathrm{d}S^\Phi + \sum_{i=1}^{N} \mu_i \mathrm{d}n_i^\Phi + \sigma\,\mathrm{d}A. \tag{4.13}$$

The analysis in [12,103] shows that all relationships between thermodynamic quantities are applicable to the systems containing grain boundaries when the interfacial characteristics – the grain boundary energy $\sigma$ – is taken into account and that these relationships are independent of the thickness of the grain boundary.

Combination of (4.6) with the Gibbs–Duhem condition $\Sigma_i n_i \mu_i = 0$ results for a binary system ($i = M, I$) in an interesting relationship for adsorption $\Gamma_i^\Phi$

$$\frac{n_I}{n_M}\Gamma_M^\Phi - \Gamma_I^\Phi = \left(\frac{\partial \sigma}{\partial \mu_I}\right)_{T,P}. \tag{4.14}$$

Equation (4.14) can be simplified for case of $\Gamma_M^\Phi = 0$ as

$$\Gamma_{I,M}^\Phi = -\left(\frac{\partial \sigma}{\partial \mu_I}\right)_{T,P}, \tag{4.15}$$

where $\Gamma^{\Phi}_{I,M}$ indicates the above condition $\Gamma^{\Phi}_{M} = 0$ [303]. It follows from (4.6) that the element, which causes a decrease of the grain boundary energy with increasing chemical potential, accumulates (segregates) at the grain boundary [12].

Substituting the chemical potential $\mu_i$ by the activity $a_i$ according to the expression [104]

$$\mu_i = \mu_i^0 + RT \ln a_i, \tag{4.16}$$

where $\mu_i^0$ is the standard chemical potential of element $i$ in the bulk and $R$ is the universal gas constant, we obtain [138, 303]

$$\Gamma^{\Phi}_{I,M} = -\frac{1}{RT}\left(\frac{\partial \sigma}{\partial \ln a_I}\right)_{P,T}. \tag{4.17}$$

In the dilute approximation, which is that of general interest, for a bulk solute molar concentration $X_I \ll 1$, in which $a_I = \gamma_I X_I$ [304] with the activity coefficient $\gamma_i \approx 1$, this becomes

$$\Gamma^{\Phi}_{I,M} = -\frac{1}{RT}\left(\frac{\partial \sigma}{\partial \ln X_I}\right)_{P,T}. \tag{4.18}$$

Equation (4.18) provides us thus with a simple relation between adsorption $\Gamma^{\Phi}_{I,M}$ on the one hand and the change of the interfacial energy $\sigma$ with the molar fraction $X_I$ of the solute in bulk on the other hand. Equation (4.18) is the most useful form of the *Gibbs adsorption isotherm* for a dilute binary system. This basic form has been used experimentally to determine the interface density of the solute $I$ from the changes of the grain boundary energy with changing the bulk composition for all interfaces in solids [13].

The Gibbs adsorption isotherm was successfully applied to quantify the first experiments on grain boundary segregation, for example on phosphorus in $\gamma$-iron [305] and tin [306, 307], silicon and sulphur in $\delta$-iron [308]. However, the measurement of the surface energy as a function of both the bulk concentration and the temperature is rather difficult. Therefore, many attempts have been made to develop alternative phenomenological models of grain boundary segregation based on simple relationships between interfacial and bulk compositions.

$\Gamma^{\Phi}_{I,M}$ is also frequently used to represent grain boundary segregation measured by APFIM methods (e.g. [231, 239, 303, 309]). In order to describe interfacial chemistry in binary as well as multi-component systems, it is useful to generalise the measured quantities. Therefore, the *grain boundary enrichment ratio*, $\beta^{\Phi}_I$, is defined as the ratio between the grain boundary and the bulk solute concentrations, $X_I{}^{\Phi}$ and $X_I$, respectively $\left(X_I^{\Phi} \gg X_I\right)$ [13],

$$\beta^{\Phi}_I = \frac{X^{\Phi}_I}{X^0 X_I} = \frac{\Gamma^{\Phi}_I}{\Gamma^0 X_I}, \tag{4.19}$$

where $\Gamma^0$ is the maximum amount of solute constituting a close packed monatomic layer of unit area and $X^0$ is the maximum grain boundary atomic concentration of the solute.

## 4.3 Langmuir–McLean Types of Segregation Isotherm

This approach is based on equality of chemical potentials (Gibbs energy) of the components in equilibrium. There exist several ways to derive the segregation isotherms based on this approach; however, some of them are thermodynamically inconsistent although all of them provide very similar results. We will follow the way presented by duPlessis and van Wyk [302, 310, 311].

Let us consider a closed system containing the grain boundary inside the crystalline surrounding. The grain boundary is considered as a region of finite thickness, while the bulk of the crystal is of infinite size. Grain boundary segregation is then defined as a redistribution of solute atoms between the crystal and the interface: From this point of view, both parts of the system are considered as open enabling an exchange of the solutes. In equilibrium, the system as a whole must possess minimum energy. The variations of the total internal energy $\delta U$ of the closed system is given by

$$\delta U = \sum_{\nu} \delta U^{\nu} = \sum_{\nu} (T^{\nu}\delta S^{\nu} - P^{\nu}\delta V^{\nu} + \delta G), \tag{4.20}$$

where $T^{\nu}$ is the temperature, $S^{\nu}$ is the entropy, $P^{\nu}$ is the pressure and $V^{\nu}$ is the volume of the structural component $\nu$ of the system (i.e. of the crystal volume and the grain boundary). $G$ is the Gibbs energy of the system. If the temperature and pressure are the same in all these structural components and the equilibrium state is reached, (4.20) reduces to [104]

$$(\delta U)_{n_j} = (\delta G)_{n_j} = 0, \tag{4.21}$$

where $n_j$ is the number of moles of the solute $j$. If the crystal is divided into $N + 1$ open subsystems ($N$ in bulk, one for the grain boundary) and the redistribution of the solutes occurs, duPlessis and van Wyk showed in very detail that the Gibbs energy of the system is given by

$$G = G_{\mathrm{B}} + \sum_{i=1}^{M} n_i^{\Phi}\left(\mu_i^{\Phi} - \mu_i^{B}\right), \tag{4.22}$$

where $n_i^{\Phi}$ is the number of moles of solute $i$ in the grain boundary, and $\mu_i^{\Phi}$ and $\mu_i^{B}$ are the chemical potentials of the $i$th solute in the grain boundary and in the bulk. The number of the solutes is $M$. In (4.22), the Gibbs energy $G_{\mathrm{B}}$ (equal to the

Gibbs energy of the bulk before segregation) is independent of changes of $n_i^\Phi$ and the sum in (4.22) represents the product of the grain boundary energy $\sigma$ and the grain boundary area $A$ [302, 310, 311]

$$\sigma A = \sum_{i=1}^{M} n_i^\Phi \left(\mu_i^\Phi - \mu_i^B\right). \tag{4.23}$$

It follows from (4.22) and (4.23) that

$$G = G_\mathrm{B} + \sigma A. \tag{4.24}$$

Since $G_\mathrm{B}$ is constant in respect to $n_i^\Phi$, the equilibrium condition $\partial G/\partial n_i^\Phi = 0$ can be written as

$$\frac{\partial \sigma}{\partial n_i^\Phi} = 0. \tag{4.25}$$

Equation (4.25) shows that *in equilibrium, the grain boundary energy possesses the minimum value* [302, 310, 311].

The equilibrium condition can be expressed as

$$\frac{\partial \sigma}{\partial n_1^\Phi} = \frac{\partial \sigma}{\partial n_2^\Phi} = \cdots = \frac{\partial \sigma}{\partial n_M^\Phi} = 0. \tag{4.26}$$

However, the conditions (4.26) are not independent because

$$n_1^\Phi + n_2^\Phi + \cdots + n_M^\Phi = n^\Phi, \tag{4.27}$$

that is the number of moles $n^\Phi$ in the finite grain boundary region is constant. Supposing $M$ is referred to the matrix element and $n_M^\Phi = 1 - \sum_{i=1}^{M-1} n_i^\Phi$, the basic condition for chemical equilibrium between the grain boundary $\Phi$ and the volume $B$ is [302, 310, 311]

$$\Delta G = \left(\mu_i^\Phi - \mu_i^B\right) - \left(\mu_M^\Phi - \mu_M^B\right) = 0 \tag{4.28}$$

for each component $i = 1, 2, \ldots, M - 1$ despite of their number.

Combining (4.16) and (4.28), the general form of the segregation equation can be written as

$$\frac{a_I^\Phi}{a_M^\Phi} = \frac{a_I}{a_M} \exp\left(-\frac{\Delta G_I^0}{RT}\right). \tag{4.29}$$

In (4.29) the *standard* molar Gibbs energy of segregation,

$$\Delta G_I^0 = \left(\mu_M^0 + \mu_{I(M)}^{0,\Phi}\right) - \left(\mu_{I(M)}^0 + \mu_M^{0,\Phi}\right) \neq 0 \tag{4.30}$$

is defined as a combination of the standard chemical potentials of the elements $I$ and $M$ in pure state at the grain boundary and in the bulk, $\mu_i{}^{0,\Phi}$ and $\mu_i^0$, respectively, at the temperature and pressure of the system and *in the structure of the matrix element* (as is indicated by index $(M)$ in brackets). Since generally,

$$a_i = \gamma_i X_i, \tag{4.31}$$

where $\gamma_i$ is the activity coefficient of an element $i$, [304], we can write (4.29) using $X_M = 1 - \sum_{J=1}^{M-1} X_J$ as

$$\frac{X_I^\Phi}{1 - \sum\limits_{J=1}^{M-1} X_I^\Phi} = \frac{X_I}{1 - \sum\limits_{J=1}^{M-1} X_I} \exp\left(-\frac{\Delta G_I^0 + \Delta G_I^{\mathrm{E}}}{RT}\right). \tag{4.32}$$

In (4.32),

$$\Delta G_I^{\mathrm{E}} = RT \ln\left(\frac{\gamma_I^\Phi \gamma_M}{\gamma_I \gamma_M^\Phi}\right) \tag{4.33}$$

$\Delta G_I^{\mathrm{E}}$ is the *excess* molar Gibbs energy of segregation. Since (4.32) was derived without any assumption about the character of the system, t represents the *general form of the segregation isotherm.*

The thermodynamic representation of the grain boundary segregation is schematically depicted in Fig. 4.1. In Fig. 4.1, the concentration dependence of the Gibbs energy of the crystal volume (*bulk*) and of the grain boundary ($\Phi$) is shown. Since the grain boundary represents the defect of crystal structure, it has to possess higher Gibbs energy comparing to the bulk in the whole concentration range of the binary system. Thus, the concentration dependence of the Gibbs energy of the grain boundary is "shifted" to higher values in comparison to a similar dependence of the Gibbs energy of the bulk. Because the grain boundary does not represent another phase, the equilibrium between the boundary and the bulk is not given by a common tangent as in phase equilibrium. To fulfil condition (4.28), the grain boundary concentration, $X_I^\Phi$, is defined by the tangent to the Gibbs energy of the grain boundary, $G^\Phi(X)$, parallel to that for the bulk, $G^{\mathrm{bulk}}$, at the concentration $X_{\mathrm{bulk}}$. On the other hand, $\Delta G_I^0 = \left(\Delta\mu_{I(M)}^{0,\Phi} - \Delta\mu_{I(M)}^0\right) - \left(\Delta\mu_M^{0,\Phi} - \Delta\mu_M^0\right) \neq 0$ (cf. Fig. 4.1). It directly follows from (4.30) and Fig. 4.1 that $\Delta G_I^0$ as the sum of the standard chemical potentials of pure substances is *independent of composition* of the system, and further, that $\mu_i^{0,\Phi}$ and $\mu_i^0$ represent respective molar Gibbs energies, that is they are composed of the terms of both the standard enthalpy and the standard entropy of pure elements in the chosen standard states.

The above derivation differs from that presented sometimes in other papers (e.g. [20, 301, 310, 312–316]). There, the chemical potential for the interface is

$$\mu_i^{0\Phi} = \xi_i^{0\Phi} + RT \ln a_i^\Phi - \sigma A_i. \tag{4.34}$$

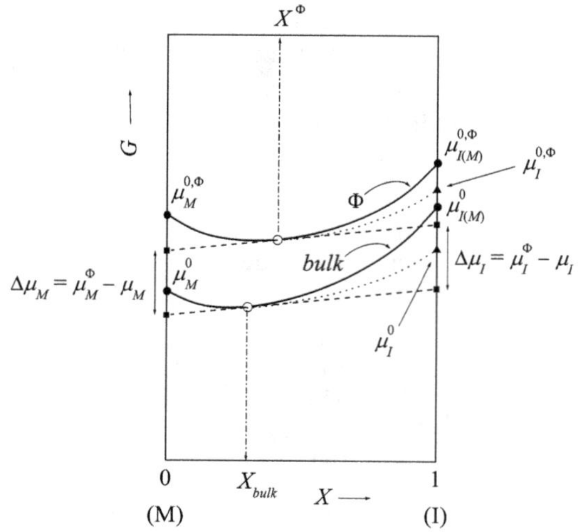


**Fig. 4.1** Schematic depiction of the concentration dependence of the Gibbs energy of the volume (*bulk*) and the grain boundary (Φ) leading to representation of the Gibbs energy of segregation. $\Delta G_I^0$ is defined by (4.30). $\mu_{I(M)}^{0,\Phi}$ and $\mu_{I(M)}^{0}$ correspond to the respective states of pure $I$ in the structure of the interface and bulk of $M$ and are higher than the respective values of $\mu_i^{0,\Phi}$ and $\mu_I^0$, corresponding to the structures of pure $I$. The equilibrium concentration $X_I^{\Phi}$ is defined by the tangent to $G^{\Phi}$ parallel to that of $G^{\text{bulk}}$ constructed in $X_I$. This conserves the condition $\Delta G = \Delta\mu_I - \Delta\mu_M = 0$ for each pair $X_I$ and $X_I^{\Phi}$ [113, 312]

In (4.34), $A_i$ is the partial molar area of species $i$ and $\xi_i$ is the partial molar Helmholz energy of species $i$ in the interface. Further treatment assuming the constant value of $\sigma$ leads to the expressions

$$\frac{X_I^{\Phi}}{(X_M^{\Phi})^{A_I/A_M}} = \frac{X_I}{(X_M)^{A_I/A_M}} \exp\left(-\frac{\Delta G_I^0 + \Delta G_I^{\mathrm{E}}}{RT}\right) \tag{4.35}$$

and

$$\Delta G_I^{\mathrm{E}} = RT\left(\ln\frac{\gamma_I^{\Phi}}{\gamma_I} - \frac{A_I}{A_M}\ln\frac{\gamma_M^{\Phi}}{\gamma_M}\right). \tag{4.36}$$

Since the values of partial molar areas are known only rarely, it is usually assumed that $A_I = A_M$, and then, (4.32) and (4.26) are obtained. This approach seems to accept too many assumptions to obtain the same result as the former derivation based on the treatment of the Gibbs energy. The former derivation is thus thermodynamically clearer; we only have to keep in mind that the grain boundary energy contribution is involved in corresponding values of the standard chemical potentials.

It is evident from (4.28) that the chemical potentials of both the solute and the solvent have to be considered in the description of thermodynamic equilibrium. Therefore, the description of solute segregation analogous to the condition for phase transformations, that is considering only the chemical potentials of the solutes ($\mu_I^{\text{bulk}} = \mu_I^{\Phi}$, e.g. [315, 317]) is incorrect, and can provide misleading results. We must keep in mind that the grain boundary does not represent "a different phase" but only the defect of the crystal lattice. This is also reflected by the fact that the Gibbs energy of the grain boundary concentration dependence is higher in the whole concentration range than that of the bulk. This is also compatible with the approach of John Cahn (cf. Chap. 2) [13, 103].

Let us mention that one can meet different notation in literature that can bring extended confusion. Sometimes the Zangwill approach [318] is referred in literature considering the formal separation of the total Gibbs energy of segregation, $\Delta G = \left(\mu_I^{\Phi} - \mu_I^{B}\right) - \left(\mu_M^{\Phi} - \mu_M^{B}\right)$, which is entirely equal to zero in equilibrium (cf. (4.28)), into two parts, *configurational* and *remaining* ones. Frequently, however, another terminology is used for these two terms (e.g. [303, 319]), the *ideal* term, $\Delta G^{\text{id}} = -T\Delta S^{\text{id}} = -RT \ln \left(X_I^{\Phi} X_M / X_M^{\Phi} X_I\right)$, and an "*excess*" term, $\Delta G^{\text{xs}}$,

$$\Delta G = \Delta G^{\text{id}} + \Delta G^{\text{xs}}. \tag{4.37}$$

Then – for a binary system – we can write

$$\frac{X_I^{\Phi}}{1 - X_I^{\Phi}} = \frac{X_I}{1 - X_I} \exp\left(-\frac{\Delta G^{\text{xs}}}{RT}\right) \tag{4.38}$$

or – supposing that not each grain boundary site is available for segregation and thus, a saturation $X^{0\Phi}$ of the boundary occurs [308],

$$\frac{X_I^{\Phi}}{X^{0\Phi} - X_I^{\Phi}} = \frac{X_I}{1 - X_I} \exp\left(-\frac{\Delta G^{\text{xs}}}{RT}\right). \tag{4.39}$$

It is evident from comparison of (4.32) and (4.38) that

$$\Delta G^{\text{xs}} = \Delta G_I^0 + \Delta G_I^{\text{E}} = \Delta G_I, \tag{4.40}$$

where two "excess" terms, $\Delta G^{\text{xs}}$ and $\Delta G_I^{\text{E}}$, appear.

It is obvious from the above considerations that the excess term $\Delta G_I^{\text{E}}$ appearing in (4.32) and (4.33) describes the *deviations between ideal and real behaviour* of any thermodynamic system. The adjective excess in this sense was introduced at beginning of the twentieth century and its usage was made common by the famous book of Lewis and Randall [304] published originally in 1923. For example, the Gibbs energy of mixing of a system is defined as $\Delta_{\text{m}}G = \Delta_{\text{m}}G^* + \Delta_{\text{m}}G^{\text{E}}$. It means that $\Delta_{\text{m}}G$ is composed of two contributions, the ideal Gibbs energy of mixing, $\Delta_{\text{m}}G^* = RT\Sigma_i X_i \ln X_i$, and the excess Gibbs energy of mixing, $\Delta_{\text{m}}G^{\text{E}} = RT\Sigma_i X_i \ln \gamma_i$ [302]. This elegant approach was chosen to conserve all relationships, which were originally derived for ideal systems, if the concentration (atom fraction) $X_i$ is replaced by the activity, $a_i$, (4.31) in real systems. It is fully and systematically implemented in thermodynamics and has been successfully used for many decades (e.g. [104,303,320,321]). The adjective excess characterising the difference between the real and ideal behaviour is well established and expresses the well-defined addition to the ideal thermodynamic state function to yield the total function (Gibbs energy, internal energy, enthalpy, entropy or volume) of the real state. The term excess thermodynamic function has no other additional adjective and thus, the adjective excess represents its necessary and sufficient specification. This term was introduced at first to thermodynamic terminology without causing any controversy.

The idea of dividing $\Delta G$ into two parts in a different way, as indicated in (4.37), is principally understandable but resulted into a controversy with the above-mentioned common thermodynamic terminology. The term $\Delta G^{\mathrm{id}}$ might be thought as "ideal", since it consists of the atomic fractions. However, the "excess" term $\Delta G^{\mathrm{xs}} = \Delta G_I$ contains both the term $-RT \ln \left(\gamma_I^{\Phi}\gamma_M/\gamma_M^{\Phi}\gamma_I\right)$, which is equal to $\Delta G_I^{\mathrm{E}}$ (4.33), and also the term $\Delta G_I^0$, which itself is ideal because it is a combination of the standard chemical potentials (4.22). Due to the presence of the latter contribution, $\Delta G^{\mathrm{xs}}$ is not the "excess" term in the sense of the discussion in the preceding paragraph, and the original Zangwill proposal [318] to call $\Delta G^{\mathrm{xs}}$ "remaining" term is more cautious.

The usage of the term "excess" for the Gibbs energy of interfacial segregation (4.40) has probably further roots. In surface physics, the adjective "excess" was introduced to denote the surface or interfacial contribution to the thermodynamic state function $Y^{\mathrm{s}}$ as compared to the bulk [303]. When considering an interface between two phases in a hypothetical system, the thermodynamic function $Y^{\mathrm{s}}$ (internal energy, entropy, ...) referring to a surface or interface s is defined as

$$Y^{\mathrm{s}} = Y - Y' - Y'', \tag{4.41}$$

where $Y$ is the "property (function) characterising the real system" [12,303], $Y'$ and $Y''$ are the functions of the phases $'$ and $''$, respectively. The thermodynamic functions $Y^{\mathrm{s}}$ are generally called surface or interfacial excess properties (i.e. functions) (e.g. [12, 303]) "that is we assign to the interface any excess of the thermodynamic functions of the real system over those of the hypothetical system" [303]. Besides these functions, so-called "excess" quantities (i.e. functions) per unit area of the surface/interface are often used, which are marked $Y^{\mathrm{xs}}$ (e.g. [303, 319]). As we have seen in the above discussion, this terminology is sometimes also used to specify characteristic thermodynamic functions of segregation, the Gibbs energy, $\Delta G^{\mathrm{xs}}$, the enthalpy, $\Delta H^{\mathrm{xs}}$ and the entropy, $\Delta H^{\mathrm{xs}}$ [303, 319].

The terminology using the complex adjective "surface/interfacial excess" is relatively new since it has never been used in surface or interfacial terminology before McLean [19], and Gleiter and Chalmers [21], and probably also not before John Cahn [102]. The adjective "excess" in the sense of characterising the surface and interfacial functions is a completely redundant over-determination because the term "surface/interfacial" itself already suggests a characteristic difference of a thermodynamic function from that of the bulk. It was introduced inconsistently without considering the established meaning of the adjective "excess" in the thermodynamics of solutions and paves the way to misunderstanding and misinterpretation. Based on the above-mentioned arguments, there is no reason to use the term "excess" to additionally characterise thermodynamic state functions for surfaces and interfaces. To avoid any confusion, it is recommended to use the simple and sufficient term *surface/interfacial thermodynamic function* to characterise the properties of the surfaces and interfaces $Y^{\mathrm{s}}$ (4.41), as used by Lewis and Randall [304], and to use the adjective *excess* exclusively to describe the differences between real and ideal behaviour [113].

## 4.3.1 Physical Meaning of Thermodynamic State Functions Appearing in Segregation Isotherms

Due to the above-mentioned ambiguity in the thermodynamic terminology of interfaces, severe misunderstanding sometimes exists in the use and interpretation of the thermodynamic state functions of segregation (cf. [113, 303, 322, 323]). Therefore, it is necessary to point out clearly the physical meaning of individual types of Gibbs energy appearing in the segregation isotherms of the Langmuir–McLean type [(4.29), (4.32), (4.35) and (4.38)] (a) the Gibbs energy of segregation, $\Delta G_I$; (b) the standard Gibbs energy of segregation, $\Delta G_I^0$; (c) the excess Gibbs energy of segregation, $\Delta G_I^{\mathrm{E}}$, and their enthalpy and entropy counterparts, as well as (d) their averaged effective functions.

### 4.3.1.1 Gibbs Energy, Enthalpy and Entropy of Interfacial Segregation

The physical meaning of the *Gibbs energy of interfacial segregation*, $\Delta G_I$, of the solute $I$ in the system $M-I$, that is considered as molar to preserve its intensive character, results from the definition given by (4.38) and (4.40). There is no doubt that $\Delta G_I$ completely determines the atomic fraction of the solute $I$ at the interface (site) $\Phi$, $X_I^{\Phi}$, at a given temperature $T$ and a given system composition $X_I$. As schematically shown in Fig. 4.2a, at constant temperature the value of $\Delta G_I$ principally changes with changing bulk composition $X_I$ and thus with changing interfacial composition $X_I^{\Phi}$ due to mutual interaction of the species in bulk and/or interface. The corresponding enthalpy, $\Delta H_I$, and entropy, $\Delta S_I$, of segregation are related to $\Delta G_I$ in sense of the basic definition [101, 104, 304, 321]

$$G = H - TS. \tag{4.42}$$

It is obvious that $\Delta H_I$ and $\Delta S_I$ are identical to $\Delta H^{\mathrm{xs}}$ and $\Delta S^{\mathrm{xs}}$ (or sometimes also to $\Delta H_{\mathrm{seg}}$ and $\Delta S_{\mathrm{seg}}$, cf. [303]), respectively, which are frequently used in the literature [319]. There is no doubt that $\Delta H_I$ and $\Delta S_I$ depend on temperature and on concentration as is obvious from many theoretical approaches. For example, the

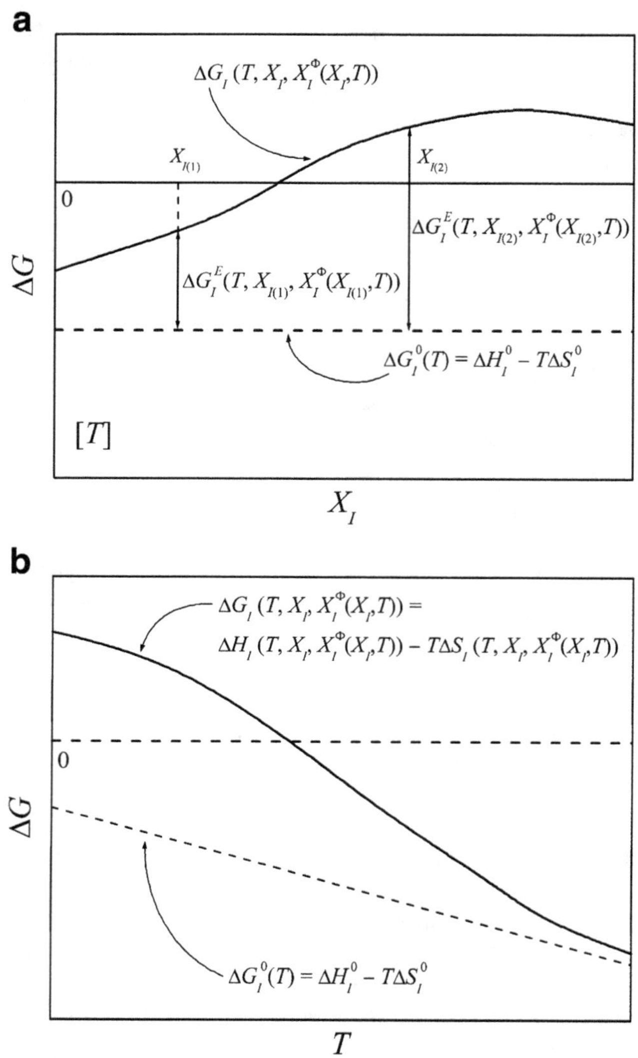


**Fig. 4.2** Schematic representation of individual types of the Gibbs energy of grain boundary segregation in a binary system appearing in (4.40) in dependence on (**a**) bulk concentration, (**b**) temperature. According to (4.40), an enrichment of the interface by the solute $I$ occurs at the bulk concentration $X_{I(1)}$ but the same interface is depleted by $I$ at $X_{I(2)}$. When the bulk concentration of $I$ is that specified by the value of $\Delta G_I = 0$, interfacial concentration of $I$ is identical with its bulk concentration. See the text for the meaning of individual types of the Gibbs energy of segregation [113]

famous model of Wynblatt and Ku [303, 324, 325] gives for $\Delta H_I$

$$\Delta H_I = (\sigma_I - \sigma_M)A^{\Phi} + 2\omega\left[z^{\mathrm{lat}}\left(X_I - X_I^{\Phi}\right) + z^{v}(X - 0.5)\right] - \frac{24\pi K_I G_M r_I (r_M - r_I)^2}{3K_I r_I + 4G_m r_M}, \quad (4.43)$$

where $\sigma_i$ is the interfacial energy of the pure component $i$ $(i = I, M)$, $A^{\Phi}$ is the molar area of the monolayer interface, $\omega$ is the regular solution parameter for the $M-I$ solution, $z^{\mathrm{lat}}$ and $z^{v}$ are the in-plane and out-of-plane co-ordination numbers, respectively, $X_I$ and $X_I^{\Phi}$ are the atom fractions of $I$ in bulk and at the interface, respectively, $K_I$ is the bulk modulus of the solute, $G_M$ is the shear modulus of the solvent (note that both are temperature dependent) and $r_i$ are the atomic radii of the pure solute $I$ and solvent $M$ atoms [303].

Another formula for the Gibbs energy of segregation (also here inaccurately called "excess") was derived on the basis of the mean-field theory by Rabkin [326].

Segregation in real systems is affected by mutual interaction of solute and solvent atoms and by presence of third elements. Under some circumstances, interfacial depletion of the solute can be observed instead of enrichment [327]. In fact, each point in the $T-X_I$ space can be characterised by different values of $\Delta G_I$ ($\Delta H_I$, $\Delta S_I$) as partially visualised in Fig. 4.2. These values describe the actual segregation at the boundary (boundary site) at a fixed point on the line drawn in Fig. 4.2a and change from one state (1) to another one (2). Because each change of temperature in turn changes $X_I^{\Phi}$ and thus changes both $\Delta H_I$ and $\Delta S_I$, these values can hardly be attributed to any general information, for example about the nature of the grain boundary and the anisotropy of the grain boundary segregation, because any orientation dependence of $\Delta H_I$ and $\Delta S_I$ varies in a complex way with temperature and composition, an example of which is shown in Fig. 4.3.

#### 4.3.1.2 Standard Thermodynamic Functions of Interfacial Segregation

The *standard (molar) Gibbs energy of interfacial segregation*, $\Delta G_I^0$, is defined as combination of the standard chemical potentials of the elements $I$ and $M$ at the grain boundary $\Phi$ and in the bulk according to (4.30). The chemical potentials related to the pure solute $I$, $\mu_{I(M)}^{0,\Phi}$ and $\mu_{I(M)}^{0}$, correspond to the standard states of the pure solute $I$ at a chosen temperature and pressure of the system but in the structure of the host material $M$, in both bulk and interface, respectively. Note that the composition dependence of $G$ for the bulk and the interface will be different when we exchange the meaning of $M$ and $I$, that is when $I$ will be considered as solvent and $M$ as solute (cf. Fig. 4.1). It clearly follows from (4.30) that $\Delta G_I^0$ (and consequently the corresponding $\Delta H_I^0$ and $\Delta S_I^0$) are principally independent of concentration (see Figs. 4.1 and 4.2a). In other words, $\Delta G_I^0 = \Delta G_I$ exclusively when the activity coefficients of all components are $\gamma_I = 1$, that is when $\Delta G_I^{\mathrm{E}} = 0$ [(4.32) and (4.33)], that is in an ideal system. Despite this obviously

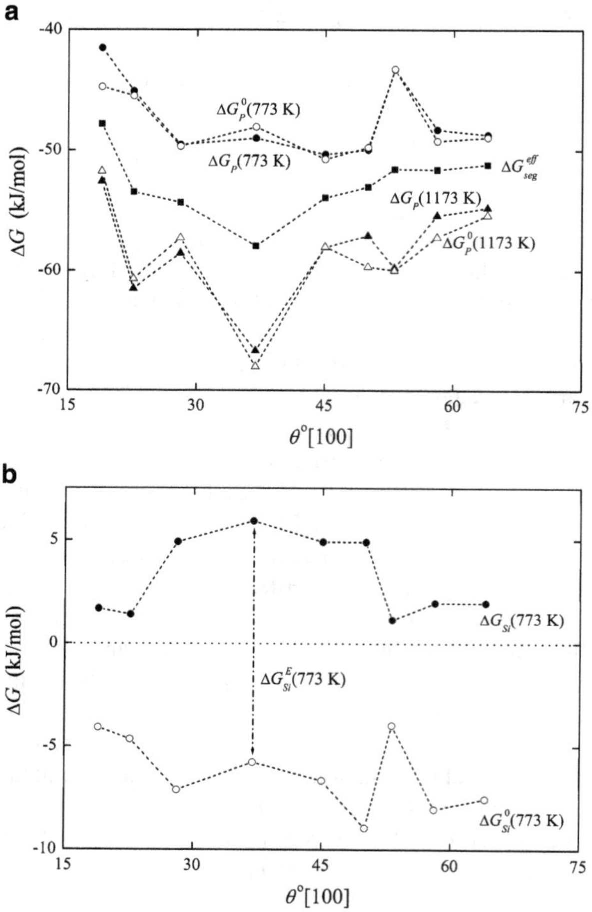


**Fig. 4.3** Anisotropy of solute segregation at [100] symmetrical tilt grain boundaries in an Fe–3.55at.%Si–0.0089at.%P–0.014at.%C alloy [19] represented by orientation dependence of different thermodynamic functions. (**a**) Gibbs energies, $\Delta G_{\mathrm{P}}^{0}$ (*empty symbols*), $\Delta G_{\mathrm{P}}$ (*solid symbols*) of phosphorus segregation at 773 K and 1,173 K. $\Delta G_{\mathrm{seg}}^{\mathrm{eff}}$ (*squares*) represents the value of $\Delta G_{\mathrm{P}}^{0}$ averaged over the same temperature range; (**b**) Gibbs energies, $\Delta G_{\mathrm{Si}}^{0}$ (*empty symbols*) and $\Delta G_{\mathrm{Si}}$ (*solid symbols*) of silicon segregation at 773 K (the value of $\Delta G_{\mathrm{Si}}^{\mathrm{E}}$ is also shown) [113]

strong limitation, it was shown in numerous papers that many systems behave practically ideal (e.g. [328, 329]). Moreover, the ideal behaviour can be reached in an infinitesimally diluted solid solution, that is when the amount of interfacial solute enrichment is very low. Therefore, $\Delta G_{I}^{0}$ characterises the segregation of an element $I$ at the most advantageous site of an interface $\Phi$ in the ideal system. The standard (molar) enthalpy, $\Delta H_{I}^{0}$, and entropy, $\Delta S_{I}^{0}$, of interfacial segregation are defined in analogy to (4.42) and have a corresponding physical meaning.

According to (4.29), constant values of $\Delta G_I^0$ and thus $\Delta H_I^0$ and $\Delta S_I^0$ define the relation between the interface activities, $a_M^\Phi$ and $a_I^\Phi$, and the bulk activities, $a_M$, and $a_I$, of solvent $M$ and solute $I$, respectively, in the whole concentration range of a binary $M$–$I$ system (see Fig. 4.3a).

In analogy to (4.30) and (4.42), the standard molar enthalpy and entropy of segregation are given by

$$\Delta H_I^0 = \left(H_{I(M)}^{0,\Phi} - H_M^{0,\Phi}\right) - \left(H_{I(M)}^0 - H_M^0\right) \tag{4.44}$$

and

$$\Delta S_I^0 = \left(S_{I(M)}^{0,\Phi} - S_M^{0,\Phi}\right) - \left(S_{I(M)}^0 - S_M^0\right) \tag{4.45}$$

which are the respective combinations of the enthalpies and entropies[1] of the pure elements $i = I, M$, related to the interface and to the bulk. According to the choice of the standard state in which both the solute and the solvent possess identical structure of the bulk and the grain boundary at a chosen $T$ and $P$, the standard thermal capacity

$$\Delta c_{p,I}^0 = \left(c_{p,I(M)}^{0,\Phi} - c_{p,I(M)}^0\right) - \left(c_{p,M}^{0,\Phi} - c_{p,M}^0\right) \equiv 0. \tag{4.46}$$

Because

$$\Delta c_{p,I}^0 = \left(\partial \Delta H_I^0 / \partial T\right) = 0 \quad \text{and} \quad \Delta c_{p,I}^0 = \left(\partial \Delta S_I^0 / \partial \ln T\right) = 0, \tag{4.47}$$

$\Delta H_I^0$ and $\Delta S_I^0$ are independent of temperature. This independence is *principal* property of $\Delta H_I^0$ and $\Delta S_I^0$: it is not the result of any averaging. This is a very important conclusion of the definition of the standard thermodynamic functions of interfacial segregation. Note that $\Delta H_I^0$ and $\Delta S_I^0$ substantially differ from any effective enthalpy and entropy of segregation, $\Delta H_{\text{seg}}^{\text{eff}}$ and $\Delta S_{\text{seg}}^{\text{eff}}$, the apparent temperature and/or concentration independence of which is frequently obtained by inappropriate averaging (see below).

For completeness, let us mention the pressure dependence. Correspondingly to the definition of $\Delta H_I^0$ and $\Delta S_I^0$ (4.44) and (4.45), the standard molar volume of segregation, $\Delta V_I^0$,

$$\Delta V_I^0 = \left(V_{I(M)}^{0,\Phi} - V_{I(M)}^0\right) - \left(V_M^{0,\Phi} - V_M^0\right). \tag{4.48}$$

According to the choice of the standard states – pure elements $I$ and $M$ at the interface and/or in the bulk at the temperature, pressure and structure of the solvent $M$, it is evident that $V_{I(M)}^{0,\Phi} = V_M^{0,\Phi}$ and $V_{I(M)}^0 = V_M^0$, and therefore, despite (4.10), $\Delta V_I^0 \equiv 0$. Since

[1] Note that $\Delta S_I^0$ includes all entropy contributions of the standard states (i.e. resulting from $\mu_i^0$), that is excluding the configuration term.

$$\left(\frac{\partial \Delta G_I^0}{\partial P}\right)_{T,n_j} = \Delta V_I^0 = 0, \tag{4.49}$$

the standard Gibbs energy of segregation is independent of pressure. Let us mention that in real systems, $\Delta V_I \neq 0$ (cf. e.g. [330]). According to definition,

$$\mathrm{d}\Delta G_I^0 = \Delta V_I^0 \mathrm{d}P - \Delta S_I^0 \mathrm{d}T + \sum_{\eta=\Phi,\nu} \sum_{i=I,M} \mu_i^\eta \mathrm{d}n_i^\eta. \tag{4.50}$$

It follows from (4.50) that

$$\left(\frac{\partial^2 \Delta G_I^0}{\partial P \partial T}\right)_{n_i} = \left(\frac{\partial \Delta V_I^0}{\partial T}\right)_{P,n_i} = -\left(\frac{\partial \Delta S_I^0}{\partial P}\right)_{T,n_i} = 0 \tag{4.51}$$

showing that $\Delta G_I^0$ and $\Delta S_I^0$ are independent of pressure. According to (4.42), it is evident that also $\Delta H_I^0$ does not depend on pressure.

Let us add for completeness that the standard Gibbs and Helmholtz energies of the segregation are identical, $\Delta G_I^0 \equiv \Delta F_I^0$. This statement follows from the basic relationship between these two thermodynamic functions, $\Delta G_I^0 = \Delta F_I^0 + P\Delta V_I^0$, where $\Delta V_I^0 \equiv 0$ as shown above.

Similarly to $\Delta G_I^0$, the standard functions $\Delta H_I^0$ and $\Delta S_I^0$ characterise interfacial segregation in an infinitesimally diluted binary solid solution. In this case, the interfacial solute enrichment is rather low. Due to their principal temperature, pressure and composition independence, $\Delta H_I^0$ and $\Delta S_I^0$ are simple but clearly defined thermodynamic functions that characterise *the tendency of the solute I to segregate at the most advantageous site of an individual grain boundary of the matrix M in an ideal system.* The actual amount of segregation at each $T-X_I$ state in a real system may then be simply determined from these values using the correction $\Delta G_I^E$ [(4.33), see below for details]. The values of $\Delta H_I^0$ and $\Delta S_I^0$ change exclusively with the structure of the grain boundary or grain boundary site (i.e. with the energy of the grain boundary (site)) and with the nature of the solvent and solute atoms. This is the great advantage of $\Delta H_I^0$ and $\Delta S_I^0$. Consequently, $\Delta H_I^0$ and $\Delta S_I^0$ can be used for general purposes, for example to characterise the general anisotropy of grain boundary segregation, which is directly related to the grain boundary classification scheme (see Chap. 2) [94].

#### 4.3.1.3 Excess Thermodynamic Functions of Interfacial Segregation

According to (4.40), the excess (molar) Gibbs energy of segregation, $\Delta G_I^E$, represents *the difference between the real behaviour and the ideal behaviour with respect to interfacial segregation.* $\Delta G_I^E$ is thermodynamically exactly defined using the activity coefficients given by (4.33). However, information about the values of the activity coefficients of the involved components in real systems is usually unknown and hardly measurable. Therefore, numerous attempts were made to find analytical expressions for this term on the basis of simplified models, which will be discussed in Sects. 4.3.3 and 4.3.4 in detail.

The role of the excess Gibbs energy of segregation, $\Delta G_I^{\mathrm{E}} = \Delta G_I - \Delta G_I^0$, as the difference between the Gibbs energy of segregation, $\Delta G_I$, and the standard Gibbs energy of segregation, $\Delta G_I^0$, is schematically shown in Fig. 4.2. In analogy, we can understand the physical meaning of the corresponding enthalpy and entropy terms. Figure 4.3 represents the experimental data of the orientation dependence of solute segregation of phosphorus and silicon at [100] symmetrical tilt grain boundaries in a multi-component Fe–Si–P–C-based alloy [20, 111] expressed in $\Delta G_I^0$ and in $\Delta G_I$. Although $\Delta G_I^{\mathrm{E}}$ is relatively small for phosphorus segregation at [100] symmetrical tilt grain boundaries (Fig. 4.3a), a pronounced effect is evident in case of silicon (Fig. 4.3b). In fact, a high value of $\Delta G_{\mathrm{Si}}^{\mathrm{E}}$ is responsible for positive values of $\Delta G_{\mathrm{Si}}$ for all grain boundaries at 773 K suggesting *depletion* of the interfaces by silicon in the multi-component system studied, that is caused by other, more interfacial active elements (P, C) in the alloy [20, 111]. However, the standard Gibbs energy of silicon segregation, $\Delta G_{\mathrm{Si}}^0$, is negative in all cases indicating that silicon principally segregates at these interfaces (when its segregation is not prevented by other solutes and/or impurities): An experimental evidence of the enrichment of the grain boundaries with silicon in pure Fe–Si-based alloys is given in [331]. As depicted in Fig. 4.3a, due to the small effect of $\Delta G_{\mathrm{P}}^{\mathrm{E}}$ on the segregation behaviour of phosphorus, the orientation dependence of both $\Delta G_{\mathrm{P}}^0$ and $\Delta G_{\mathrm{P}}$ is similar, showing on the one hand, local maxima for the $\Sigma = 5$, 36.9°[100], {013} and 53.1°[100], {012} special grain boundaries at 773 K and on the other hand, minima of $\Delta G_{\mathrm{P}}^0$ as well as of $\Delta G_P$ for these grain boundaries at 1,173 K. At the higher temperature, minima of $\Delta G_{\mathrm{P}}^0$ and $\Delta G_{\mathrm{P}}$ were also observed at the $\Sigma = 13$, 22.6°[100], {015} grain boundary (Fig. 4.3a). All these findings are in excellent agreement with the orientation dependence of $\Delta H_I^0$ (see Fig. 4.4 for silicon), which is used to characterise the basic segregation behaviour of individual grain boundaries: The {015}, {013} and {012} grain boundaries in α-iron are considered as special with principally low tendency to solute segregation (singular maxima of $\Delta H_I^0$, that is minima of the absolute values $|\Delta H_I^0|$) [94]: Actually, $\Delta H_I^0$ is the most characteristic parameter for description of the anisotropy of grain boundary segregation (Fig. 4.4) [20]. $\Delta G_{\mathrm{Si}}^0$ exhibits maxima for the 36.9°[100], {013} and 53.1°[100], {012} special grain boundaries at 773 K, while $\Delta G_{\mathrm{Si}}$ shows a minimum for the 53.1°[100], {012} but a maximum for the 36.9°[100], {013} interface (Fig. 4.3b). This result clearly indicates that the standard Gibbs energy of segregation, $\Delta G_I^0$, reflects the classification of grain boundaries while $\Delta G_I$ does not.

#### 4.3.1.4 Effective Thermodynamic Functions of Interfacial Segregation

Although the thermodynamic functions of interfacial segregation are clearly characterised, as reviewed above, some published experimental data (and sometimes

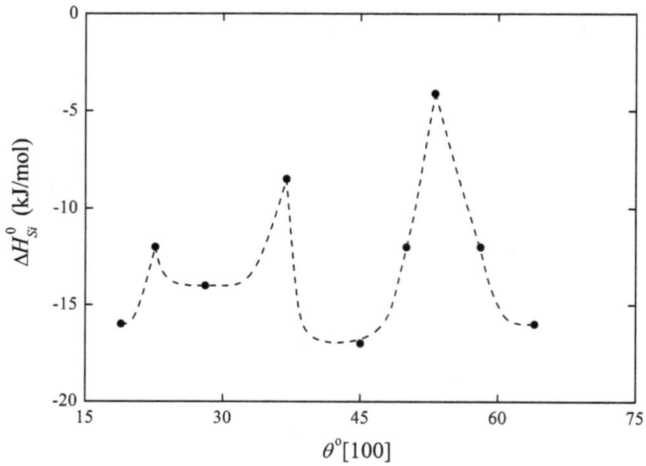


**Fig. 4.4** Anisotropy of the standard enthalpy of silicon segregation in α-iron, $\Delta H^0_{\mathrm{Si}}$. Data from [20]

also theoretical work) are not consistent with their definitions. Data of this kind are usually obtained under additional assumptions (sometimes insufficiently specified or even not specified at all) and represent averages of one of the above-mentioned types of the Gibbs energy (enthalpy, entropy) of interfacial segregation. These data may be called *effective* Gibbs energy, enthalpy and entropy of segregation, $\Delta G^{\mathrm{eff}}_{\mathrm{seg}}$, $\Delta H^{\mathrm{eff}}_{\mathrm{seg}}$ and $\Delta S^{\mathrm{eff}}_{\mathrm{seg}}$ [113, 303].

There are several ways of averaging the data to obtain effective thermodynamic functions of interfacial segregation:

1. Determination of $\Delta G^{\mathrm{eff}}_{\mathrm{seg}}$ from the averaged chemical composition of the interfaces (i.e. neglecting the spatial distribution of segregated species) according to the segregation isotherms [(4.32) and (4.34)]
2. Averaging the values of $\Delta G_I$ for differently concentrated alloys of the same system using the ideal approximation for real systems
3. Averaging the values of $\Delta G_I$ ($\Delta H_I$ or $\Delta S_I$) or $\Delta G^0_I$ over a temperature range (i.e. neglecting the entropy term)
4. Averaging the values of $\Delta G_I$ or $\Delta G^0_I$ over various interfaces/sites (i.e. neglecting the anisotropy of interfacial segregation)

All the above-listed types result in averaging the values of segregation enthalpy and entropy, thus providing effective functions, $\Delta H^{\mathrm{eff}}_{\mathrm{seg}}$ and $\Delta S^{\mathrm{eff}}_{\mathrm{seg}}$ [303]. It is obvious that – due to the randomly chosen averaging (cf. Fig. 4.5) – all these thermodynamic parameters have no clear physical meaning as has been frequently emphasised in literature [113, 303, 332]. Owing to this averaging, these ill-defined thermodynamic functions apparently possess similar properties as the standard thermodynamic functions $\Delta H^0_I$ and $\Delta S^0_I$ – especially the independence of concentration and/or

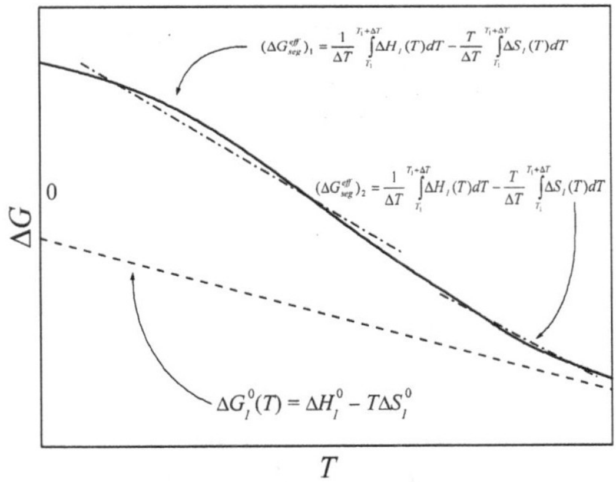


**Fig. 4.5** Schematic comparison of the standard Gibbs energy of segregation, and effective Gibbs energy of segregation showing apparently similar behaviour (i.e. apparently constant $\Delta H^{\text{eff}}_{\text{seg}}$ and $\Delta S^{\text{eff}}_{\text{seg}}$) but obviously their different values

temperature. Because of this apparent but artificial similarity, these two different types of thermodynamic functions are sometimes used in place of each other and, therefore, they are misinterpreted. The main differences between the effective and standard Gibbs energies, enthalpies and entropies of segregation are discussed in the following according to the above-mentioned items.

1. *Determination of $\Delta G^{\text{eff}}_{\text{seg}}$ from averaged chemical composition of the interfaces (i.e. neglecting the spatial distribution of segregated species) according to the segregation isotherms* [(4.32) and (4.38)]. This problem of averaging is closely connected with the way of obtaining the experimental data on grain boundary concentration. As discussed in Chap. 3, the experimental data can be basically obtained in two major ways according to the analytical tools employed, namely by means of (a) microscopic techniques and (b) surface analysis techniques. In contrast to very localised and sensitive microscopic techniques the optimum lateral and depth resolutions (typically of about 10 nm and 1 nm, respectively) as well as the analytical sensitivity (0.1–1 at.%) in AES [119] are worse, and only allow to obtain a convolution of the chemical composition of typically the top 3–10 atomic layers at the surface, and averaged over a relatively large area ($\geq 10 \times 10\,\text{nm}^2$). When such "effective" concentrations are used automatically to determining the values of the thermodynamic functions of segregation, the above-mentioned effective values of $\Delta H^{\text{eff}}_{\text{seg}}$ and $\Delta S^{\text{eff}}_{\text{seg}}$ are obtained, which are averaged over an interface volume in which the segregation is confined. Indeed, the significance of these values is low if any at all.

However, there already exist techniques such as angle-resolved AES and XPS [333] and sputter depth profiling, which enable careful deconvolution of the data [333, 334], thus providing additional information about the extent of the segregation and its confinement to the chemical composition of the top interface layer (e.g. [111]). From such kind of data, the true values of $\Delta G_I$ (4.38) and also

$\Delta S_I = -(\partial \Delta G_I/\partial T)$ and $\Delta H_I = \Delta G_I + T\Delta S_I$ are obtained that represent the correct characterisation of each point in the $T-X_I$ space in agreement with the above analysis (see Fig. 4.2). With an appropriate correlation of the term, $\Delta G_I^{\mathrm{E}}$ (see Sects. 4.3.3 and 4.3.4), it is possible to extract the values of $\Delta G_I^0$ at individual temperatures and thus to determine the values of $\Delta H_I^0$ and $\Delta S_I^0$ from measurements of grain boundary segregation including relatively higher solute concentrations or even in multi-component systems [94, 111]. If the measurements of the solute segregation are performed at a single grain boundary of a bicrystal in a dilute system with low level of $X_I^{\Phi}$ and if the quantification is performed in a correct way, the values of $\Delta H_I^0$ and $\Delta S_I^0$ are obtained describing the segregation at the energetically most favourable site of that boundary (see above) [94, 111]. The data obtained in this way are the *true* values of $\Delta H_I^0$ and $\Delta S_I^0$. Their values are no averages and they are principally independent of concentration and temperature.

2. *Averaging the values of $\Delta G_I$ for differently concentrated alloys of the same system (i.e. using the ideal approximation for real systems).* As mentioned above, substantial difference can exist between the values of $\Delta G_I$ and $\Delta G_I^0$ that is related to the excess contribution, $\Delta G_I^{\mathrm{E}}$ (4.40). In some cases, the measured experimental data are incorrectly interpreted by the authors who average the values of $\Delta G_I$ and take them for $\Delta G_I^0$. As an example, let us address this case in a study of grain boundary segregation of antimony in iron [335]. The temperature dependence of grain boundary composition was measured in three Fe–Sb alloys containing 0.0055at.%, 0.0225at.% and 0.0414at.% of antimony. The summary data on temperature dependence were correlated by a single pair of the segregation enthalpy and entropy, $\Delta H_{\mathrm{Sb}} = -19\,\mathrm{kJ/mol}$ and $\Delta S_{\mathrm{Sb}} = +28\,\mathrm{J/(mol\,K)}$, respectively. These values differ substantially from those of phosphorus, $\Delta H_{\mathrm{P}}^0 = -34.3\,\mathrm{kJ/mol}$ and $\Delta S_{\mathrm{P}}^0 = +21.2\,\mathrm{J/(mol\,K)}$, measured by Erhart and Grabke [328]. Although antimony and phosphorus exhibit similar solid solubility in α-iron (3.3at.%Sb and 4.2at.%P), the above data suggest much lower segregation of antimony as compared to phosphorus. Based on this discrepancy, Briant [336] casts doubts on the general validity of the inverse relationship between the extent of interfacial segregation of an element and its solid solubility as proposed by Seah [138] (see also Chap. 5 [337]). However, a detailed analysis [338] of the data on antimony segregation [335] revealed that the values of the above-given enthalpy and entropy of segregation are $\Delta H_{\mathrm{seg}}^{\mathrm{eff}}$ and $\Delta S_{\mathrm{seg}}^{\mathrm{eff}}$, because the grain boundary composition of each Fe–Sb alloy exhibits its own temperature dependence (see Fig. 4.6a). To determine the values of $\Delta H_{\mathrm{Sb}}^0$ and $\Delta S_{\mathrm{Sb}}^0$ that do not depend on concentration and temperature, a limited amount of the grain boundary sites has to be taken into account. That means $X^{0\Phi} < 1$ in (4.39). As seen in Fig. 4.6b, the re-evaluation of the experimental data using the value $X^{0\Phi} = 0.27$ as the best-fit [338] provides an optimum correlation of original data [335]. The corresponding values of the standard enthalpy and entropy of grain boundary segregation of antimony are then $\Delta H_{\mathrm{Sb}}^0 = -23\,\mathrm{kJ/mol}$ and $\Delta S_{\mathrm{Sb}}^0 = +37\,\mathrm{J/(mol\,K)}$ [338].

With the above correction, the value of $\Delta H_{\mathrm{Sb}}^0$ is closer to $\Delta H_{\mathrm{P}}^0$ although quantitative agreement is still not good enough. However, we have to consider that AES

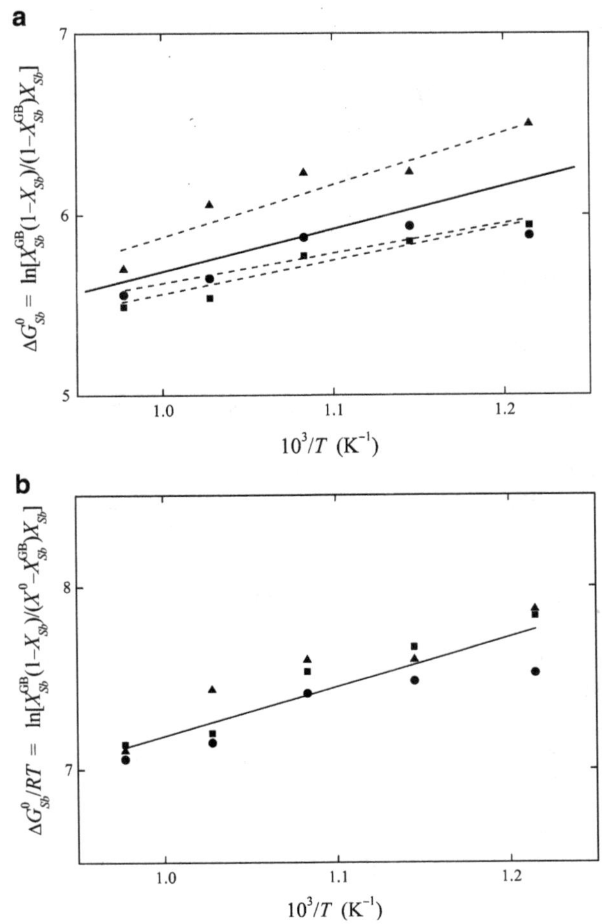


**Fig. 4.6** Plot of experimental data on grain boundary segregation of antimony in polycrystalline bcc iron. Bulk concentration of antimony: 0.0055at.%Sb (triangles), 0.0225at.%Sb (circles), 0.0431at.%Sb (squares). (**a**) original data [335] correlated (*solid line*) with $X^0 = 1$, (**b**) re-evaluated data using $X^0 = 0.27$ showing independence of $\Delta G^0_I$ of bulk concentration [338]

measurements were done at Fe–Sb polycrystals with unknown character of the fractured grain boundaries (see below): If the fracture surface contains a substantial part of special interfaces (see Chap. 2), the value of $\Delta H^0_{Sb}$ reflects this fact and its absolute value is reduced as compared to the case where only general grain boundaries are present on the fracture surface. (Frankly speaking, they also represent the effective values...) The good fit of the pair $\Delta H^0_{Sb}$ and $\Delta S^0_{Sb}$ in the compensation effect (see Chap. 5) supports the assumption of a considerable amount of special interfaces in the experiment [335]. This result substantially weakens the Briant arguments [336] against the inverse relationship between the strength of interfacial segregation and the bulk solid solubility.

3. *Averaging the values of $\Delta G_I(\Delta H_I$ or $\Delta S_I)$ or $\Delta G_I^0$ over a temperature range (i.e. neglecting the entropy term).* In some papers, the values of $\Delta H_I$ and $\Delta S_I$ or even the values of $\Delta G_I$ are averaged as [113]

$$\Delta G_{\mathrm{seg}}^{\mathrm{eff}} = \frac{1}{\Delta T} \int_{T}^{T+\Delta T} \Delta G^{\mathrm{xs}}(T)\mathrm{d}T = \mathrm{const.} \tag{4.52}$$

This averaging strongly simplifies the application of the segregation isotherms because single constant parameters are used to determine the chemical composition of interfaces at different temperatures. However, the resulting values of the effective thermodynamic functions of segregation have no physical meaning. If the Gibbs energy is averaged, the result should be considered to be the effective enthalpy of segregation, because the segregation entropy is completely neglected. For example, the segregation of phosphorus [339, 340], carbon [339], as well as silicon [340] in α-iron was evaluated in this way. As seen in Fig. 4.3a, the value of $\Delta G_{\mathrm{seg}}^{\mathrm{eff}}$ for phosphorus, averaged over the temperature range 773–1,173 K differs substantially from the values of $\Delta G_{\mathrm{P}}^0$ and $\Delta G_{\mathrm{P}}$. Let us repeat that in general, $\Delta H_I$ and $\Delta S_I$ depend on temperature and concentration. To overcome this complication, some authors average $\Delta H_I$ and $\Delta S_I$ over temperature (e.g. [341]) and/or concentration ranges in various ways analogously to $\Delta G_{\mathrm{seg}}^{\mathrm{eff}}$ in (4.52). Then, the segregation isotherm (4.32) for a binary system transforms into

$$\frac{X_I^{\Phi}}{1 - X_I^{\Phi}} = \frac{X_I}{1 - X_I} \exp\left(-\frac{\Delta H_{\mathrm{seg}}^{\mathrm{eff}} - T\Delta S_{\mathrm{seg}}^{\mathrm{eff}}}{RT}\right). \tag{4.53}$$

It is clear that the effective values of $\Delta H_{\mathrm{seg}}^{\mathrm{eff}}$ and $\Delta S_{\mathrm{seg}}^{\mathrm{eff}}$ are physically meaningless [113, 303] and (4.53) only represent an empirical correlation of the data. Nevertheless, their constant values are apparently similar to those of $\Delta H_I^0$ and $\Delta S_I^0$ although a fundamental difference exists between these two pairs of the thermodynamic functions as is clearly seen from their schematic comparison in Fig. 4.5. Despite this fact, these functions are sometimes misinterpreted. This is the important misunderstanding, which appears in literature (cf. [303]).

4. *Averaging the values of $\Delta G_I$ or $\Delta G_I^0$ over various interfaces/sites (i.e. neglecting anisotropy of interfacial segregation).* Frequently, the averaging of the values of $\Delta G_I$ or $\Delta G^0$ over numerous (usually non-specified) interfaces is applied, because many thermodynamic data on grain boundary segregation are extracted from AES measurements on polycrystalline samples (cf. [342]). In this case, the composition of different grain boundaries is measured in different samples at different temperatures, and the values of the segregation enthalpy and entropy (although considered as $\Delta H_I^0$ and $\Delta S_I^0$) cannot be ascribed to a single grain boundary or even site.

Let us briefly analyse the relation between the true values of the standard enthalpy and entropy with their effective counterparts. In a polycrystalline aggregate, we can express the segregation to each grain boundary in analogy to (4.32) in the simplest case of the binary ideal system as

$$\frac{X_{Ik}^{\Phi}}{X_{Mk}^{\Phi}} = \frac{X_I}{X_M} \exp\left(-\frac{\Delta G_{Ik}^0}{RT}\right), \tag{4.54}$$

where index $k$ denotes the particular grain boundary. For the averaged composition, $X_{I,\text{eff}}^{\Phi}$, over all measured grain boundaries, we can write

$$\frac{X_{I,\text{eff}}^{\Phi}}{X_{M,\text{eff}}^{\Phi}} = \frac{\sum_{k=1}^{m} X_{Ik}^{\Phi}}{\sum_{k=1}^{m} X_{Mk}^{\Phi}} = \frac{X_I}{X_M} \exp\left(-\frac{\Delta G_{I,\text{eff}}^0}{RT}\right) \tag{4.55}$$

with the effective Gibbs energy of segregation, $\Delta G_{I,\text{eff}}^0$. We can write (4.55) as

$$\Delta G_{Ik}^0 = -RT \ln \frac{X_{Ik}^{\Phi}}{X_{Mk}^{\Phi}} + RT \ln \frac{X_I}{X_M}. \tag{4.56}$$

Then

$$\Delta G_{I,\text{eff}}^0 = -RT \ln \frac{X_{I,\text{eff}}^{\Phi}}{X_{M,\text{eff}}^{\Phi}} + RT \ln \frac{X_I}{X_M}. \tag{4.57}$$

In (4.57),

$$\frac{X_{I,\text{eff}}^{\Phi}}{X_{M,\text{eff}}^{\Phi}} = \frac{\prod_{k=1}^{m} X_{Ik}^{\Phi}}{\prod_{k=1}^{m} X_{Mk}^{\Phi}} = \frac{X_I}{X_M} \exp\left(-\frac{\Delta G_{I,\text{eff}}^0}{RT}\right). \tag{4.58}$$

To get the single value of $\Delta G_{I,\text{eff}}^0$, (4.55) and (4.58) should be identical. This condition is only fulfilled when the arithmetic and geometric averages of the interfacial concentrations over the studied grain boundaries are identical. It occurs when (a) a single grain boundary (in a bicrystal) is measured, or (b) the grain boundaries have identical composition and, consequently, are characterised by single value (or very close values) of $\Delta G_{Ik}^0$ for all $k$ [342]. The latter condition seems to be rather restrictive. Nevertheless, it can be approximately met even in polycrystalline materials when studying fracture surfaces with, for example AES. Because in general, decohesion is increasing with the amount of interfacial solute enrichment, it is most probable that the highly enriched general grain boundaries are opened during the fracture. The segregation levels at these boundaries at a particular temperature are very similar and their values of $\Delta G_{Ik}^0$ ($\Delta H_{Ik}^0$ and $\Delta S_{Ik}^0$) are close to each other (e.g. [94]). Therefore, the thermodynamic functions obtained from AES measurements on polycrystalline samples, which are definitely of effective character, can serve as a rough estimate of the standard enthalpy and entropy of segregation at the general grain boundaries. Nevertheless, it is necessary to keep in mind all the assumptions leading to this conclusion.

A final critical remark seems to be necessary for this section. In a few publications, the data for $\Delta G_I$, $\Delta H_I$ and $\Delta S_I$ are replaced by temperature and concentration independent "effective" values (e.g. [335,339,343,344]), here denoted as $\Delta G_{\mathrm{seg}}^{\mathrm{eff}}$, $\Delta H_{\mathrm{seg}}^{\mathrm{eff}}$ and $\Delta S_{\mathrm{seg}}^{\mathrm{eff}}$ (cf. [303]). Because their derivation is based on generally wrong assumptions or illicit simplifications, the effective values have no physical meaning as correctly pointed out in [303]. Fortunately, many of the published data on segregation represent the well-defined thermodynamic standard functions of interfacial segregation, $\Delta G_I^0$, $\Delta H_I^0$ and $\Delta S_I^0$. The standard enthalpy, $\Delta H_I^0$, and entropy, $\Delta S_I^0$, of solute segregation are *principally* independent of temperature and concentration. The reader should pay attention to the real meaning of each set of published thermodynamic data on interfacial segregation and let us repeat it here again, should not mistake the fundamental thermodynamic standard functions $\Delta H_I^0$ and $\Delta S_I^0$ for the meaningless $\Delta H_{\mathrm{seg}}^{\mathrm{eff}}$ and $\Delta S_{\mathrm{seg}}^{\mathrm{eff}}$ with apparently similar properties (i.e. concentration and temperature independence, artificially obtained by averaging). An obvious lack of distinction between these two types of thermodynamic functions led some authors to the recommendation that any constant (i.e. temperature and concentration independent) values of segregation enthalpy and entropy should be avoided in interpreting measurements of anisotropy of interfacial segregation because of missing physical significance (e.g. [303]). This erroneous conclusion has to be principally rejected. During the past 20 years, careful experimental work resulted in numerous publications on anisotropy of interfacial segregation based on cautiously determined values of thermodynamic standard state functions of segregation, $\Delta H_I^0$ and $\Delta S_I^0$ (e.g. [20, 94, 111]).

### 4.3.2 *Langmuir–McLean Model*

The original derivation of the simplest form of the segregation isotherm was made by McLean [19]. In his classical approach, $C$ solute atoms are distributed randomly among $N$ lattice sites and $c$ solute atoms are also randomly distributed among $n$ independent grain boundary substitution sites. The internal energy $U$ of this binary ideal system containing the solute atoms is [138]

$$U = cu_1 + CU_1 - kT\,[\ln n!N! - \ln(n-c)!c!(N-C)!C!]\,, \tag{4.59}$$

where $U_1$ and $u_1$ are the internal energies of the solute atom in the lattice and in the grain boundary, respectively, and $k$ is the Boltzman constant. The term at the end of the right-hand side of (4.59) is related to the configurational entropy of the random arrangement of the solute atoms in the bulk and grain boundary. The equilibrium state of the system occurs at the minimum value of $U$. This may be determined by differentiating $U$ with respect to $c$, noting that the sum of $c + C$ remains constant. The grain boundary analogue of the Langmuir adsorption to the free surfaces is obtained,

$$\frac{X^{\Phi}}{1 - X^{\Phi}} = \frac{X}{1 - X}\exp\left(-\frac{\Delta u}{kT}\right). \tag{4.60}$$

In (4.60), it is supposed that $c \ll n$ and $C \ll N$. Then $X^{\Phi} = c/n$ and $X = C/N$. $\Delta u = u_1 - U_1$ is the molar internal energy of segregation. Let us stress out that according to (4.60), the segregation is driven by changes of the internal energy while in the binary real analogue, (4.32), this term is replaced by the Gibbs energy of segregation. It means that in (4.60), all entropy contributions such as anharmonic, vibration, etc. are neglected except the configurational entropy, $k \ln[(X^{\Phi}(1 - X))/(X(1 - X^{\Phi}))]$, which is used for determination of the grain boundary composition. From this point of view, the segregation isotherm expressed by (4.32) is more general, because it was derived without any additional assumption concerning the distribution of the species in the system. Equation (4.32) also considers – besides the configurational term – other contributions to the segregation entropy, which were shown in many cases to play an important role in grain boundary segregation [20, 94, 327, 341, 345].

Equation (4.60) is the well-known Langmuir–McLean segregation isotherm. In its derivation it is assumed that all grain boundary positions are substitution sites and are available for segregation. In some cases, only a submonolayer fraction may be available for grain boundary segregation in saturation. As shown by Hondros and Seah [308], the Langmuir–McLean isotherm should be rewritten as

$$\frac{X_i^{\Phi}}{X^{0\Phi} - X_i^{\Phi}} = \frac{X_i}{1 - X_i}\exp\left(-\frac{\Delta g}{kT}\right) \tag{4.61}$$

where the internal energy of segregation was already replaced by corresponding Gibbs energy. The same result can be obtained from (4.32) and (4.33) supposing a binary system ($X_I = 1 - X_M$) with ideal behaviour ($X_i = a_i$, i.e. $\gamma_i = 1$ and thus $\Delta\overline{G}_I^{\mathrm{E}} = 0$) and considering the saturation limit $X^{0\Phi}$. For example, the average saturation level, $X^{0\Phi}$, of grain boundaries in copper for antimony segregation was found to vary between 0.38 and 0.65 of Sb monolayer [346] and 0.27 for segregation of antimony in bcc iron [338]. The Langmuir–McLean segregation isotherm was successfully applied to the description of the equilibrium composition of the grain boundaries in a spectrum of binary systems such as bcc iron with molybdenum and niobium [231].

Besides the substitutional sites, however, the grain boundary also contains interstitial sites. These sites can be understood as the sites, which cannot be occupied by the matrix atoms in equilibrium but may accept the atoms of other solutes. Due to the atomic structure of these interfaces, the free volume among equilibrium atom positions is large enough to accept – besides typical interstitial atoms – even much larger atoms that occupy the substitutional sites in regular crystal lattice. For example, phosphorus, tin and antimony in bcc iron and bismuth in copper are substitutional solutes, however, they segregate at interstitial positions. To describe the interstitial segregation, we will assume that the boundary in a binary system is composed of two sublattices. The "substitution" sublattice is completely filled by the atoms of the base element. The interstitial positions in matrix material, $M_{\square}$, are partially filled by the atoms of the solute $I$ so that there occurs an exchange $I + M_{\square}^{\Phi} \leftrightarrow I^{\Phi} + M_{\square}$. Defining the chemical potentials and activities for individual components including the unoccupied interstitial positions $M_{\square}$ in both, the grain boundary and the bulk, we can write an expression principally analogous to (4.29),

$$\frac{a_I^{\Phi}}{a_{M_{\square}}^{\Phi}} = \frac{a_I}{a_{M_{\square}}} \exp\left(-\frac{\Delta G_I^0}{RT}\right). \tag{4.62}$$

In this case, $\Delta G_I^0$ the standard molar Gibbs energy of segregation is defined as $\Delta G_I^0 = \left(\mu_I^{0,\Phi} + \mu_{M_{\square}}^0\right) - \left(\mu_I^0 + \mu_{M_{\square}}^{0,\Phi}\right)$. $\mu_{M_{\square}}^{0,\Phi}$ and $\mu_{M_{\square}}^0$ are the respective chemical potentials of the grain boundary vacancies $M_{\square}^{\Phi}$ and of the volume vacancy or substitutional atom depending on whether the solute is substitutional or interstitial in volume [347].

In case of interstitial segregation, however, not all grain boundary positions can be occupied by the solute segregating at interstitial positions: the substitutional positions in the grain boundary always remain occupied by the matrix atoms. The interstitial segregation is completed when all allowed interstitial positions are occupied by the segregated species. Thus, the "usual" segregation occurs in the interstitial sublattice,

$$\frac{\theta_I^{\Phi}}{\theta_{M_{\square}}^{\Phi}} = \frac{X_I}{X_{M_{\square}}} \exp\left(-\frac{\Delta G_I^0 + \Delta G_I^{\mathrm{E}}}{RT}\right), \tag{4.63}$$

where $\theta_i^{\Phi}$ is the atomic concentration of segregated interstitial solute $I$ or fraction of free interstitial positions $M_{\square}$ in the interstitial sublattice. $\theta_I^{\Phi}$ is related to the total concentration at the grain boundary by

$$\theta_I^{\Phi} = \frac{X_I^{\Phi}}{X_{\mathrm{int}}^{0\Phi}} \tag{4.64}$$

with $X_{\mathrm{int}}^{0\Phi}$ being the fraction of the interstitial positions available for solute segregation at the grain boundary. We can then rewrite (4.63) as

$$\frac{X_I^{\Phi}}{X_{int}^{0\Phi} - X_I^{\Phi}} = \frac{X_I}{1 - X_I} \exp\left(-\frac{\Delta G_I^0 + \Delta G_I^{\mathrm{E}}}{RT}\right), \tag{4.65}$$

which is similar to the Hondros and Seah formulation of the segregation isotherm (4.61) [173, 308] supposing $\Delta G_I^{\mathrm{E}} = 0$ (i.e. ideal behaviour) [316].

To describe the grain boundary segregation rigorously in an ideal multicomponent system, (4.32) can be used with $\Delta G_I^{\mathrm{E}} = 0$,

$$\frac{X_I^{\Phi}}{1 - \sum\limits_{i=1}^{M-1} X_i^{\Phi}} = \frac{X_I}{1 - \sum\limits_{i=1}^{M-1} X_i} \exp\left(-\frac{\Delta G_I^0}{RT}\right). \qquad (4.66)$$

Expression (4.66) is often used to describe so-called *site competition* of segregating solutes: The site competition effect is considered to be involved in the sum $\sum_{i=1}^{M-1} X_i^{\xi}$. The term $1 - \sum_{i=1}^{M-1} X_i^{\xi}$, however, only denotes the concentration of the matrix element at the grain boundary ($\xi = \Phi$) or in the volume. However, in some cases, the segregation of an element is unaffected by other present elements at all. This is, for example, for elements that segregate at different boundary sites (interstitial and substitution sites, see Sect. 4.3.3). Then the system is considered as "pseudobinary" and (4.60) or (4.61) can be applied to describe the concentration of the segregated impurity. Such case was found, for example for phosphorus segregation in a low-alloy steel [329]. In real systems, however, the interaction between the solute atoms may play a role and two solutes can be mutually repulsed or attracted despite they segregate at substitution or interstitial positions [348]. Sometimes, it is not easy to distinguish between repulsive interaction and site competition [349]: only in the case when the atoms of both solutes occupy different grain boundary sites (for example silicon and phosphorus), the repulsive interaction can be unambiguously proved [350]. For more details see Sects. 4.3.3 and 4.3.4.

Grain boundary "site competition" has been reported between nitrogen and sulphur in bcc iron [351] but not under all conditions [352]. The most important example of the site competition is the replacement of phosphorus by carbon at grain boundaries in bcc iron [328, 339, 353–356]. Further published examples of this phenomenon are listed in Table 4.1. However, in some cases such as Ni–C the site competition is hardly acceptable when taking into account that both elements segregate at different sites in bcc iron matrix (nickel segregates at substitutional sites while carbon at interstitial positions) [98, 375].

Langmuir–McLean segregation isotherms represent historically the first model for the description of equilibrium segregation at solid interfaces in dependence on the bulk composition and temperature. In principle, it should be successfully applied to describe the grain boundary segregation in all binary and pseudobinary systems dilute enough to exhibit ideal behaviour. It has been frequently used to characterise grain boundary segregation in such systems, for example, it was used to correlate grain boundary segregation of phosphorus [329], tin [123] and sulphur [376] in bcc iron, phosphorus, carbon and boron [344], manganese and phosphorus [377], manganese, phosphorus and boron [378] and phosphorus [379] in fcc-based iron systems including 17Cr–12Ni stainless steels, indium and sulphur in nickel [334] and platinum, palladium and rhodium in gold [380].

**Table 4.1** Examples of site-competitive grain boundary segregation

| Matrix | Competitively segregating elements | Reference |
|---|---|---|
| α-Fe | P–S | [357–361] |
| | P–C | [328, 339, 353, 354, 360, 362–367] |
| | P–N | [360, 363] |
| | P–B | [367] |
| | P–Sb | [357] |
| | N–C | [360] |
| | S–C | [351, 354, 360, 368, 369] |
| | S–N | [351, 360, 368] |
| | Sb–C | [370] |
| | Sn–S | [367, 371] |
| Mo | C–O | [130] |
| W | P–C | [372] |
| γ-Fe | P–N | [373] |
| Ni | P–S, Sb–S | [374] |

Phosphorus is probably the most frequently examined element in respect to the grain boundary segregation in α-iron. This interest was evoked by its large embrittling potency in iron and steels. As mentioned above, the first quantitative study of the grain boundary segregation by AES was focused just on phosphorus segregation in low-alloy ferrite steels [121]. In the first thorough quantitative study of grain boundary segregation of phosphorus in α-iron, Erhart and Grabke [328] studied chemical composition of the grain boundaries in polycrystalline iron containing various volume concentrations of phosphorus ranging from 0.003 to 0.33 mass%, at temperatures between 400 and 800°C. Using the Langmuir–McLean segregation isotherm (4.60) to correlate the experimental data, they evaluated $\Delta H_P^0 = -34.4\,\mathrm{kJ/mol}$ and $\Delta S_P^0 = +21.5\,\mathrm{J/(mol\,K)}$. Argon ion sputtering of the fracture surfaces as well as FIM studies confirmed that the segregation effects are confined in a single or merely few monatomic layers along the grain boundary [198].

Similarly to the Fe–P system, a detail study was performed to evaluate the sulphur grain boundary segregation in polycrystalline α-iron by Briant on the alloys ranging from 0.0035 to 0.013 at.%S in volume at temperatures 400, 480 and 550°C. It was shown that maximum grain boundary concentration of sulphur appears at intermediate temperature. It reflects complex effect of the segregation tendencies and strong decrease of the bulk solid solubility of sulphur with decreasing temperature. At lower temperatures, the amount of sulphur in bulk ferrite solution is substantially reduced: it reflects in the amount of this element at the grain boundaries. The enthalpy of sulphur grain boundary segregation in α-iron was determined to be $\Delta H_S^0 = -51.5\,\mathrm{kJ/mol}$ [381]. However, its bulk content is generally strongly suppressed by presence of manganese and thus, by precipitation of MnS resulting from strong attractive interaction [313]. Thus, the bulk concentration of sulphur in steels is so low that no extended sulphur grain boundary segregation is observed [123]. Similar example of the "scavenging" of soluble impurities in iron is the precipitation of iron oxides due to the low solubility of oxygen in iron: then the amount of soluble oxygen is too small to segregate in iron [313, 382].

Grain boundary segregation of tin was found to be saturated well below a monolayer [123], although it exhibits a large scatter of measured values. This was ascribed to the strong effect of anisotropy of grain boundary segregation (see Chap. 7). The values of the thermodynamic parameters of the grain boundary segregation were determined on basis of the Langmuir–McLean segregation isotherm, (4.66), as $\Delta H^0_{\mathrm{Sn}} = -22.5$ kJ/mol and $\Delta S^0_{\mathrm{Sn}} = +26.1$ J/(mol K) [123]. These data are in good agreement with the measurements of Seah and Hondros [307] providing the value $\Delta G^0_{\mathrm{Sn}}$ (823 K) $= -44$ kJ/mol. However, tin exhibits large tendency to surface segregation and it is considered that this effect is responsible for temper embrittlement of low alloy steels by tin. Tin is allowed to segregate at the surfaces of the voids and cavities inside the material upon creep as well as in the intergranular crack front at intermediate temperatures during dynamic embrittlement similarly to copper-based alloys [383] (see also Chap. 7).

An interesting example of the limited sites allowed for segregation is the grain boundary chemistry in ordered intermetallics. Principally, the solutes and impurities segregate at the grain boundaries in ordered alloys in very low concentrations [127, 347, 384]. Therefore, we can well assume that the segregation level is controlled by ordering of the system and then, we can use (4.61) to describe this kind of grain boundary segregation. Similarly to free surfaces [385], grain boundaries in ordered systems may also exhibit extended or less extended ordering tendencies. Let us assume that no segregation is allowed at the completely ordered grain boundary while the segregation at the disordered parts exhibits the Langmuir–McLean behaviour. Accepting that the portion of the ordered grain boundary region can be described by the grain boundary order parameter $m^{\Phi}(T)$, the average grain boundary concentration $X^{\Phi}_I$ can be expressed as [179]

$$X^{\Phi}_I(T) = X^{\Phi\mathrm{D}}_I(T)[1 - m^{\Phi}(T)] + X_I m^{\Phi}(T), \tag{4.67}$$

where $X^{\Phi\mathrm{D}}_I$ is the grain boundary concentration of a solute $I$ in conditions of disordered solid solution (i.e. the Langmuir–McLean behaviour). Combining (4.61) and (4.67) and introducing a general concentration

$$Y^{\Phi}_I \equiv X^{\Phi}_I - X^{\Phi}_{Im}(T), \tag{4.68}$$

we obtain

$$\frac{Y^{\Phi}_I}{X^{0\Phi} - Y^{\Phi}_I} = \frac{X_I}{1 - X_I}\exp\left(-\frac{\Delta G^0_I}{RT}\right), \tag{4.69}$$

where

$$X^{0\Phi} = 1 - m^{\Phi}(T). \tag{4.70}$$

If the product $X_i m^{\Phi}(T)$ is small in comparison to $X_I^{\Phi}$, (4.69) transforms onto (4.61). It means that for solute segregation at partially ordered grain boundaries, $X^{0\Phi}$ is equal to the portion of the disordering of the grain boundary at a given temperature, characterised by the grain boundary order parameter $m^{\Phi}(T)$. In other words, the ordered material can be considered as a chemical compound with limited free species as well as positions available for segregation.

The parameter, $m^{\Phi}(T)$, can be evaluated according to the expression [385, 386]

$$m^{\Phi}(T) = \left(\frac{T_c^{\Phi} - T}{T_c^{\Phi}}\right)^{\beta^{\Phi}}, \tag{4.71}$$

where $T_c^{\Phi}$ is the grain boundary ordering temperature, and $\beta^{\Phi}$ is the grain boundary exponent. The value of $\beta^{\Phi}$ lies between those of the bulk, $\beta^{\text{bulk}} = 0.3$, and of the surface, $\beta^{\text{surface}} = 0.8$ [179, 385]. We can well suppose – considering the number of the broken bonds of the atoms in various surroundings – that $\beta^{\Phi}$ will possess an intermediate value, probably closer to that of the bulk. The value of the grain boundary ordering temperature, $T_c^{\Phi}$, can differ from that of the bulk, $T_c$ [385]. In fact, both relationships, $T_c^{\Phi} > T_c$ and $T_c^{\Phi} < T_c$ are possible. Generally, the value of $T_c^{\Phi}$ may change with grain boundary type and orientation. The values are expected to be closer to $T_c$ in the case of special grain boundaries as compared to general ones. Since the values of $T_c^{\Phi}$ representing a non-clearly-defined average value over all interfaces in a polycrystal, are not known, this term represents a fitting parameter. However, it has to keep its physical meaning as the temperature: Therefore, only $T_c^{\Phi}$ values comprised in the range (0, $T_m$), where $T_m$ is the melting temperature, are accepted.

Let us now point an interesting consequence of existence of different saturation limits. Supposing the limited number of positions for segregation, $X_1^0$, as described for example by (4.65), we can add a further portion of attainable positions, $X_2^{0\Phi}$, as limited by another source. Then, the left-hand side of (4.65) will change to $\theta_I^{\Phi}/\left(X_2^{\Phi} - \theta_I^{\Phi}\right)$. According to the definition of $\theta_I^{\Phi} = X_I^{\Phi}/X_1^0$ (e.g. (4.64)), we can write it as $X_I^{\Phi}/\left(X_1^{0\Phi} X_2^{0\Phi} - X_I^{\Phi}\right)$ and, in general, it results in

$$X_{\text{total}}^{0\Phi} = X_1^{0\Phi} \times X_2^{0\Phi} \times \cdots \times X_k^{0\Phi} = \prod_{i=1}^{k} X_i^{0\Phi}. \tag{4.72}$$

In both the stoichiometric $Ni_3Al$-based intermetallics and in the Al-rich $Ni_3Al$, no segregation effects have been observed [387, 388]. In Ni-rich $Ni_3Al$ alloys, boron segregation is observed accompanied probably by nickel co-segregation. Maximum reported amount of 14at.% of boron was detected at grain boundaries of a Ni–24at.%Al(0.1at.%B) alloy at 1,273 K [389]. Boron segregation in Ni-rich $Ni_3Al$ was correlated by $\Delta H_B = -11$ kJ/mol and $\Delta S_B = 22$ J/(mol K) for the alloy containing 0.048 at.% of boron, while $\Delta H_B = -10$ kJ/mol and $\Delta S_B = 9$ J/(mol K) were found for the material with 0.48at.% of boron. In all these measurements, very low levels of boron segregation of about 1–2 at.% were found [390]. The level of enrichment of the grain boundaries in Ni–49.7at%Al–0.7at.%Mo–0.4at.%Zr did also not exceed 1.5 at.% [391]. Ni-rich $Ni_3Si$ also exhibits boron segregation with the maximum reported level of about 5 at.% at 873 K [392].

### 4.3.3 Guttmann Model of Grain Boundary Segregation in Interacting Multi-component Systems

In the above description, we concentrated mainly on ideal (infinitesimally diluted) binary (pseudobinary) systems where the activities can simply be replaced by concentrations, that is where $\Delta G_I^{\mathrm{E}}$ in (4.32) and (4.35) can be neglected. To study the grain boundary segregation at higher concentration levels, this term has to be taken into account because the activity of the solute differs from the concentration due to the solute–solute interactions. Then, the activities (activity coefficients) should be evaluated to obtain $\Delta G_I^0$. However, this evaluation is not easy at all and therefore, various types of empirical or semi-empirical correlation are used to evaluate the excess Gibbs energy of segregation.

The most common model, which has not been overcome till now, is based on the regular solid solution according to the "zero-order quasichemical approximation". This model supposes randomly distributed solute and solvent atoms among the fixed number of equivalent sites both in the bulk and in the grain boundary and constant pair interaction energy between the nearest neighbours [313,314,320,393]. The interaction coefficients $\alpha_{ij}$ in a regular solution are related to the excess Gibbs energy, $\Delta G_I^{\mathrm{E}}$, which is equal to the enthalpy of mixing, $\Delta_m H$,

$$\Delta G_I^{\mathrm{E}} = \Delta_m H = \sum_{i<j} \alpha_{ij} X_i X_j \tag{4.73}$$

with

$$\alpha_{ij} = ZN_0 \left[ \varepsilon_{ij} - \frac{\varepsilon_{ii} + \varepsilon_{jj}}{2} \right], \tag{4.74}$$

where $Z$ is the coordination number in the crystal (boundary), $N_0$ is the Avogadro's number and $\varepsilon_{ij}$ are energies of the $i-j$ bonds [393]. Combining (4.73) and (4.74) with the condition $X_M = 1 - \sum_{J=1}^{M-1} X_J$, we obtain

$$\Delta G_I^{\mathrm{E}} = -2\left(\alpha_{IM}^{\Phi} X_I^{\Phi} - \alpha_{IM} X_I\right) + \sum_{J \neq I,M} \left(\alpha_{IJ}'^{\Phi} X_J^{\Phi} - \alpha_{IJ}' X_J\right), \tag{4.75}$$

where $\alpha_{IJ}' = \alpha_{IJ} - \alpha_{IM} - \alpha_{JM}$ is the net interaction between solutes $I$ and $J$, $\alpha_{IJ}$, with respect to their interactions with the matrix element $M$, $\alpha_{IM}$ and $\alpha_{JM}$. Although the interaction coefficients can principally be different for the grain boundary and for the bulk due to different co-ordination numbers and atomic distances in the respective structures, the role of chemical interactions should be the same if it is assumed that $\alpha_{IJ}^{\Phi} \cong \alpha_{IJ}$ [313]. Then we can write

$$\Delta G_I^{\mathrm{E}} = -2\alpha_{IM}^{\Phi}\left(X_I^{\Phi} - X_I\right) + \sum_{J \neq I,M} \alpha_{IJ}'^{\Phi}\left(X_J^{\Phi} - X_J\right). \tag{4.76}$$

Combination of (4.32) and (4.75) or (4.76) represents the well-known form of the Guttmann type of the segregation isotherm for the case of solute segregation in non-ideal (regular) multicomponent system [313]. Let us notice that in the present notation, repulsive interaction is characterised by the values $\alpha_{IJ}'^{\Phi} > 0$ while $\alpha_{IJ}'^{\Phi} < 0$ describes attractive interaction. The interacting elements in various matrices are listed in Table 4.2.

The Guttmann model of interactive segregation in multicomponent systems was successfully used to describe the grain boundary segregation in various systems. The character of the ternary interaction was found to be repulsive for Si–P ($\alpha'_{\mathrm{SiP}} = 92$ kJ/mol) and for P–C ($\alpha'_{\mathrm{CP}} = 7$ kJ/mol), and attractive in case of Si–C ($\alpha'_{\mathrm{SiC}} = -3$ kJ/mol) pairs [111]. The strong P–Si repulsive interaction was directly

**Table 4.2** Examples of interaction pairs during grain boundary segregation

| Matrix | Type of interaction | Interacting elements | Reference |
|---|---|---|---|
| α-Fe | Repulsive | P–B, P–S, C–Sn | [348, 394] |
| | | P–Si | [348, 350, 395–397] |
| | | C–Si | [347, 349] |
| | | Si–N | [398] |
| | | Si–Sb | [398] |
| | | Si–B | [241] |
| | | P–Nb | [399] |
| | | Sn–S | [366] |
| | Attractive | P–Ni, P–Mn, Sb–Mn, Sb–Cr, Sb–Mo | [348, 400] |
| | | Sb–Ni | [348, 364, 401] |
| | | P–Cr, P–Mo | [348, 364, 366, 402] |
| | | Ni–Sn | [366, 367] |
| | | C–P | [349, 351] |
| | | V–P | [313, 363, 402] |
| | | Sb–Ti | [355, 367] |
| | | P–Ti | [355, 403] |
| | | S–Mn | [404] |
| | | S–H, C–H | [405] |
| | | Ti–H | [406] |
| | | Cr–N | [366] |
| Ir | Repulsive | Th–Si | [407] |
| W | Repulsive | Fe–C, Fe–O | [407] |
| γ-Fe | Repulsive | P–S | [408, 409] |
| | | P–B | [410] |
| | | P–C | [411] |
| | | P–Mo | [412] |
| | Attractive | Cu–Sn, Cu–Sb | [401] |
| Ni | Attractive | Mo–P, Nb–P | [126] |
| | Repulsive | S–N | [144] |

proved by scanning tunnelling microscopy at low-index free surfaces of bcc iron showing a depletion of silicon in vicinity of islands of segregated phosphorus atoms [413]. Zhang et al. showed a pronouncedly increasing repulsive interaction between phosphorus and sulphur with decreasing temperature (54 kJ/mol at 973 K while 79 kJ/mol at 773 K) [394]. Attractive interactions were detected between phosphorus and various transition metals, P–Cr ($\alpha'_{CrP} = -3$ kJ/mol), P–Mo ($\alpha'_{MoP} = -23$ kJ/mol) and P–V ($\alpha'_{VP} = -11$ kJ/mol [394] or $-144$ kJ/mol [313]). Attractive interaction was also indicated for phosphorus with molybdenum and niobium in IN 718 nickel-based superalloys [127]. From strong repulsive interaction of tin and silicon on the (4.46) surface of bcc iron ($\alpha'_{SnSi} = -50$ kJ/mol), it is deduced that similar repulsion will exist at the grain boundaries [414]. The finding that silicon segregates at a special grain boundary of $\alpha$-iron, while boron segregates at general grain boundaries suggests repulsion between these two kinds of atoms [241].

Equation (4.76) is the simplest correlation of $\Delta G_I^{\mathrm{E}}$ for the regular system containing the substitution atoms only (model 1). This model was extended to more complex cases supposing regular solutions of both substitution and interstitial solutes without competition (model 2), quasimolecular behaviour with site competition taking into account formation of an $I_x J_y$ compound (model 3), quasimolecular non-competitive behaviour in two distinct (substitution and interstitial) sublattices considering the formation of an $I_x J_y$ compound (model 4), and – last but not the least – formation of a two-dimensional ternary compound (model 5) [313]. These models were developed by Guttmann and McLean [313] and thoroughly discussed by Militzer and Wieting [415]. In general, the segregation equations corresponding to the models 1–5 can be displayed (by neglecting the binary interaction parameters and supposing dilute bulk solid solution) as

$$\frac{Y_I^{\Phi}}{Y_M^{\Phi}} \cong X_I \exp\left(-\frac{\Delta G_I}{RT}\right) \tag{4.77}$$

with

$$\Delta G_I = \Delta G_I^0 + \sum_{J \neq I,M} Q_{IJ} Y_J^{\Phi}, \tag{4.78}$$

where $Y_I^{\Phi}$ and $Q_{IJ}$ are the generalised terms for interfacial concentration and ternary solute interaction, respectively. The meaning of these quantities is given in Table 4.3 for individual models.

For classical example of the mutual enhancement of grain boundary segregation of two solutes, there may serve the behaviour of nickel and antimony in bcc iron. It was found that presence of nickel increases grain boundary segregation of antimony and vice versa [349, 370, 393, 416–418]. Guttmann et al. quantitatively interpret this kind of segregation behaviour as a consequence of an attractive interaction between the atoms of nickel and antimony [175, 393, 416] (Fig. 4.7). This conclusion was frequently criticised on basis of other experimental findings showing that changes in antimony concentration do not affect nickel segregation [419–422]. The main objection following from these experiments is that the effect is caused by interaction of

**Table 4.3** Parameters of individual models 1–5 of grain boundary segregation in multicomponent alloys [313] [(4.77) and (4.78)]

| Model | $Y_I^\Phi$ | $Y_M^\Phi$ | $Q_{IJ}$ | Note |
|---|---|---|---|---|
| 1 | $X_I^\Phi$ | $1-\Sigma_{J\neq M}X_J^\Phi$ | $\alpha'_{IJ}$ | |
| 2 | $\dfrac{X_I^\Phi}{1-\sum_{\text{int}}X_J^\Phi}$ | $1-\dfrac{\sum_{\text{subst}\neq M}X_J-(b/c)\sum_{\text{int}}X_J^\Phi}{1-\sum_{\text{int}}X_J^\Phi}$ | $b\alpha'_{IJ}$ | $I=$ subst(itutional) |
| | | | $c\alpha'_{IJ}$ | $J=$ int(erstitial) |
| 3 | $X_I^\Phi$ | $1-\Sigma_{J\neq M}X_J^\Phi$ | → | Complex functions of $Y_I^\Phi$ and $Y_M^\Phi$ |
| 4 | $\dfrac{X_I^\Phi}{1-\sum_{\text{int}}X_J^\Phi}$ | $1-\dfrac{\sum_{\text{subst}\neq M}X_J-(b/c)\sum_{\text{int}}X_J^\Phi}{1-\sum_{\text{int}}X_J^\Phi}$ | → | Complex functions of $Y_I^\Phi$ and $Y_M^\Phi$ |
| 5 | $\dfrac{X_I^\Phi}{1-\sum_{\text{int}}X_J^\Phi}$ | $1-\dfrac{\sum_{\text{subst}\neq M}X_J-(b/c)\sum_{\text{int}}X_J^\Phi}{1-\sum_{\text{int}}X_J^\Phi}$ | $c\alpha'_{IJ}$ | |

$b$ and $c$ are the fractions of substitution and interstitial lattice sites, respectively ($b+c=1$)

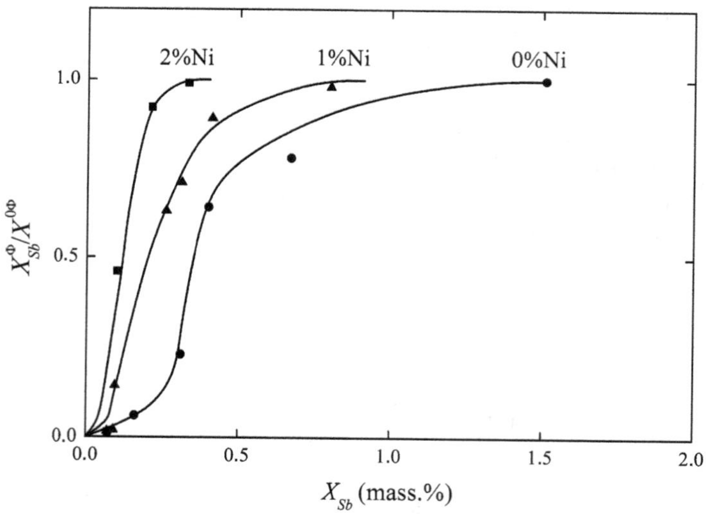


**Fig. 4.7** Isotherms of the grain boundary segregation of antimony in Fe–Ni–Sb alloys as a function of the bulk antimony concentration for different content of nickel in the alloy. According to [416]

these elements with carbon as the trace impurity in the material used. Grabke and Briant proposed to explain the segregation behaviour in multicomponent systems on basis of both the changed solubility and activity of the solutes in iron due to presence of another impurity and the site competition. In the above case of the segregation in an Fe–Ni–Sb alloy, the increased segregation of antimony with increasing content of nickel is explained as reduction of the bulk concentration of carbon as an impurity in the system. Thus, carbon is less effective in site competition with antimony and, therefore, the grain boundary concentration of antimony may increase [419–422].

**Table 4.4** Examples of scavenging or gettering effects influencing grain boundary segregation

| Matrix | Scavenging/gettering effect | Reference |
|---|---|---|
| α-Fe | Fe + O → oxides | [423] |
| | Mn,Ca,Cr,Ce + S → sulphides | [367,423–425] |
| | Fe,Ni,Mo,Nb,Ti,V,Cr,Mn + C → carbides | [363,400,420,423,424,426] |
| | Mo,Nb,Ti + P → phosphides | [123,355,381,400,424,427,428] |
| | Cr,Ti,Mo + N → nitrides | [123,408,429,430] |
| | Mn,Fe,Ni,Ti,Nb,Cr + Sb → antimonides | [422,431] |
| | Ti + H → hydride | [405] |
| γ-Fe | Nb,Mn + S → sulphides | [432] |
| | Fe,Nb,Cr,Mo,Ni + C → carbides | [373,377,432–434] |
| Ir | Ir,Th + Si → silicides | [406] |
| Ni | Ti,Nb,Ca,Mg,Y,Zr + S → sulphides | [125,435] |
| | Ti,Nb + N → nitrides | [125] |
| | Ti + C + N → carbidonitrides | [435,436] |
| | Cr + C → carbides | [436] |

This situation is rather general because carbon exerts a large tendency to precipitate as carbides with many other solutes such as molybdenum, vanadium, titanium or tantalum [349, 417, 420] (cf. Table 4.4). This fact serves as a general explanation of reduced competition with other segregating elements such as phosphorus [417]. However, the effect of site competition should be of little power in dilute alloys and the interactive segregation seems still to be the most reasonable explanation for observed behaviour [336]. Let us state that site competition and repulsive interaction do not compete each to the other: Both effects can synergistically be used to correlate the data more precisely [313, 437] as was shown, for example of phosphorus, silicon and carbon grain boundary segregation in iron [350].

Let us conclude that the segregation behaviour in a multicomponent alloy reflects various effects that should be considered in full interpretation of the data [423]. We should consider (a) the basic tendency of the solutes to segregate at the grain boundaries, (b) the changes in the bulk composition due to the scavenging and gettering effects and following precipitation, (c) the site competition and (d) the mutual interaction of different atoms. However, if we want to study the grain boundary composition in a multicomponent alloy, we should avoid all effects that are not connected with segregation. Primarily, we should only consider the true concentrations of dissolved solutes and impurities in the matrix phase that are available for segregation, and not to use the nominal composition of the system. In this way, the problems with possible scavenging and gettering do not come into discussion and the situation becomes clearer. Another effect, which should be avoided, is the saturation of the boundaries: When a boundary is saturated, fine details of co-segregation may be invisible [418]. Then, we can simply consider the changes of the activities of the system that can be correlated by various approaches (although presently, the only available approach is the interaction model of Guttmann (4.83) and (4.84) and relate the grain boundary concentration of a segregated element to the grain boundary concentration of the matrix element (site competition, (4.66)) to describe the segregation behaviour in the system under consideration. The Guttmann model may seem to be too simplified but – as already mentioned above – till now no other alternative has been proposed to describe interfacial segregation in real systems.

### 4.3.4 Fowler Model

In a binary system, the ternary interaction coefficients $\alpha'_{IJ}$ will be neglected and (4.76) reduces into

$$\Delta G_I^{\mathrm{E}} = -2\alpha_I^{\Phi}\left(X_I^{\Phi} - X_I\right). \tag{4.79}$$

Expression (4.79) is very similar to the Fowler interaction term [438]

$$\Delta G_I^{\mathrm{E}} = 2Z_1\omega X_I^{\Phi}/X^{0\Phi}, \tag{4.80}$$

where $Z_1$ is the number of the nearest neighbours in the interface (i.e. lateral co-ordination number) and $\omega$ is the pair interaction energy of the $I$ atoms. The combination of (4.32) with (4.79) or (4.80) represents the well-known Fowler segregation isotherm [13,438].

Analysis of the Fowler isotherm results in an interesting consequence. In case of $\alpha_I^{\Phi} < 0$ (i.e. repulsive *I–I* interaction), the absolute value of the exponential term in (4.32) is reduced, and therefore, the extent of segregation is lowered progressively as $X_I^{\Phi}$ grows. On the other hand, if the presence of a segregated atom $I$ enhances the probability of occupation of the nearest neighbour sites by the same kind of the atoms (i.e. positive values of $\alpha_I^{\Phi}$ corresponding to attractive *I–I* interaction), the exponential term and the extent of segregation is enhanced with increasing $X_I^{\Phi}$. As $\alpha_I^{\Phi}$ becomes more positive, the segregation increases more strongly with decreasing temperature until the rise in segregation becomes eventually discontinuous at the value of $\alpha_I^{\Phi} > 2RT$. This can be documented, for example of selenium and tellurium segregation in iron (Fig. 4.8) [13,307,439] and antimony in iron [416]. It is also well documented in Fig. 4.8, for example of phosphorus segregation in iron [328], that the Langmuir–McLean segregation isotherm (4.65) represents the special case of the Fowler isotherm with $\alpha_I^{\Phi} = 0$. For example, the Fowler type of grain boundary segregation with $\alpha_I^{\Phi} = 2.5RT$ was observed for bismuth in copper [440].

Grain boundary concentration of tellurium at 700°C was found to increase with increasing bulk concentration up to about 100 ppm [441,442]. This bulk concentration represents probably the limit of solid solubility and beyond this limit, the same amount of tellurium remains dissolved in ferrite solid solution and thus available for segregation while the remaining tellurium precipitates as the second phase. A wide scatter of measured grain boundary concentrations is ascribed to strong anisotropy of grain boundary segregation. Large differences in grain boundary segregation and thus in grain boundary energy result in faceting of grain boundaries, which then exhibit large difference in tellurium content [442].

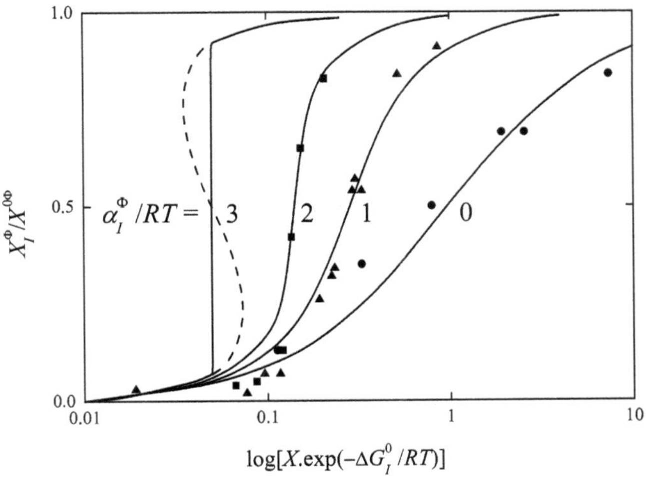


**Fig. 4.8** Correlation of experimental data on grain boundary of phosphorus (*circles*) [328], selenium (*triangles*) and tellurium (*squares*) [439] in iron using the Fowler isotherms (according to [308]). The value of the Fowler parameter $\alpha_I^\Phi/RT$ (4.71) is marked at individual curves

Fowler model was also used to describe low level of boron grain boundary segregation in ordered polycrystalline Fe–40at.%Al intermetallic alloys [128, 443]. This correlation resulted in the values of $\Delta G_B^0$ ranging from $-30$ to $-34$ kJ/mol and $\alpha_I^\Phi$ ranging from $-110$ to 160 kJ/mol. These values suggest strong repulsive interaction between boron atoms. However, we can alternatively fit these experimental data according to the Langmuir–McLean model considering the saturation of the grain boundary. In this case, the ordering of the system and the grain boundary segregation in interstitial positions has also to be taken into account, $X_{tot}^{0\Phi} = X_{ord}^{0\Phi} \times X_{int}^{0\Phi}$. The best fit of the experimental data was obtained using $X_{tot}^{0\Phi} = 0.058$, which corresponds to the values $X_{ord}^{0\Phi} = 0.23$ (determined according to (4.70) and (4.71) with $m^\Phi(T = 673\ \text{K}) = 0.77$ using $T_c^\Phi = 1{,}354$ K and $\beta^\Phi \approx 0.5$ [179, 385]) and to $X_{int}^{0\Phi} = 0.6$ (the fraction of octahedral interstitial positions in bulk bcc lattice), provided the best correlation supposing that about one interstitial positions are occupied per unit cell. The corresponding value of the standard Gibbs energy of segregation was found to be $\Delta G_I^0 = -48$ kJ/mol. As it is apparent from Fig. 4.9, these data fit with the experimental values very well.

Let us compare now the correlation of the segregation data using both the Fowler approach and the site competitive segregation without interaction. It is apparent from Fig. 4.9, that both types of correlation provide us with very similar results [347]. If we compare both types of correlation (i.e. (4.65) with $\Delta G_I^E = 0$ and $X^{0\Phi} \neq 1$ on the one hand, and (4.65) with $X^{0\Phi} = 1$ and $\Delta G_I^E$ given by (4.79) on the other hand) and suppose $X_I^\Phi \gg X_I$, we obtain

$$X^{0\Phi} - X_I^\Phi = \left(1 - X_I^\Phi\right) \exp\left(-\Psi X_I^\Phi\right), \tag{4.81}$$

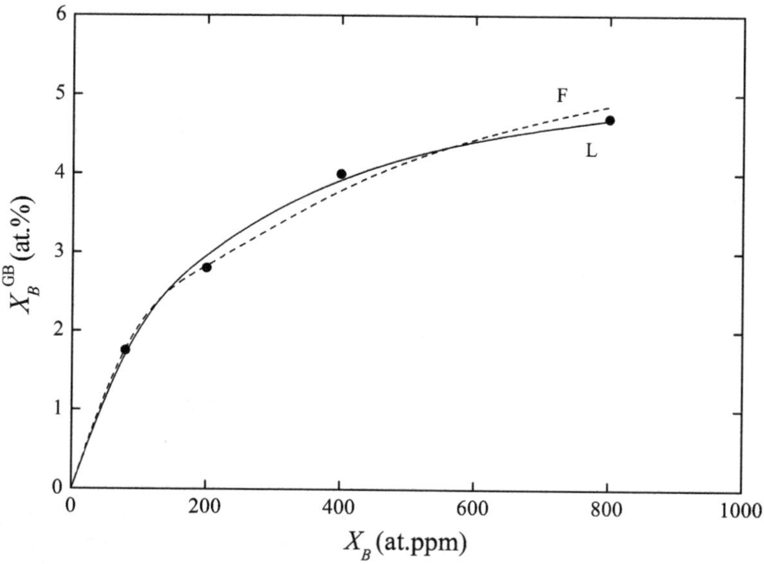


**Fig. 4.9** Correlation of the concentration dependence of equilibrium boron segregation at grain boundaries of a polycrystalline Fe–40at. %Al alloy at 673 K (*solid circles*) according to the Langmuir–McLean model with limited number of positions available for segregation $(X^{\Phi} = 0.058,\ \Delta G_B^0 = -48\,\text{kJ/mol})$ (L, *full line*) and the Fowler model $(\alpha_B^{\Phi} = -114\,\text{kJ/mol},\ \Delta G_B^0 = -34\,\text{kJ/mol})$ (F, *dashed line*) [347]

where

$$\Psi = -2\alpha_I^{\Phi}/RT. \tag{4.82}$$

Differentiation of (4.81) by $X_I^{\Phi}$ results in

$$\exp\left(-\Psi X_I^{\Phi}\right) = \frac{1}{1 + \left(1 - X_I^{\Phi}\right)\Psi} \tag{4.83}$$

and its substitution back in (4.81) provides us with a general expression assuming only $X^{0\Phi} \neq 1$,

$$\frac{1}{\Psi} = \frac{\left(1 - X_I^{\Phi}\right)\left(X^{0\Phi} - X_I^{\Phi}\right)}{1 - X^{0\Phi}}. \tag{4.84}$$

It clearly follows from (4.84) that the Fowler interaction parameter, represented by $\Psi$, is closely related to the saturation limit: While the negative value of $\Psi$ between segregating atoms corresponds to the solute segregation in more than one monolayer as, for example in the case of bismuth segregation in copper [105] or tin in iron [91], positive values of $\Psi$ are well correlated with a saturation limit for segregation. It is clearly obvious from (4.65) that the value $X^{0\Phi} = 1$ corresponds to ideal Langmuir–McLean behaviour ($\Psi = 0$).

Equation (4.84) suggests that the interaction parameter $\Psi$ should depend on interfacial concentration although it is concentration independent. In fact, however, the value of $\Psi$ changes with the concentration range of the data used for the correlation as can be tested for any experimental or model data. Therefore, the mean value of $\Psi$ for the whole concentration range should be determined as

$$\frac{1}{\Psi^{\text{mean}}} = \frac{\int\limits_0^{X^{0\Phi}} \frac{1}{\Psi} \mathrm{d}X_I^{\Psi}}{\int\limits_0^{X^{0\Phi}} \mathrm{d}X_I^{\Psi}}. \tag{4.85}$$

Substituting (4.84) into (4.85) and integrating it, we obtain

$$\frac{1}{\Psi^{\text{mean}}} = \frac{X^{0\Phi}(3 - X^{0\Phi})}{6(1 - X^{0\Phi})}. \tag{4.86}$$

Supposing low saturation limit for interfacial segregation, that is $X^{0\Phi} \ll 1$ (that is typical for ordered alloys and intermetallics [347]), (4.86) reduces to

$$X^{0\Phi} = \frac{2}{\Psi^{\text{mean}}}. \tag{4.87}$$

Comparing the values of $X^{0\Phi} = 0.06$ for saturated boron segregation in FeAl-based alloys [125,443], and of $2/\Psi^{\text{mean}} = 0.05$ obtained using $\alpha_I^{\Phi} = -114\,\text{kJ/mol}$, exhibits a very good agreement [347].

A close relationship between $X^{0\Phi}$ and $\alpha_I^{\Phi}$ (cf. (4.87) and (4.82)) can be well understood because the repulsive interaction between the segregating species prevents the segregation of the same atoms in close vicinity: This fact also can be interpreted as a limitation of the number of the sites accessible for segregation. This close relationship can also be documented by successful correlation of antimony segregation in iron on the basis of both the Fowler model [417] and the Langmuir–McLean model considering the grain boundary saturation [338].

### 4.3.5 Other Models for Grain Boundary Segregation

In this section, there are various approaches listed, which are used to describe the interfacial segregation by phenomenological models based on the chemical potentials. Many of them were developed for surface chemistry, however, due to similarity of the behaviour of free surfaces and internal interfaces they may also be applied to the grain boundaries.

#### 4.3.5.1 Modern Thermodynamic Calculation of Interfacial Properties

Mezey and Giber [444–446] combined (4.34)–(4.36) and obtained an expression

$$\kappa_I = \frac{X_I^{\Phi}/X_I^{\text{bulk}}}{\left[(1-X_I^{\Phi})/(1-X_I^{\text{bulk}})\right]^{A_I/A_M}} = \exp\left\{\left(\frac{A_I}{A_M}\right)\left[\frac{\mu_M-\mu_M^{\text{E}}}{RT}\right]-\left[\frac{\mu_I-\mu_I^{\text{E}}}{RT}\right]\right\}, \tag{4.88}$$

where the parameter $\kappa_I$ is a modification of the enrichment ratio $\beta_I$ (cf. (4.19)). Supposing $A_i = A_i^0$ for pure $i$ $(i = I, M)$, $\mu_i^{\text{E}} = \sigma_i^0 A_i^0 + \tilde{\mu}_i^{\text{E}}$, $\mu_i \approx \tilde{\mu}_i^{\text{E}}$, and neglecting lattice distortion effects in the interface for the solvent ($\mu_M(d) \approx 0$), (4.88) can be rewritten as

$$\kappa_I = \exp\left\{\frac{(\sigma_M^0-\sigma_I^0)\,A_I^0}{RT}+\frac{\mu_I(d)}{RT}\right\}. \tag{4.89}$$

A detailed analysis showed that the solute $I$ segregates if $(\sigma_M^0-\sigma_I^0)/\sigma_I^0 \geq 0.05$ or if $|(\sigma_M^0-\sigma_I^0)/\sigma_I^0| < 0.05$ and simultaneously $|(r_M-r_I)/r_M| \geq 0.15$, where $r_i$ are the atomic radii of the component elements. A qualitative agreement was found in the case of surface segregation in many systems [446]. This method was also modified to include the effects of the anisotropy of grain boundary energy and the lattice distortion caused by a solute and its release at the interface [447–450]. In this modification, the interfacial concentration $X_I^{\Phi}$ of a solute $I$ in a binary system $M$–$I$ can be expressed as

$$X_I^{\Phi} = X_I C_m C_n C_r, \tag{4.90}$$

where $C_m$ is the configuration term

$$C_m = \left[\left(1-X_I^{\Phi}\right)/(1-X_I)\right]^{q_M}, \tag{4.91}$$

$C_n$ is the dangling bond term

$$C_n = \exp\left[\frac{q_M\alpha_I^0 G_I^a-\alpha_M^0 G_M^a}{RT}\right] = \exp\left[\frac{\left(\sigma_I^0-\sigma_M^0\right)A_M^0}{RT}\right], \tag{4.92}$$

and $C_r$ is the real mixture term

$$C_r = \exp\left[\frac{q_M\Delta\mu_I^{\text{E}}(m)-\Delta\mu_M^{\text{E}}(m)}{RT}\right], \tag{4.93}$$

where

$$\Delta\mu_i^{\text{E}}(m) = -\alpha\mu_i^{\text{E}\Phi}X_i^{\Phi}+(1-2\alpha)\left[\mu_i^{\text{E}}(X_i)-\mu_i^{\text{E}\Phi}\left(X_i^{\Phi}\right)\right]. \tag{4.94}$$

In (4.91)–(4.94), $q_M$ is the molar internal free energy of atomisation, and $\alpha_i^0$ and $\alpha$ are the parts of $G_i^a$ related to the neighbours missing in the interface, and of $\mu_i^{\mathrm{E}}$ related to the interactions of the atoms above and below the interface, respectively. The values of individual parameters in the above expressions can be estimated on basis of corresponding values of interfacial energy and molar enthalpy of mixing.

As mentioned above, this method was primarily used for description of surface segregation, especially for determination of the anisotropy of platinum surface segregation on nickel [448–450].

#### 4.3.5.2 Model of Luthra and Briant

Another expression for segregation isotherm was proposed by Luthra and Briant [315],

$$\frac{X_I^{\Phi}}{X_I^{\mathrm{bulk}}} = \left(\gamma_I^{\mathrm{bulk}}\right)^{1-f_1} \exp\left[\frac{(\sigma_M f_2 - \sigma_I)A_I^0}{RT}\right], \tag{4.95}$$

where $\gamma_I^{\mathrm{bulk}}$ is the activity coefficient of the solute $I$ in bulk, $f_1$ and $f_2$ are the parameters reflecting the ratio of the Gibbs energies of the grain boundary and of the bulk, and the ratio of the partial molar surface area of component $I$ in a solution and in pure substance, respectively. $\sigma_M$ and $\sigma_I$ are the grain boundary interfacial energies of the matrix element $M$ and of the segregating element $I$, respectively, and $A_I^0$ is the standard molar grain boundary area of the component $I$ [315]. Similarly to the Guttmann model, this approach tries to correlate the non-ideal behaviour of the segregating system by considering the activity coefficients of the components. However, the values of some of the above parameters are not known and therefore, are used as fitting parameters. In fact, it is an analogous approach to that used in the Guttmann model [313], where the non-ideal behaviour of the system is correlated by the regular (quasichemical) approximation. Guttmann model allows simple prediction of both segregation enthalpy and entropy as the fundamental characteristics of interfacial segregation, together with the ternary interaction coefficients that are clearly defined [375]. In fact, the interaction parameters in Guttmann approach on one hand and $\gamma_i$ and $f_i$ in Luthra and Briant model are not known and represent exclusively the correlation of the non-measurable parameter $\Delta G_I^{\mathrm{E}}$, that is from this point of view, both models are very similar. The main drawback of the approach of Luthra and Briant [315] is the physically incorrect basic assumption $\mu_I^{\mathrm{bulk}} = \mu_I^{\Phi}$ ([310], see above), in which the presence of the matrix element in segregating system is completely ignored. This assumption should have a strict consequence that the grain boundary enrichment ratio $\beta^{\Phi}$ (4.19) is independent of the nature of the matrix element. However, the measured values of $\beta^{\Phi}$ for a solute in different matrices exhibit large differences, for example in case of phosphorus, $\beta_{\mathrm{P}}^{\Phi} \approx 3 \times 10^4$ in tungsten while $\beta_{\mathrm{P}}^{\Phi} \approx 200$ in bcc iron [168] and in case of boron, $\beta_{\mathrm{B}}^{\Phi} \approx 5 \times 10^3$ in bcc iron [138] while $\beta_{\mathrm{B}}^{\Phi} \approx 10$ in $Ni_3Al$ [425]. Proper derivation based on realistic starting conditions should result in more realistic formulae, which can be used to describe grain boundary segregation in multicomponent alloys. Equation (4.95) could be applied, however, for dilute systems with negligible bulk as well as interfacial concentration of the solute $I$.

### *4.3.6 Models for Different Segregation Sites*

As it was mentioned in Chap. 2, different grain boundaries possess different structures as characterised by the structural units of individual interfaces. Each structural unit consists of several atoms with different binding energies [275]. In consequence, the Gibbs energy of segregation will also differ for various grain boundary positions of the solute atoms. Similarly, different values of $\Delta G_I^0$ will be found for different grain boundaries. In polycrystalline materials, a spectrum of the segregation Gibbs energies thus exists determining the equilibrium composition of individual grain boundaries. However, only a single value of $\Delta G_I^0$ is involved in the segregation isotherms mentioned above. The only physically meaningful value of $\Delta G_I^0$ is that corresponding to the solute segregation at a particular position of a given grain boundary [451]. Such localised segregation, however, is hardly experimentally measurable at present state-of-art of the detection methods (cf. Chap. 3) and therefore, there is a lack of information about the local environment of the segregated atom [44]. The above given segregation isotherms can thus correctly be applied only to the determination of characteristic thermodynamic parameters of segregation of an element at a chosen grain boundary in bicrystal from the temperature dependence of its chemical composition providing low-grain boundary concentrations. Only in this case, one may expect that the solute atoms segregate at identical grain boundary sites (causing the highest reduction of the Gibbs energy of the system) [20]. An application of the above-outlined segregation isotherms for description of the grain boundary segregation in polycrystals only provides the values of the characteristic parameters averaged over the large spectrum of grain boundaries and grain boundary sites with little physical meaning [342]. On the other hand, there were developed several models for describing the segregation at different grain boundary sites.

#### 4.3.6.1 BET Isotherm

In case of free surfaces, there is a well-known BET theory [452] describing multilayer surface gas adsorption accounting for different energetic states of atoms in individual layers. Let us assume that the surface is filled by solute atoms as shown in Fig. 4.10. There are $n_1$ sites on the surface occupied by one atom of solute, $n_2$ sites occupied by two atoms, etc. It is assumed that the Gibbs energy of adsorption in the first layer, $\Delta G_I^1$, has a particular value while its value in higher layers, $\Delta G_I^i$, is equal to the Gibbs energy of condensation [453]. In case of interfacial segregation, we can consider it as two (or more) different values of the Gibbs energy of segregation. It is also supposed that the adsorption/desorption processes occur exclusively

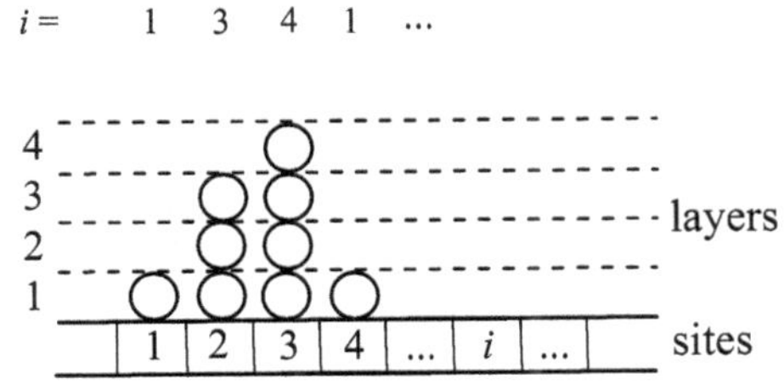


**Fig. 4.10** Schematic depiction of formation of multilayer segregation in BET model

between the interface and the bulk, and the atoms are not allowed to move from one layer to another. In equilibrium, the rates of adsorption and desorption for each layer must be identical. Based on these assumptions, the BET isotherm is formulated as

$$\theta^{\Phi} = \frac{X_I \exp\left[-\left(\Delta G_I^i - \Delta G_I^1\right)/RT\right]}{(1 - X_I)\left\{1 - X_I + X_I \exp\left[-\left(\Delta G_I^i - \Delta G_I^1\right)/RT\right]\right\}}. \tag{4.96}$$

It is apparent that for low bulk concentrations $X_I$, the Langmuir-McLean isotherm (4.61) is obtained [452]. Principally, (4.96) can be extended to involve several different values of $\Delta G_I^i$ and be thus able to describe the segregation to different interfacial positions.

BET approach was applied to describe the grain boundary segregation of tin in bcc iron [306, 307], and to interpret sulphur, phosphorus and antimony segregation in iron and nickel [454]. Multilayer segregation was also detected in other systems such as tellurium and selenium in bcc iron, phosphorus in tungsten and bismuth in copper (cf. [92]).

#### 4.3.6.2 Individual Site Models

Assuming a grain boundary containing $N$ distinct sites of different values of the standard Gibbs energy of segregation, $\Delta G_{I,k}{}^0 (k = 1, \ldots, N)$, White et al. extended McLean treatment for a binary *M–I* system [455, 456]. In their model, the interfacial concentration of the solute $I$, $X_{I,k}^{\Phi}$, at position $k$ is given by

$$X_{I,k}^{\Phi} = \frac{X_I \exp\left(-\Delta G_{I,k}^0/RT\right)}{1 + X_I \left[\exp\left(-\Delta G_{I,k}^0/RT\right) - 1\right]}. \tag{4.97}$$

The total experimentally observable atom fraction of the solute at the interface, $X_I^{\Phi}$, would then be a weighted average summed over all interfacial sites with the fraction $F_k$ of the total sites at the interface having the same value of the standard Gibbs energy of segregation,

$$X_I^{\Phi} = \sum_k^N F_k X_{I,k}^{\Phi}. \tag{4.98}$$

Nowicki and Biscondi [457, 458] modified this approach by normalisation of the parameters per unit area,

$$F_k = \frac{Q_k}{P'}, \tag{4.99}$$

where $P'$ is the mean surface atomic density of the solvent in the volume and $Q_k$ is the number of sites of type $k$ per unit area. According to this model, it is possible to compare the results of the calculated values of the Gibbs energy of segregation with the measurements of the grain boundary concentration at different temperatures and different grain boundaries [458]. In this way, similar values $\Delta H_{\mathrm{O}}^0 \approx 232\,\mathrm{kJ/mol}$ were determined for oxygen segregation at different sites of the 32°[100], {027} symmetrical tilt grain boundary in molybdenum [457] although such value seems to be unexpectedly high. Similar approach was also used by Rittner [219], Udler and Seidman [332] and Kirchheim [459, 460]. Equations (4.97)–(4.99) can be applied for dilute binary systems. In more concentrated systems, it is necessary to consider the solute–solute interaction and to approximate them, for example according to a simple mean field or quasichemical approach. The Fowler interaction term will then additionally appear in (4.97).

On the other hand, Suzuki [461] supposes that the average amount of the segregated species, $X_{\mathrm{av}}^{\Phi}$, must reflect the structural effect of the distribution of the Gibbs energy of segregation,

$$X_{\mathrm{av}}^{\Phi} = \int_0^{\infty} X_I^{\Phi}(\Delta G_I) f(\Delta G_I) \mathrm{d}\Delta G_I, \tag{4.100}$$

where $f(\Delta G_I)$ is the distribution function of the Gibbs energy of segregation and $X_I^{\Phi}(\Delta G_I)$ is the concentration of solute $I$ at a grain boundary with the Gibbs energy of segregation $\Delta G_I$. In the simplest case, the Gaussian distribution can be taken to represent the distribution function $f(\Delta G_I)$. A comparison of the model calculations of phosphorus segregation in α-iron with experimental data of Erhart and Grabke [328] documented the difference of the thermodynamic parameters characterised mainly by an underestimation of the segregation enthalpy in the experiments being performed on polycrystals.

## 4.4 Models for Thermodynamic Functions of Interfacial Segregation

The phenomenological theories presented above can provide us with the values of thermodynamic and/or interaction parameters of the grain boundary segregation by fitting the experimental data. Knowledge of the values of $\Delta H_I$ and $\Delta S_I$ is, however, essential for an estimate of the segregation behaviour of a chosen system. Therefore, considerable effort has been spent to develop methods, which could enable assessing the values of these functions on basis of the state of atomic bonding and structural factors at individual interfaces. Since the structure and bonding are simpler for free surfaces than for grain boundaries, the majority of the theories to predict thermodynamic parameters of solute segregation were originally developed for free surfaces. Because of similarity of the nature of free surfaces and internal interfaces and analogous description of their segregation behaviour, the theoretical models developed for free surfaces can be modified to describe the grain boundary segregation. However, their application requires further assumptions to be made, mainly with respect to structural details of the grain boundaries. Although these models have been used only rarely to evaluate the thermodynamic parameters of the grain boundary segregation, a brief description of some models is given here.

### *4.4.1 Model of Wynblatt and Ku*

Wynblatt and Ku [324,325] modified the so-called bond-breaking model of Williams and Nanson [462] based on the enthalpies of formation and mixing to calculate the enthalpy and entropy of interfacial segregation. The model considers the contribution of two different terms, the bond alteration at the interface or "chemical" contribution, $\Delta H_I^{\mathrm{chem}}$ [463], and the elastic strain energy relief, $\Delta H_I^{\mathrm{el}}$ [19, 464], which may be modified by the electronic density relaxation [465]. The total enthalpy of interfacial segregation is then

$$\Delta H_I = -\Delta H_I^{\mathrm{el}} + \Delta H_I^{\mathrm{chem}}. \tag{4.101}$$

The elastic enthalpy term arises from the mismatch between the atoms of solute $I$ and solvent $M$ and is generally expressed as [19, 138, 324, 466]

$$\Delta H_I^{\mathrm{el}} = \frac{24\pi B\mu r_I r_M (r_I - r_M)^2}{3Br_I + 4\mu r_M}, \tag{4.102}$$

where $B$ is the bulk modulus of solute $I$, $\mu$ is the shear modulus of solvent $M$ and $r_I$ and $r_M$ are the effective radii of the solute and solvent, respectively. The term $\Delta H_I^{\mathrm{chem}}$ is defined by the energy change if an atom $M$ is replaced by an atom $I$ at the interface. In the regular solution approximation,

$$\Delta H_I^{\mathrm{chem}} = (\sigma_I - \sigma_M)A^{\Phi} - \frac{2\Delta_m H}{ZX_I X_M}\left[Z_{\mathrm{L}}\left(X_I^{\Phi} - X_M\right) + Z_{\mathrm{P}}\left(X_I - \frac{1}{2}\right)\right]. \tag{4.103}$$

Here $\sigma_i$ are the interfacial energies of pure components ($i = I, M$), $A^{\Phi}$ is the interface area per atom, $\Delta_m H$ is the enthalpy of mixing of the $M$–$I$ alloy, $Z$ is the co-ordination number of mixing in bulk, and $Z_L$ and $Z_P$ are the lateral and the perpendicular co-ordination numbers, respectively, of an atom in the interface layer.

This combined model reflects three principal contributions to the interfacial segregation in metallic systems (a) the *chemical driving force* represented by the difference of the interfacial energy of pure components $M$ and $I$ (cf. (4.103)); (b) *interatomic driving force* depending on the regular solution constant (4.103) which vanishes in the case of ideal solution ($\Delta_m H = 0$) and (c) *elastic driving force* represented by the strain energy contribution reflecting the degree of the misfit of the solute in a solution (4.102) that also vanishes in ideal system [303]. $\Delta H_I$ thus involves both the ideal contribution of the grain boundary energy to the segregation enthalpy and the excess term and thus differs from the standard enthalpy of grain boundary segregation. It follows from (4.101)–(4.103) that the standard enthalpy of segregation is $\Delta H_I^0 = (\sigma_I - \sigma_M)A^\Phi$ [113,303] although the enthalpy contribution to interfacial Gibbs energies $\sigma_i$ should only be taken into account,

$$\Delta H_I^0 = \left(h_I^\Phi - h_M^\Phi\right) A^\Phi. \tag{4.104}$$

Correspondingly, the entropy of segregation, $\Delta S_I$, is also composed of one elastic and two chemical contributions [325],

$$\begin{aligned}\Delta S_I = (s_I - s_M)A^\Phi - \frac{2\Delta_m S}{Z X_I X_M}\left[Z_L\left(X_I^\Phi - X_M\right) + Z_P\left(X_I - \frac{1}{2}\right)\right] \\ + \frac{d}{dT}\left[\frac{24\pi B\mu r_I r_M (r_I - r_M)^2}{3Br_I + 4\mu r_M}\right],\end{aligned} \tag{4.105}$$

where $\Delta_m S =$ is the excess entropy of mixing of the alloy $M-I$, and $s_I$ and $s_M$ are the specific interfacial entropies of the pure components. These three terms can either be of the same sign to reinforce each other or have different signs to partially cancel out. Generally, $|\Delta S_I|$ will be large when $|\Delta H_I|$ possesses large value [303]. This is compatible with the compensation effect (see Chap. 5). Similarly to $\Delta H_I^0$, the standard entropy of grain boundary segregation should be given as

$$\Delta S_I^0 = \left(s_I^\Phi - s_M^\Phi\right) A^\Phi. \tag{4.106}$$

In metals, the grain boundary energy is relatively small and the difference $\sigma_I - \sigma_M$ is less important than for free surfaces [467]. Therefore, the term $\Delta_m H$ has comparatively larger influence. If $\Delta_m H < 0$, the segregation enthalpy is lowered, the system tends to ordering and oscillations of the concentration in deeper layers are predicted [324,467]. If $\Delta_m H > 0$, the binary system has the properties representative of a miscibility gap [303] and the segregation is enhanced, which can lead to clustering, that is to multilayer segregation [467]. Later, the theory of Ku and Wynblatt was extended to ternary alloys $M-I-J$ by extending the right side of (4.103) by the term [468]

$$\frac{1}{Z}\left[\frac{\Delta_m H_{IJ}}{X_I X_J} - \frac{\Delta_m H_{IM}}{X_I(1-X_I-X_J)} - \frac{\Delta_m H_{JM}}{X_J(1-X_I-X_J)}\right] \times\left[Z_{\mathrm{L}}\left(X_I^{\Phi}-X_I\right)+Z_{\mathrm{P}}X_I\right]. \tag{4.107}$$

In contrast to metallic materials, the following driving forces controlling grain boundary segregation in ceramic materials can be distinguished (a) elastic; (b) electrostatic and (c) dipole interactions [468,469]. The elastic term arises from interactions between solutes and the ceramic interface due to the space charge and can be expressed as

$$\Delta_{\mathrm{el}} H_I = -\frac{6B\pi r_M (r_I - r_M)^2}{1+3B/4\mu}. \tag{4.108}$$

The electrostatic term arises from interactions between solutes and the ceramic interface due to the space charge and can be written as

$$\Delta_{\mathrm{e}} H_I = -\frac{1}{2}\rho u^{\Phi}. \tag{4.109}$$

where $\rho$ is the charge density and $u^{\Phi}$ is the potential distribution in the interface. The third contribution stems from the tendency of the charged solutes to combine with the defects of the opposite charge in a ceramics, and to form electrically neutral complex with a dipole moment. It is given by

$$\Delta_{\mathrm{d}} H_I = -\frac{1}{2}Ep, \tag{4.110}$$

where $E$ is the electric field and $p$ is the dipole moment [469].

### *4.4.2 Model of Seah*

According to Seah [13, 371], $\Delta S_I^0$ generally consists of three contributions that are associated with changes of vibrational $\left(\Delta S_I^{\mathrm{vib}}\right)$, anharmonic $\left(\Delta S_I^{\mathrm{an}}\right)$ and site multiplicity $\left(\Delta S_I^{\mathrm{mult}}\right)$ entropies. In general, $\Delta S_I^{\mathrm{an}}$ and $\Delta S_I^{\mathrm{mult}}$ can be neglected in comparison to $\Delta S_I^{\mathrm{vib}}$, which is expressed as the change of the Debye temperature for a solute in the matrix and at the interface [13, 371, 470]

$$\Delta S_I^{\mathrm{vib}} = 3R\left[1+\ln\left(\frac{kT}{h\nu_I}\right)\right], \tag{4.111}$$

where $kT \gg h\nu_I$, $\nu_I$ is the Einstein frequency and $h\nu_I$ may be rewritten as $k\theta_{\mathrm{E},J}$ or $0.775k\theta_{\mathrm{D},J}$ where $\theta_{\mathrm{E},J}$ and $\theta_{\mathrm{D},J}$ are Einstein and Debye temperatures, respectively, so that

$$\Delta S_I^{\mathrm{vib}} = 3R\ln\left(\frac{\theta_{\mathrm{D},J}}{\theta^*_{\mathrm{D},J}}\right), \tag{4.112}$$

where $\theta^*_{\mathrm{D},J}$ is the Debye temperature for the solute atom at the distorted site of the interface. Often, it is supposed that the entropy is controlled by the vibrational term and that $|\Delta S_I^{\mathrm{vib}}| < 3.3RT$ [13, 299, 300, 371]. However, much higher values of segregation entropy were reported in literature, for example $\Delta S_{\mathrm{P}}^0 \approx 5.4R$ was found for the {013} grain boundary in bcc iron [111] and $\Delta S_{\mathrm{C}}^0 \approx 5.2R$ for carbon segregation in polycrystalline bcc iron [355]. High values of segregation entropy were also determined theoretically ($\Delta S_{\mathrm{Sn}}^0 \approx 5.4R$ for $T \leq 1{,}184$ K and ($\Delta S_{\mathrm{Sn}}^0 \approx 6.4R$ for $T \geq 1{,}184$ K) [371]. The origin of the high values of segregation entropy conveys with the compensation effect discussed in Chap. 5.

### 4.4.3 Model of Miedema

Solute segregation in alloys of transition metals can be determined considering the effects of (a) the heat of solution, (b) the differences in interfacial energy of pure metals and (c) the elastic size mismatch energy [465]. Miedema established the basic expression for interfacial segregation in binary $M-I$ alloy,

$$\frac{X_I^\Phi}{X_I} = \exp\left\{\left[f\Delta_{\mathrm{sol}}H_{MI} - g\left(H_I^\Phi - H_M^\Phi\right)V_I^{2/3}\right]/RT\right\}, \tag{4.113}$$

where $\Delta_{\mathrm{sol}}H_{MI}$ is the enthalpy of solution of $I$ in $M$, $H_{M,I}^\Phi$ is the interfacial enthalpy of $M$ and $I$ and $f$ and $g$ are constants. The interfacial enthalpies are obtained from an empirical theory considering two basic parameters: the electrochemical parameter, $u^*$ and the charge parameter, $\rho^{1/3}$. $\Delta u^*$ serves as the measure of the charge transfer between metals $I$ and $M$, and $\Delta\rho^{1/3}$ reflects the difference in interfacial tension between the elemental metals. The difference of the interfacial energies of the two pure metals is generally controlling the segregation behaviour [465]. This easy method offers a very realistic view on surface segregation in many systems [471].

### 4.4.4 Model of Kumar

Kumar [472] used the bond-breaking model to extend a quasichemical formulation for chemical composition at the surfaces of non-regular solid solutions. Considering only the nearest-neighbour interactions, different behaviour of individual layers parallel to the interface and the equal relaxations for all bonds, the enthalpy of segregation, $\Delta H_I$, in the interface layer in a $I_xM_y$ binary alloy can be expressed as

$$\Delta H_I = \frac{\varepsilon_{II} - \varepsilon_{MM}}{2}\left[Z - (Z_{\mathrm{L}} + Z_{\mathrm{P}})(1 + \alpha)\right], \tag{4.114}$$

where $\varepsilon_{ii}$ are the bond enthalpies for the *ii*-type nearest neighbour pair. $Z$, $Z_L$ and $Z_P$ are the total, interlayer and intralayer numbers of the nearest neighbours of an atom in the interfacial layer, respectively, and $\alpha$ is the relaxation parameter. The segregation entropy can be expressed in a similar way. A good agreement between the results obtained according to this model and experimental data was found for surface segregation in silver–gold and copper–nickel systems.

### 4.4.5 Model of Mukherjee and Morán-López

According to this model based on a simple tight-binding theory of the surface segregation in alloys of transition metals [473, 474], the equilibrium interface composition of an $I_X M_Y$ alloy can be determined by minimising the total free energy

$$F(X_0, X_1, \ldots, X_N) = \sum_{\lambda} [E_\lambda + kT(X_\lambda \ln X_\lambda + Y_\lambda \ln Y_\lambda)] - \mu N \sum_{\lambda} X_\lambda \tag{4.115}$$

with

$$E = -\frac{1}{2} W_n (1 - n) + n\varepsilon_0, \tag{4.116}$$

$$W_{MI}^2 = 12\left[XY(\varepsilon_I - \varepsilon_M)^2 + (XW_I + YW_M)^2/Z\right] \tag{4.117}$$

and

$$\begin{aligned}\left(W_{MI}^{\Phi}\right)^2 &= 12[X_0 Y_0 \left(\varepsilon_I^{\Phi} - \varepsilon_M^{\Phi}\right)^2 + Z_L(X_0 W_I + Y_0 W_M)^2/Z^2 \\ &\quad + Z_P(X_0 W_I + Y_0 W_M)(XW_I + YW_M)/Z^2].\end{aligned} \tag{4.118}$$

In (4.115)–(4.118), $X$ and $Y$ denote the bulk composition of the alloy $I_X M_Y$ and $Y_\lambda$ are the compositions of individual surface ($\lambda = 0$) and subsurface layers $(1, \ldots)$, $W$ is the bandwidth centred at the energy $\varepsilon_0$, and $n$ is the fractional occupation number. $W_{MI}$ and $W_{MI}^{\Phi}$ are the effective bandwidths in the bulk and at the surface, respectively, $\varepsilon_i^{\Phi}$ and $\varepsilon_i$ are the centres of the surface and the bulk $d$-bands of the pure metals $i$ ($i = I, M$), respectively. $Z$, $Z_L$ and $Z_P$ are the bulk, intralayer and interlayer coordination numbers, respectively. Mukherjee and Morán-López used a rectangular shape of the $d$-band density of states with the bandwidth, band centre and band filling as only input parameters and predicted the surface segregation for any binary alloy of transition metals. In contrast to the model of Miedema [465,470], the equilibrium surface concentration can be calculated for any bulk composition as a function of temperature [473].

### 4.4.6 BFS Model

Bozzolo, Ferrante and Smith developed a technique for calculating the alloy properties such as the heat of formation [475], the surface energy of binary alloys [476] and the heat of segregation of substitution impurities [477,478]. Based on their BFS model, the energy of segregation, $\Delta E_I^{\text{seg}}$, is defined as the difference between the heat of formation of a semi-infinite crystal $M$ with an impurity $I$ located at a lattice site on a plane parallel to the surface and the heat of formation of the same structure but with the atom $I$ located in a lattice site in the bulk. In this way, $\Delta E_I^{\text{seg}}$ can be expressed (analogously to the model of Wynblatt and Ku) as a sum of two contributions, the strain term, $\Delta E_I^{\text{strain}}$, and the chemical term, $\Delta E_I^{\text{chem}}$,

$$\Delta E_I^{\text{seg}} = \Delta E_I^{\text{chem}} + \Delta E_I^{\text{strain}}, \tag{4.119}$$

where

$$\Delta E_I^{\text{strain}} = e_I^{\text{strain},\Phi} - e_I^{\text{strain,bulk}} - e_M^{\text{strain},\Phi} \tag{4.120}$$

and

$$\begin{aligned}\Delta E_I^{\text{chem}} = \sum_q g_{Mq}\left[f_q^{\Phi} e_{Mq}^{\prime\text{chem}} + g_q^{\Phi} e_{Mq}^{\prime\prime\text{chem}}\right] - N_1 e_{Mb}^{\prime\text{chem}} \\ - N_2 e_{Mb}^{\prime\prime\text{chem}} + g_{I\Phi} e_I^{\text{chem},\Phi} - g_{Ib} e_I^{\text{chem},b}\end{aligned} \tag{4.121}$$

In (4.120) and (4.121), $e_i^{\text{strain}}$ and $e_i^{\text{chem}}$ are the strain and chemical energies, respectively, of atoms $i$ $(i = I, M)$ at the interface $\Phi$ and in the bulk. $G_i$ are the coupling terms of $i$ at the interface and in the bulk $b$ and $e_{Mq}^{\prime\text{chem}}$ and $e_{Mq}^{\prime\prime\text{chem}}$ are the chemical energies between atoms for the nearest and the next-nearest neighbours, respectively, at the $q$th layer. $N_1$ and $N_2$ are the total numbers of the nearest and the next-nearest neighbours, respectively, and $f_q^{\Phi}$ and $g_q^{\Phi}$ are the numbers of the nearest and the next-nearest neighbours, respectively, in layer $q$ in respect to the atom located in layer $\Phi$. Equations (4.119)–(4.121) were derived for the unrelaxed thermodynamic state. Monte Carlo calculations at zero temperature enable selected atoms to undergo relaxation and provide relaxed configurations and, thus, the segregation energies. Individual parameters of the BFS theory are determined from pure elemental data and from only two-alloy properties [478].

BFS model has successfully been tested for surface segregation [477] but can be extended to the grain boundaries after appropriate modification. The main advantage of this model is that it allows the derivation of simple approximate expressions describing the trends in segregation as well as the elucidation of driving mechanisms for these phenomena.

### 4.4.7 SMA–TBIM Approach

To study interfacial segregation, a tight-binding Ising model (TBIM) used for studies of the surface segregation [479, 480] was extended to the grain boundary segregation assuming that the energetic parameters of the Ising model are determined by atomistic simulations utilising atomic potentials derived from the second-moment approximation (SMA) of the tight-binding scheme [481, 482]. Using the tight-binding Hamiltonian and generalised perturbation method, an effective Ising Hamiltonian can be obtained. By means of this Hamiltonian, the part of the energy that depends on chemical configuration can be calculated. Supposing the simplest case, the index of the possible sublattices is omitted. Then, the concentration $X_I^p$ of solute $I$ in $p$th plane parallel to the grain boundary plane can be determined according to [481,482] as

$$\frac{X_I^p}{1 - X_I^p} = \frac{X}{1 - X}\exp\left(-\frac{\Delta G_{I,p}}{RT}\right). \tag{4.122}$$

In (4.122), $\Delta G_{I,p}$ is the Gibbs energy of segregation of $I$ at $p$th plane. The enthalpy part, $\Delta H_{I,p}$, consists of two contributions – the standard and the excess enthalpies,

$$\Delta H_{I,p} = \Delta H_{I,p}^0 + \Delta H_{I,p}^{\mathrm{E}}, \tag{4.123}$$

where the concentration-dependent excess contribution takes into account for the interaction between the segregating atoms,

$$\Delta H_{I,p}^{\mathrm{E}} = 2\sum_R \left( \sum_{p'=-q}^{p'=+q} Z_R^{p,p+p'} V_R^{p,p+p'} X_{I,p+p'} - Z_R V_R X_I \right). \tag{4.124}$$

In (4.124), $V_R = \left(V_R^{AA} + V_R^{BB} - 2V_R^{AB}\right)/2$ is the effective pair interaction energy between $R$th neighbours, $Z_R$ is the bulk co-ordination number for the $R$th shell of neighbours, $Z_R^{p+p'}$ is the number of $R$th neighbours between planes $p$ and $p'$. The index $q$ defines the number of planes that have to be considered ($2q + 1$ in bulk) consistently with the spatial extension of $V_R$. The effective pair interactions can vary at the interface. This explains the occurrence of the exponent in $V_R^{p,p+p'}$ in the term depending on $X_{I,p+p'}$ [482]. The term $\Delta H_{I,p}^0$ consists of three contributions [479,481],

$$\Delta H_{I,p}^0 = \Delta H_{I,p}^{\mathrm{size}} + \Delta H_{I,p}^{\mathrm{site}} + \Delta H_{I,p}^{\mathrm{EPI}}, \tag{4.125}$$

that is (a) of the size mismatch between solute $I$ and matrix $M$, (b) of the difference of sites on the $p$th plane between the pure constituents and (c) of the effective pair interaction, respectively [482]. The latter term is related to $V_R$

$$\Delta H_{I,p}^{\mathrm{EPI}} = \sum_{R} V_R \left( Z_R - \sum_{p'=-q}^{p'=+q} Z_R^{p,p+p'} \right). \tag{4.126}$$

The most important result of the SMA-TBIM model is the determination of the energetic parameters of (4.123)–(4.126) on basis of realistic simulations [481,483]. $N$-body atomic potentials may be derived from the SMA of the tight-binding scheme [482, 484]. Minimisation of the enthalpy with respect to atomic positions done by quenched molecular dynamics algorithm results in the value of $\Delta H_{I,p}^{\mathrm{site}}$. To calculate $\Delta H_{I,p}^{\mathrm{size}}$, the atomic potentials are the same for the *I–I, M–M* and *I–M* interactions except that fixing the lattice parameter for each metal. Thus, the size effect is separated from the site and effective pair interaction effects. $V_R$ can be obtained by considering the difference of the enthalpy of a relaxed system, $\Delta H_R$, containing two isolated solute atoms (initial state) and of two solute atoms in $R$th neighbour positions (final state), $V_R = \Delta H_R/2$ [483].

The vibrational part of the segregation entropy which is involved in $\Delta G_{I,p}$ and $\Delta S_{I,p}$ can be estimated by a recursion method [485] using the relaxed values of the atomic positions and the force constants obtained on the basis of SMA potential [482].

# Chapter 5
# Effect of Variables on Equilibrium Grain Boundary Segregation

There are many examples of both experimental and theoretical evidence of the effect of various thermodynamic and structural variables on the grain boundary segregation [20]. We did already see the complex effect of presence of several segregating elements (Chap. 4). Accepting its thermodynamic description in the sense of the thermodynamic approach [(4.32) and (4.33)], we can understand that the influence of individual parameters such as magnetic field, grain boundary structure and character of both segregating and matrix elements on grain boundary segregation is primarily involved in variations of the values of $\Delta H_I^0$ and $\Delta S_I^0$. Additionally, temperature affects the values of the whole exponential term in (4.32). Let us also mention that all the above-mentioned effects as well as pressure, and bulk and grain boundary concentrations contribute to variations of the values of $\Delta G_I^{\mathrm{E}}$. Among the intensive thermodynamic parameters affecting the segregation behaviour of individual systems, we can consider temperature, pressure, magnetic field and grain boundary energy. Grain size is an additional parameter affecting interfacial segregation.

## 5.1 Temperature and Bulk Composition

In a binary system, the value of the argument of the exponential term of segregation isotherm (4.32) approaches to negative infinity at low temperatures supposing $\Delta H_I^0 \neq 0$, and thus the exponent is equal to zero at $T = 0\,\mathrm{K}$. With increasing temperature, the absolute value of the exponent reduces causing that the grain boundary concentration decreases, as it is apparent from Fig. 5.1.

It is also seen from Fig. 5.1 that the course of this concentration decay depends on the values of both, the enthalpy and entropy of segregation, and the bulk concentration of the solute. Because practically the equilibrium is hardly reached at temperatures below 650 K, complete saturation of the boundary by the segregating element is not reached. In case of the solute with low absolute value of $\Delta H_I^0$ (Fig. 5.1b), the equilibrium concentration at such temperatures is so low that we speak in such case about less segregating element although theoretically, it should completely cover the boundary at very low temperatures. In case of the element with

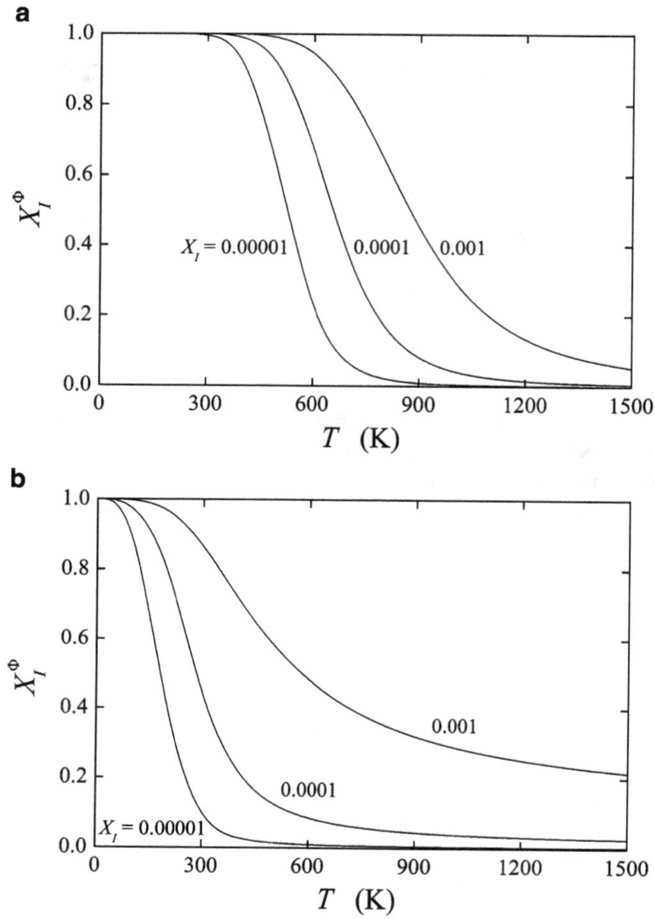


**Fig. 5.1** Temperature dependence of grain boundary concentration in model binary systems characterised by (**a**) $\Delta H_I^0 = -50$ kJ/mol, $\Delta S_I^0 = 0$, and (**b**) $\Delta H_I^0 = -10$ kJ/mol, $\Delta S_I^0 = +40$ J/(mol K), for three different bulk concentration levels, 0.1, 0.01 and 0.001at.%

rather high absolute value of $\Delta H_I^0$, the monolayer segregation can also be reached at real temperatures. It is also worth noting that solute segregation does not depend on the melting point: The segregation isotherm (cf. (4.32)) is identical for all elements and the melting temperature is not included there. Therefore, in the case of the matrix elements with low melting point (such as aluminium or indium), we may expect to observe large segregation effects just below the melting point, while in case of the matrix elements with high melting point, the segregation effects below the melting temperature are very weak.

In case of a multi-component alloy, the segregation behaviour depends on mutual interaction of the segregating elements. Assuming the segregation of the same solutes in identical matrix as shown, for example in Fig. 5.1, we can distinguish principally four different cases:

(a) Non-interactive segregation on different sites
(b) Site competitive segregation at the same sites
(c) Segregation with repulsive interaction
(d) Segregation with attractive interaction

All these cases are documented in Figs. 5.2 and 5.3. It is apparent that the character of the temperature dependence of the grain boundary concentration is different in individual cases. In case of non-interactive segregation at different sites at the

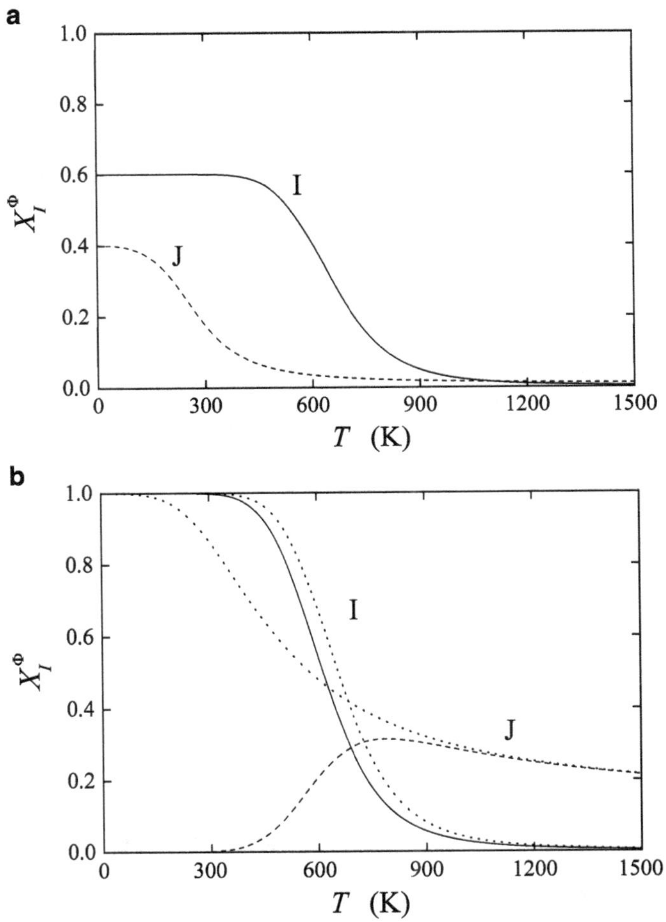


**Fig. 5.2** Temperature dependence of the grain boundary concentration in model ternary systems characterised by $\Delta H_I^0 = -50\,\text{kJ/mol}$, $\Delta S_I^0 = 0$, $\Delta H_J^0 = -10\,\text{kJ/mol}$, $\Delta S_J^0 = +40\,\text{J/(mol K)}$, $X_I = 0.0001$, $X_J = 0.001$. (**a**) Non-interactive segregation to different sites ($X_I = 0.6$, $X_J = 0.4$); (**b**) site competitive segregation at identical sites. *Dotted lines* represent the behaviour of the solutes in binary alloys (cf. Fig. 5.1)

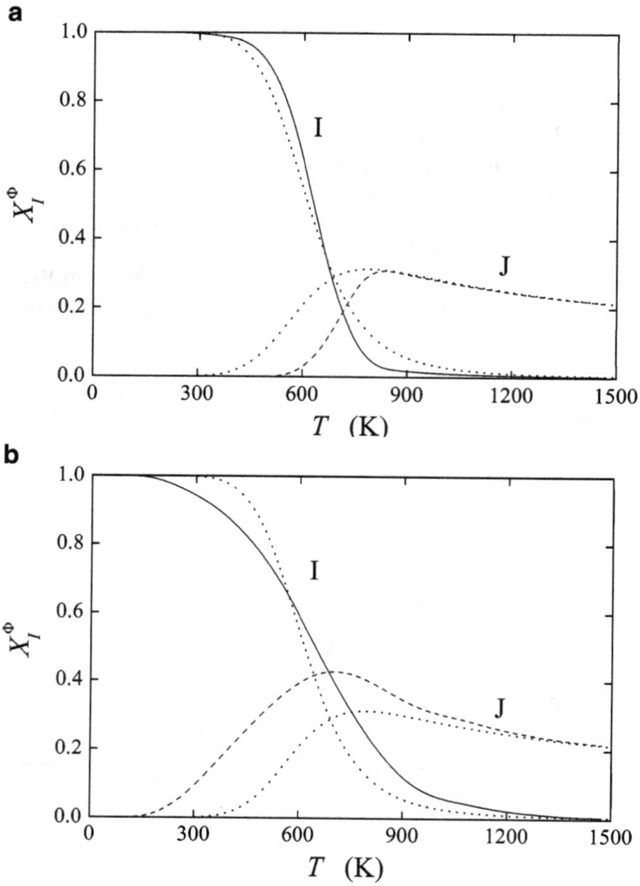


**Fig. 5.3** Temperature dependence of the grain boundary concentration in model ternary systems characterised by $\Delta H_I^0 = -50$ kJ/mol, $\Delta S_I^0 = 0$, $\Delta H_J^0 = -10$ kJ/mol, $\Delta S_J^0 = +40$ J/(mol K), $X_I = 0.0001$, $X_J = 0.001$. (**a**) Segregation with repulsive interaction, $\alpha'_{IJ} = +20$ kJ/mol; (**b**) segregation with attractive interaction, $\alpha'_{IJ} = -20$ kJ/mol. *Dotted lines* represent the behaviour of the solutes in non-interacting ternary alloys (cf. Fig. 5.2b)

interface (Fig. 5.2a), the segregation of both elements is similar to that found for a binary system, that is independent of the grain boundary concentration of the other element. In this case, we speak about the segregation in a pseudobinary system. The only difference is that the grain boundary concentrations of both solutes do not approach 1 at 0 K, but they approach the doles of the sites of corresponding type at the interface.

When both solutes segregate at the same sites, a *site competition* occurs (cf. Chap. 4). At low temperatures, it is characterised by complete occupation of the interface by the solute $I$ with a higher segregation tendency (higher $|\Delta H_I^0|$), while the less segregating solute $J$ (lower $|\Delta H_J^0|$) is completely repelled from the grain boundary. At higher temperatures, the concentration of $J$ increases but it is still $X_J^\Phi < X_J$: this situation is understood as the grain boundary *depletion* from $J$. With increasing temperature, $X_I^\Phi$ decreases similarly as it does in the binary alloy (dotted line in Fig. 5.2b) and therefore, more grain boundary sites become available for segregation of $J$. $X_J^\Phi$ then increases with increasing $T$, reaches maximum and decreases thus approaching the concentrations comparable to its equilibrium value of the grain boundary segregation in a binary alloy (dotted line in Fig. 5.2b) at the same temperature. When the solutes interact with each other, the courses of the temperature dependence of their grain boundary concentrations change quantitatively depending on the character of the interaction (Fig. 5.3). In case of the repulsive interactions (Fig. 5.3a), the segregation of the more surface-active element $I$ dominates at lower temperatures again but the decrease of its interfacial concentration at intermediate temperatures is more sharp than in a non-interacting system, that is the derivation of the temperature dependence in the inflection point is more negative under repulsive interaction. The absolute value of the slope of the temperature dependence of the grain boundary segregation of the more active solute increases with increasing value of the ternary interaction coefficient $\alpha'_{IJ}$ and at high values, it may cause an abrupt change from a very high concentration to a very low one. The segregation of the less surface-active element $J$ is reduced due to repulsive interaction and is shifted to higher temperatures.

In case of the attractive interaction between segregating solutes, the more surface-active solute also possesses higher interfacial concentration, however, due to an attractive interaction, the less surface-active solute starts to segregate at much lower temperatures than in a non-interactive system. The slopes of the curves are much lower than in system with repulsive interaction as well as in non-interacting systems.

In fact, the repulsive interaction means that both elements repeal each other from the grain boundary, while in case of the attractive interaction both elements tend to segregate more strongly compared to the non-interactive case. It can be understood in terms of the grain boundary concentration of the matrix element: At a given temperature, its interfacial concentration is higher when both solutes mutually repeal, and lower when they attract (Fig. 5.4).

The character of the temperature dependence of the grain boundary segregation in a ternary alloy also changes with changing values of the thermodynamic parameters of interfacial segregation and the bulk concentration: Under some circumstances, a qualitative change of the temperature dependence of the grain boundary segregation may even occur (e.g. a dominating segregation of the solute with lower $|\Delta H_I^0|$ when its bulk concentration is large enough) [327].

## 5.2 Pressure

It follows from (4.49) and (4.51) that $\Delta G_I^0$, $\Delta H_I^0$ and $\Delta S_I^0$ are *principally independent of pressure*. However, the studies – although only rarely described in literature – show a pronounced effect of pressure on interfacial segregation. Lee and Chiang

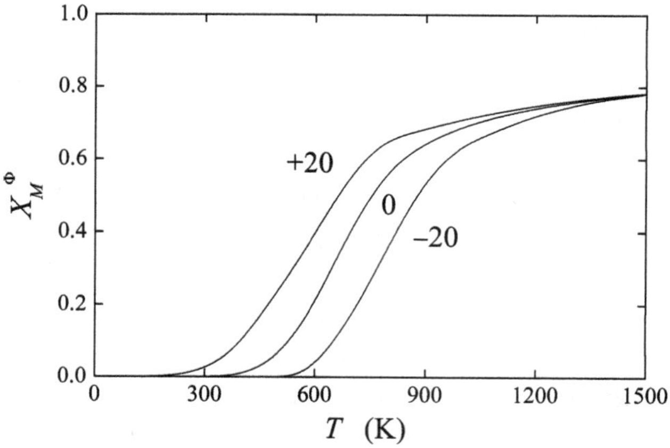


**Fig. 5.4** Temperature dependence of the grain boundary concentration of the matrix element in model ternary systems characterised by $\Delta H_I^0 = -50\,\text{kJ/mol}$, $\Delta S_I^0 = 0$, $\Delta H_J^0 = -10\,\text{kJ/mol}$, $\Delta S_J^0 = +40\,\text{J/(mol K)}$, $X_I = 0.0001$, $X_J = 0.001$ and by the values of the ternary interaction parameter, $\alpha'_{IJ} = +20\,\text{kJ/mol}$, 0 and $-20\,\text{kJ/mol}$

found that the grain boundaries of ZnO are saturated by bismuth at ambient pressure (0.1 MPa) while its segregation at 1 GPa is completely suppressed [486]. Since $\Delta G_I^0$ is independent of pressure, all changes in the chemical composition of grain boundaries are involved in the pressure dependence of the excess Gibbs energy of segregation,

$$\left(\frac{\partial \Delta G_I}{\partial P}\right)_{T,n_i} = \left(\frac{\partial \Delta G_I^{\mathrm{E}}}{\partial P}\right)_{T,n_i} \tag{5.1}$$

and, thus, in the pressure dependence of corresponding activity coefficients. This effect is so high to reduce the Gibbs energy of segregation from $-65\,\text{kJ/mol}$ at 0.1 MPa to zero at 1 GPa.

This finding is in agreement with expectations that high pressure will prevent grain boundary segregation [487]. The reason for it might also be connected with the pressure effect on the phase diagrams by shifting the solvus-line, that is by increasing the solid solubility limit at a given temperature with increased pressure.

## 5.3 Magnetic Effects

Similarly to the effect of pressure, magnetism affects segregation behaviour of individual solutes. Here, we should recognise two effects (a) the effect of the magnetic state of the material and (b) the effect of external magnetic field on grain boundary segregation.

The Gibbs energy of segregation, $\Delta G_I$, in (4.65) is characteristic for segregation in *paramagnetic* state. When the matrix is ferromagnetic, the total Gibbs energy of segregation, $\Delta G_I^{\mathrm{T}}$, used in exponent of the segregation isotherm will additionally contain a ferromagnetic term $\Delta G_I^{\mathrm{F}}$ [488],

$$\Delta G_I^{\mathrm{T}} = \Delta G_I + \Delta G_I^{\mathrm{F}}. \tag{5.2}$$

The ferromagnetic term $\Delta G_I^{\mathrm{F}}$ is temperature dependent and can be approximated by

$$\Delta G_I^{\mathrm{F}}(T) = (1 - mX_I)\left(T_{\mathrm{c}}/T_{\mathrm{c}}^0\right)\Delta G_M^{\mathrm{F}}(T_r), \tag{5.3}$$

where $\Delta G_M^{\mathrm{F}}$ is the ferromagnetic term of the Gibbs energy for ferromagnetic matrix element $M$, $T_{\mathrm{c}}^0$ and $T_{\mathrm{c}}$ are the Curie temperatures of pure matrix $M$ and $M{-}I$ solid solution, $T_r$ is the temperature term defined as $T_r = TT_{\mathrm{c}}^0/T_{\mathrm{c}}$. $m$ is the factor representing the effect of the solute element $I$ on the size of ferromagnetic term approximated by $m = 1$ for nonmagnetic solutes and $m = 0$ for magnetic solutes [488]. It follows from this treatment that

$$\Delta H_I^{\mathrm{F}}(T) = (1 - mX_I)\left(T_{\mathrm{c}}/T_{\mathrm{c}}^0\right)\Delta H_M^{\mathrm{F}}(T_r) \tag{5.4}$$

and

$$\Delta S_I^{\mathrm{F}}(T) = (1 - mX_I)\Delta S_M^{\mathrm{F}}(T_r), \tag{5.5}$$

where the meaning of the enthalpy, $\Delta H_M^{\mathrm{F}}$, and entropy, $\Delta S_M^{\mathrm{F}}$, corresponds to that of Gibbs energy, $\Delta G_M^{\mathrm{F}}$. It means in fact, that the dependence of $G^{\Phi}$ in ferromagnetic region of the concentration dependence of the Gibbs energy (Fig. 4.1) is shifted to higher values. The equilibrium concentration, $X_I^{\Phi}$, corresponding to the parallel tangents of the shifted curve will also be higher as compared to the paramagnetic state. Detailed study of sulphur and antimony segregation in CoNi resulted in the values of $\Delta H_{\mathrm{Sb}}^0 = -5\,\mathrm{kJ/mol}$, $\Delta S_{\mathrm{Sb}}^0 = 38\,\mathrm{J/(mol\,K)}$, $\Delta S_{\mathrm{S}}^0 = -32\,\mathrm{kJ/mol}$ and $\Delta S_{\mathrm{S}}^0 = 47\,\mathrm{J/(mol\,K)}$ [488]. An analysis of the data on phosphorus segregation in bcc iron from the viewpoint of the magnetic effect provided us with the values of $\Delta H_{\mathrm{P}} = -15\,\mathrm{kJ/mol}$ and $\Delta S_{\mathrm{P}} = 33\,\mathrm{J/(mol\,K)}$ [488]. In fact, the correlation of experimental data on phosphorus grain boundary concentration [328] in the ferromagnetic region fully corresponds with that using the values of $\Delta H_{\mathrm{P}}^0 = -34.3\,\mathrm{kJ/mol}$ and $\Delta S_{\mathrm{P}}^0 = 21.5\,\mathrm{J/(mol\,K)}$ [328] but differs slightly in paramagnetic region. In fact, the data on solute segregation at grain boundaries given in Appendix A should be considered as corresponding to ferromagnetic state. Further conclusion about the effect of ferromagnetism is that it strengthens the segregation effects.

External magnetic field may affect the grain boundary segregation of both magnetic and surprisingly even non-magnetic materials. An increase of the grain boundary energy in the Fe–0.8at.%Sn alloy by 15 rel.% with increasing external magnetic field from 0 to 6 T was found during annealing the samples at 973 K [489, 490]. Based on findings of Seah and Hondros [307], it suggests that tin grain boundary concentration decreases with increasing effect of external magnetic field. This deduction was proved by direct measurement of grain boundary chemistry after

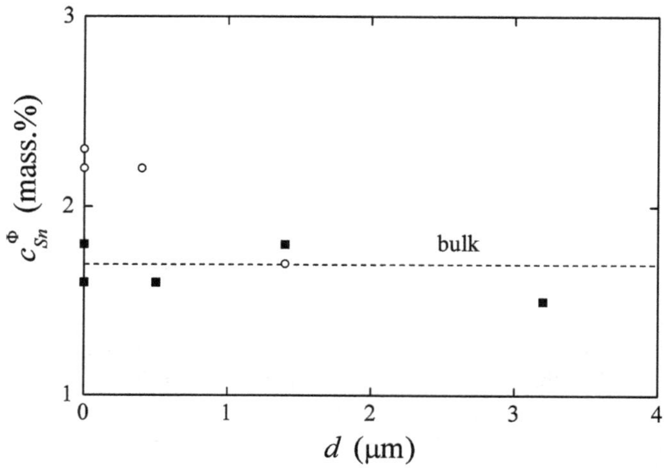


**Fig. 5.5** Concentration of tin at grain boundaries of iron in depth $d$ under the grain boundary after annealing at 973 K for 6 h without application of magnetic field (*empty circles*) and after annealing in magnetic field of 3 T (*solid squares*) (according to [490])

"magnetic annealing" at 973 K: While tin enrichment of the grain boundaries by factor 1.5 was found after annealing without application of magnetic field, no segregation was detected after annealing in magnetic field of 3 T [489] (Fig. 5.5). In fact, annealing of the samples in magnetic field is known for nearly 100 years as providing finer and more homogeneous structures of materials than conventional annealing without magnetic field [490].

Tsurekawa et al. [491] consider two sources for the effect of magnetic field on interfacial segregation (a) magnetic Gibbs energy and (b) grain boundary magnetism.

Magnetic Gibbs energy in ferromagnetic materials related to unit volume is generally given as [491]

$$\Delta G_I^{\mathrm{F}} = -\mu_0 \left( H - \frac{NM_{\mathrm{s}}}{2} \right) M_{\mathrm{s}}, \tag{5.6}$$

where $\mu_0$ is the magnetic permeability of vacuum, $M_{\mathrm{s}}$ is the saturation magnetism, $H$ is the strength of magnetic field and $N$ is the demagnetising factor. In dia- and paramagnetic materials, the magnetic Gibbs energy per unit volume is

$$\Delta G_I^{\mathrm{D}} = \Delta G_I^{\mathrm{P}} = -\frac{1}{2}\mu_0 \chi (1 - \chi N) H^2, \tag{5.7}$$

where $\chi$ is the susceptibility. Paramagnetic tin atoms segregated at the grain boundaries of iron may be considered as "clouds," thus enhancing $\Delta G_I^{\mathrm{F}}$ of the system in applied magnetic field despite the fact that these "clouds" do not affect the Curie temperature [492]. The reason for this enhancement is much higher magnetic free energy of the system at the "cloud" than in the grain volume in magnetic field, that is $\Delta G_I^{\mathrm{P}} \gg \Delta G_I^{\mathrm{F}}$ due to extremely low susceptibility of tin ($2.7 \times 10^{-8}$ [491]). An estimate of the energy difference $\Delta G_I = \Delta G_I^{\mathrm{F}} - \Delta G_I^{\mathrm{P}}$ according to (5.6) and (5.7) assuming $N = 0$ gives $-6 \times 10^6\,\mathrm{J/m^3}$ for $H = 6\,\mathrm{T}$. Comparing this value with $\Delta G_{\mathrm{Sn}} = -6.5 \times 10^5\,\mathrm{J/m^3}$ reported by Seah and Lea for tin segregation in α-iron at 973 K [371], one can conclude that the magnetic Gibbs energy is one order of magnitude lower than the Gibbs energy of grain boundary segregation. As a consequence, application of a magnetic field will repeal tin atoms from the grain boundaries, because this process results in lowering the magnetic Gibbs energy and, thus, the Gibbs energy of the system [491]. It can also be assumed that the demagnetisation factor for the "cloud" depends on the direction of the magnetic field. It will be much lower for magnetic field being parallel to the grain boundary compared to the perpendicular case. Thus, the series $\Delta G_I^{\mathrm{P}}(\perp) > \Delta G_I^{\mathrm{P}}(||) \gg \Delta G_I^{\mathrm{F}}$ or $\Delta G_I^{\mathrm{D}}(\perp) > \Delta G_I^{\mathrm{P}}(||) \gg \Delta G_I^{\mathrm{F}}$ should be valid for segregated elements in paramagnetic state (e.g. tin) or diamagnetic state (e.g. copper), respectively.

Increased grain boundary magnetism is another possible explanation of the effect of magnetic field on grain boundary segregation. It was shown experimentally [493, 494] as well as theoretically [495] that the magnetic moment at the grain boundary is different from that of the grain volume and that it increases with decreasing atomic density. In case of tin in α-iron it can be well assumed that both the magnetisation and the Curie temperature are lower for segregated grain boundary than for the volume. Simultaneously, it can be expected that the difference in magnetisation of the grain boundary and the volume in a sample annealed at a temperature close to the Curie point will intensify the grain boundary segregation of tin. An external magnetic field will then evoke the field-induced magnetisation resulting in decrease of the difference in magnetisation of the grain boundary and the volume thus suppressing the grain boundary segregation caused by the magnetic effect [491].

## 5.4 Grain Boundary Energy

Changes in grain boundary energy affect the values of the chemical potentials of the solute and matrix elements at the grain boundary and thus the standard molar Gibbs energy of segregation (cf. Chap. 4, (4.30)). It results in *anisotropy of grain boundary segregation*.

Grain boundary energy exhibits a more-or-less pronounced dependence on the structure of the interface. Such dependence is shown, for example in Fig. 5.6. Due to the fact that grain boundary energy substantially contributes to the value of the chemical potentials (4.23) and according to the definition of thermodynamic quantities of grain boundary segregation as combination of chemical potentials (4.28) and (4.30), anisotropy of the thermodynamic quantities of grain boundary segregation must exist. Similarly, anisotropy of grain boundary composition may also

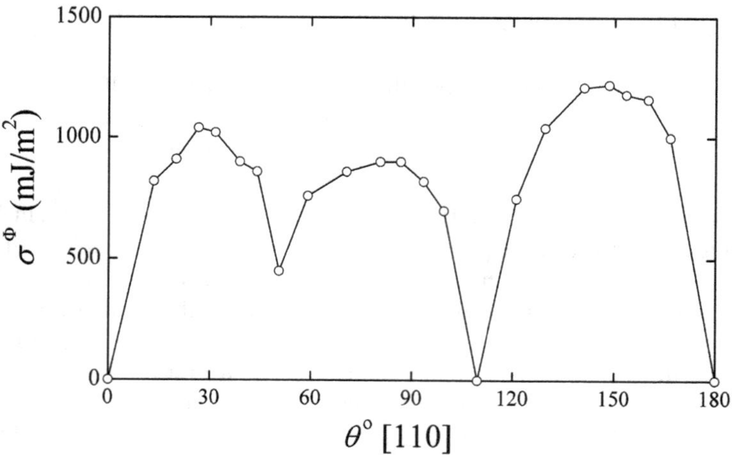


**Fig. 5.6** Dependence of energy of symmetrical tilt grain boundaries, $\sigma^{\Phi}$, on [110] misorientation angle $\theta$ in Ni–4at.%Pd alloy at 0 K (according to [341])

appear, however, due to its additional dependence on temperature, pressure and bulk composition of the alloy, it is not so straightforward as the anisotropy of the standard thermodynamic quantities.

Anisotropy of grain boundary segregation was experimentally detected for the first time from a scatter of data on chemical composition of individual "facets" on intergranular fracture surfaces, that is opened grain boundaries in polycrystalline materials albeit crystallographically unspecified. For example, variations of about ±15 rel.% were found for tin and sulphur segregation in α-iron [307] and ±45 rel.% for phosphorus segregation in a Cr–Mo low-alloy steel [173]. Grain boundaries in bismuth-doped copper were either fully decorated by bismuth or completely free of it [496]. Indications of non-homogeneous grain boundary segregation with scatters ranging from ±30 to ±60 rel.% were also reported in many other systems: for segregation of phosphorus [67, 166, 349, 497, 498], carbon [166, 430, 499], antimony [67,419,424,500,501], nickel [428,500], manganese [67,498], tin [500], molybdenum [426, 497], and chromium and oxygen [166, 428] in α-iron, and ferrite-based iron alloys and steels. Other examples are grain boundary segregation of antimony, sulphur and phosphorus in nickel [374,454], of boron in $Ni_3Al$ [502], of phosphorus in tungsten [370] or of iron in magnesia [192]. Generally, the majority of the literature data are confined in a broad region limited by the Maxwell distribution with scatter ±60 rel.% for 50% of the Maxwell curve [20].

Experimental data on chemistry of a single grain boundary measured by methods of surface analysis on fracture surfaces can exhibit a scatter. This scatter up to about ±10 rel.% can be caused by deviations of the fracture path from the interface, inhomogeneities of bulk chemical composition and lack of equilibrium. A larger scatter then reflects anisotropy of grain boundary segregation [67, 503]. Theoretical analysis of the bond strength suggests that intergranular fracture passes through a metallic material by breaking the weakened metal–metal bonds in the neighbourhood of the strong metal–metalloid bonds [503–505]. The fracture path zigzags along the boundary and the segregated species are distributed between the two newly created fracture surfaces. It is usually supposed that this distribution is homogeneous, that is that on average, there is always one-half of the amounts of the segregated element(s) on each fracture surface. Some measurements of grain boundary composition on both matching fracture surfaces support this idea, e.g. phosphorus concentration on fracture surfaces of polycrystalline tungsten [372], Cr–Ni ferrite steel [67] and α-iron [343, 354]. Careful studies of composition of both matching fracture surfaces of well-characterised single grain boundaries in bicrystals brought progress in understanding this distribution. Markedly different phosphorus concentrations were found on both sides of a separated symmetrical {013} tilt grain boundary in a Fe–Si alloy [139]. Model calculations suggested that the fracture did not zigzag through the grain boundary core but passed parallel to the grain boundary at a distance of one atomic layer from the core over a relatively large area. After meeting, a defect at the grain boundary such as deformation twin occurring in front of the crack during breaking the bicrystal, the fracture can jump onto the corresponding path on the other side of this boundary. However, once the equilibrium grain boundary segregation is established, the fracture runs just in the grain boundary core and distributes the segregated species homogeneously to both fracture surfaces in the case of this grain boundary and in the case of other symmetrical interfaces [111]. On the other hand, systematic differences of composition between the two fracture surfaces found for asymmetrical grain boundaries suggest that fracture unevenly distributes the segregated species between the fracture surfaces [111]. The difference between chemical composition of both fracture surfaces can reach as much as ±30 rel.% [94, 506, 507].

Li and Williams [508] studied phosphorus segregation at individual grain boundaries in rapidly solidified Fe–0.6mass%P alloy. Despite a strong non-equilibrium character of such interfaces, they found heterogeneous distribution of phosphorus at the grain boundaries ranging from 0.05 to 0.4 monolayer with significantly lower segregation level at low-angle and low-$\Sigma$ grain boundaries comparing to high-angle and high-$\Sigma$ ones. During this AEM study, different phosphorus concentrations were detected on matching planes of the grain boundary.

To elucidate the relationship between the structure and chemical composition of individual grain boundaries, the measurements of interfacial segregation should be performed on well-characterised grain boundaries in bicrystals. Numerous experiments of this kind were performed in the past decades. Very low segregation has usually been observed at coherent twin grain boundaries, for example in case of copper in lead [509], phosphorus in an Fe–Si alloy [134, 510], bismuth in copper [511] and silicon and titanium in alumina [512]. It is worth noting that coherent twin boundaries are often found free of segregation [511]. Other grain boundaries exhibit higher or lower levels of segregation. Lower segregation was measured for antimony at 15°[110] and 100°[110] symmetrical tilt grain boundaries of copper than at 54°[110] and 67°[110] ones. Highly coincident grain boundaries did not show any segregation of antimony. Variation of antimony concentration along facetted grain boundaries indicates that grain boundary plane orientation plays an important role in impurity segregation [511]. This conclusion is supported by the finding of weak silicon and titanium segregation at grain boundaries parallel to at least one dense lattice plane of alumina such as (0001) [512]. Three times higher sulphur concentration was detected at the 20°[100] tilt grain boundary in nickel as compared to the 40°[100] one [513]. Low levels of carbon and oxygen segregation were detected at the 5°[100] low-angle grain boundary in tungsten than at the 50°[100] and 48°[110] high-angle grain boundaries [514]. These findings may suggest that the tendency to grain boundary segregation increases with increasing misorientation angle of the two adjoining grains. Slightly higher segregation of calcium and silicon was detected at the $\Sigma = 17, 28.5°[001]$ twist grain boundary in magnesia than at the $\Sigma = 5, 36.9°[001]$ one: much higher levels of segregation were detected at general grain boundaries [515]. In Mn–Zn ferrites, low calcium segregation was detected at highly coincident grain boundaries such as $\Sigma = 13, \{015\}$ and asymmetrical grain boundaries having one of the boundary planes close to a low-index one, for example (110) and (111) in comparison to low-coincidence grain boundaries. Segregation of calcium was accompanied by a depletion of $Fe^{2+}$ ions relatively to $Fe^{3+}$ ions [516]. Similarly, highly coincident [100] grain boundaries ($\Sigma = 3, 5$ and 13) in tetragonal $YBa_2Cu_3O_{7-\delta}$ structure were found free of segregation while general grain boundaries often contained vitreous phase indicating strong segregation effects [517]. Low-angle [518] and low-$\Sigma$ [519, 520] grain boundaries in $YBa_2Cu_3O_{7-\delta}$ exhibit identical composition as the grain interior. On the other hand, molybdenum, zirconium and strontium were found to segregate at less coincident grain boundaries [518].

Grain boundary structure also seems to have an effect on character of segregation in multi-component alloys. For example, silicon was found to segregate at both, $\Sigma = 9, (011)$ twist grain boundary [241] and $\Sigma = 9, [011]$ tilt grain boundary [521], whereas boron segregates at general grain boundaries in an Fe–Si alloy [240,521]. In both cases, carbon segregation at the same level was measured at the grain boundary fracture surfaces [240, 521], however, it is not clear whether its presence is due to segregation or surface contamination. A comparison of solute segregation at tilt and twist interfaces showed larger absorptive capacity of twist interfaces [522, 523].

Grain boundary concentration of a solute is affected by many factors discussed above, such as temperature, bulk concentration of segregating element and presence of other components in the system. For general description, the values of the enthalpy and entropy of segregation are necessary. There are several attempts to evaluate the values of the thermodynamic functions of segregation for individual grain boundaries. Surprisingly high values of $\Delta H_I$ were determined for niobium and iron segregation to stacking faults (i.e. twin grain boundary) in cobalt, $\Delta H_{Nb} = -60$ kJ/mol, $\Delta H_{Fe} = -30$ kJ/mol [522, 523]. This may result from inadequate application of the regular solution model to evaluate the thermodynamic data, because the driving force for segregation originates from third nearest neighbours and higher interactions [523]. The values $\Delta H^0_{In} = -38$ kJ/mol and $\Delta S^0_{In} = -0.05$ J/(mol K), and $\Delta H^0_{In} = -39$ kJ/mol and $\Delta S^0_{In} = -0.05$ J/(mol K) were determined for indium segregation at {115} and {1 1 10} grain boundaries of nickel, respectively,

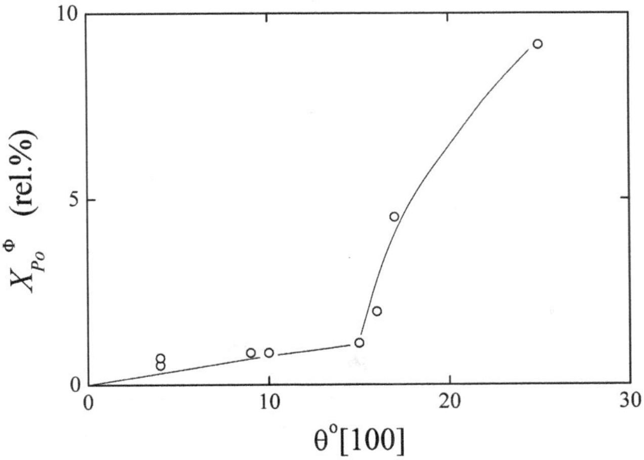


**Fig. 5.7** Dependence of relative concentration of polonium at [100] symmetrical tilt grain boundaries of lead [525]

using the Langmuir–McLean segregation isotherm (4.61) [524]. Since the entropy term has only a very small effect on $\Delta G^0_{\text{In}}$, we may accept $\Delta G^0_{\text{In}} \approx \Delta H^0_{\text{In}}$. The value $\Delta G^0_{\text{In}} \approx -40$ kJ/mol is lower than $-50$ kJ/mol measured for general grain boundaries in polycrystalline material. Thus, the above mentioned well-characterised grain boundaries cannot be considered as general [524] but rather as vicinal.

To disclose characteristic differences in behaviour of individual interfaces, the detail systematic measurements of orientation dependence of solute segregation on a set of grain boundaries is necessary. Due to large variety of grain boundaries (5 DOFs, cf. Chap. 2), such systematic study is very complicated. Usually, a cut of the 6-dimensional space is used, where only one DOF is changing or that cut is well characterised. The dependence of solute segregation on the misorientation angle $\theta$ for symmetrical tilt grain boundaries can serve, for example of the last-mentioned case [20]. The first systematic study of structural dependence of this type was reported for polonium segregation at [100] symmetrical tilt grain boundaries of lead by autoradiography [525]. It showed an increase of grain boundary segregation with increasing misorientation angle. An abrupt rise of polonium concentration appears for grain boundaries above $\theta = 15°$ (Fig. 5.7). This change of the slope of the orientation dependence is very probably caused by the change of the grain boundary structure from low-angle dislocation one to that formed by structural units in case of the high-angle grain boundaries.

The dependence of the Auger peak-to-peak heights ratios $r_I$ (which is a measure of the grain boundary concentration) on the misorientation angle of tilt grain boundaries without specifying the other DOFs was constructed for tin and silicon grain boundary segregation in bcc iron [526,527] (Fig. 5.8). The data were fitted by a monotonic dependence (full lines in Fig. 5.8) and concluded that no increase of silicon concentration exists for low-angle grain boundaries $\theta < 15°$, while a very

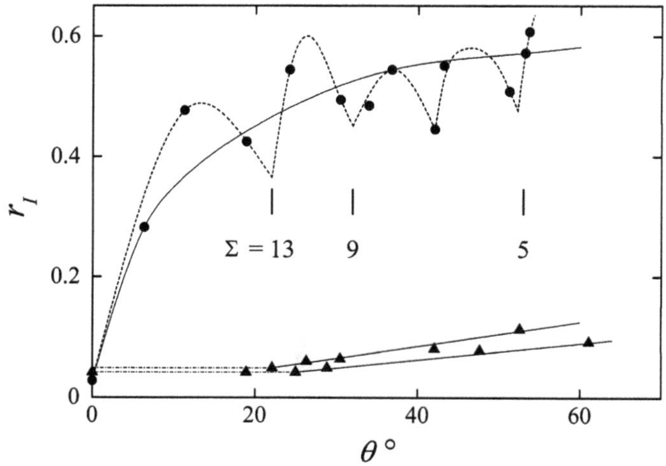


**Fig. 5.8** Dependence of relative Auger peak-to-peak height ratios, $r_I$, of silicon (*triangles*) [527] and tin (*circles*) [526] on tilt angle of rotation as presented by Watanabe et al. (*full lines*). *Dotted line* shows an alternative interpretation of the data taking into account possible low segregation at special (highly coincident) grain boundaries [20]

slight increase can be deduced for high-angle ones (15° > $\theta$ > 60°). On the other hand, tin segregation is more pronounced and exhibits steep increase for low-angle grain boundaries and "saturation" of high-angle grain boundaries [526]. No systematic trend was observed in silicon segregation at twist grain boundaries [527] while a week increase of tin concentration was apparent for tin segregation at twist grain boundaries characterised by $\theta < 4°$ [526]. These studies of anisotropy of grain boundary segregation are pioneering and very instructive. Nevertheless, the studied systems were not fully characterised [526, 527]. First, not all DOFs of individual grain boundaries were fully characterised, which does not enable to construct an unambiguous structural dependence of grain boundary segregation. Second, the samples were slowly (furnace) cooled after annealing at high temperatures (1,523 K for an Fe–Si alloy and 1,670 K for an Fe–Sn alloy): It may induce an apparent solute segregation at lower temperatures so that the segregation cannot be considered as equilibrium. Third, there were other trace elements (carbon, phosphorus, sulphur) present in particular binary alloys that are able to segregate at the interfaces. These elements then affect the amount of silicon and tin at the grain boundaries and, in addition, the fracture of the samples necessary to open the grain boundaries for the AES analysis. Additionally, the fracturing was performed on air which certainly caused an intensive contamination of the fracture surfaces by oxygen and carbon. Supposing that lower segregation occurs at special grain boundaries, which may be – for simplicity – characterised by low values of $\Sigma$ as compared to general ones [527], the experimental data [526, 527] allow such a construction (dashed line in Fig. 5.8) [20]. However, no significant dependence of oxygen, nitrogen and carbon segregation was observed in molybdenum [528].

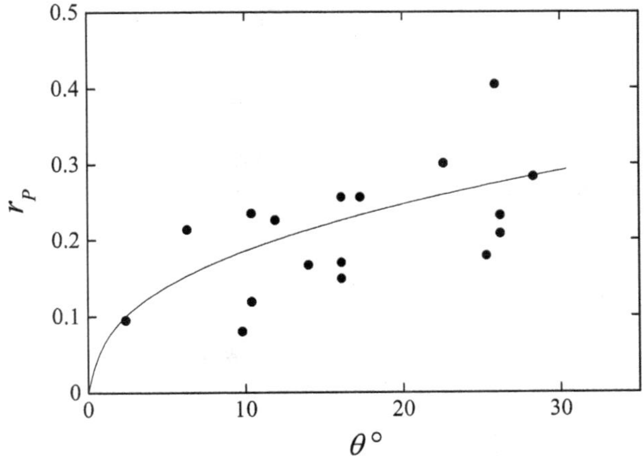


**Fig. 5.9** Relation between Auger peak-to-peak height of phosphorus $r_P$ measured on the grain boundary fracture surface of iron, and the tilt misorientation angle $\theta$ [343]

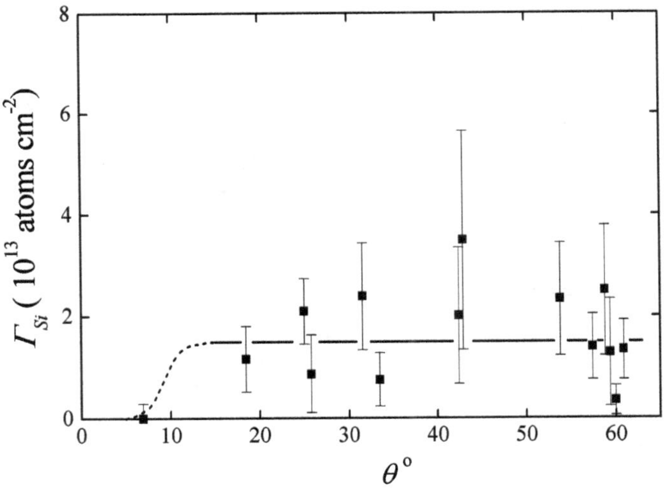


**Fig. 5.10** Dependence of grain boundary concentration of silicon $\Gamma_{Si}$ on the twist misorientation angle in bcc iron [241]

Phosphorus segregation in iron was found to increase with increasing tilt misorientation angle [343] (Fig. 5.9). Additionally, the level of silicon segregation at high-angle grain boundaries of bcc iron was interpreted to be completely independent of misorientation angle [241] (Fig. 5.10). These results qualitatively fit with the above-mentioned findings on silicon and tin segregation in bcc iron [526, 527] (Fig. 5.8). In case of silicon segregation in bcc iron [241] (Fig. 5.10), the summary dependence of grain boundary composition was plotted regardless the rotation axis and the grain boundary characteristics. In addition, the tendency of silicon to segregate at grain boundaries of bcc iron is low, so, no large variations of its concentration can be expected at different grain boundaries [20]. Small amount of phosphorus was observed at low-angle and twin grain boundaries of fcc Fe–Ni–Cr alloy in comparison with general grain boundaries [529]. In case of general grain boundaries, phosphorus was found to segregate less to those oriented closely to low-index planes of one of the grains. Such boundaries were also found straight and non-facetted [529]. Similar results were also reported for phosphorus segregation in bcc iron [530].

The above-mentioned examples of orientation dependence of grain boundary segregation only qualitatively show an increase of grain boundary concentration of a solute with increasing misorientation angle of two adjacent grains. However, they say nothing instructive about the character of the anisotropy of grain boundary segregation. Increasing oxygen segregation with increasing misorientation angle of low-angle grain boundaries and its saturation was also reported for high-angle grain boundaries in molybdenum [100] tilt bicrystals [528] (Fig. 5.11a). Similarly to tin segregation in iron (Fig. 5.8), oxygen segregation was interpreted as independent of the structure of grain boundaries (full line in Fig. 5.11a). However, very few measured data only do not exclude existence of anisotropy of oxygen segregation with minima at special grain boundaries (dotted line in Fig. 5.11a) [20]. In case of [110] grain boundaries, however, cusps were already measured corresponding to the 50.5°[110], {113} and 70.5°[110], {112} grain boundaries (Fig. 5.11b).

Two maxima of bismuth segregation were found at [100] symmetrical tilt grain boundaries in copper, one at about 30°[100], the other at about 60°[100] [531]. It can also be interpreted as the dependence with a single broad minimum at about 45°[100] (Fig. 5.12). However, a detail analysis of this quite large amount of experimental data provided two sharp minima at about 37°[100] and 50°[100] misorientation angles (dotted line in Fig. 5.12) [20]. This fit could also be correlated to the orientation dependence of grain boundary energy of copper [36]. In contrast to these results, interfacial segregation of antimony in copper exhibits minima at 28.1°[100], {015}, 36.9°[100], {013}, 53.1°[100], {012} and 61.9°[100], {023} symmetrical tilt grain boundaries. These grain boundaries are also more resistant to the brittle fracture compared to the other interfaces [532].

A surprisingly reversed course of orientation dependence of the grain boundary concentration was observed in case of silicon segregation at [100] symmetrical tilt grain boundaries in 17Cr–13Ni austenitic stainless steel [533]. As shown in Fig. 5.13, the highest levels of silicon segregation were found at 36.9°[100] ($\Sigma = 5$), 53.1°[100] ($\Sigma = 5$) and 67.4°[100] ($\Sigma = 13$) grain boundaries although – with respect to the atomic structure of these low-periodicity interfaces – an opposite result should be expected. Although this result seems to be strange, it may represent an exemplar consequence of the *compensation effect* (see Sect. 5.5.3).

Another important effect on the segregation properties is the orientation of the grain boundary plane. Symmetrical {111} and {112} facets of the twin grain boundary ($\Sigma = 3$) in nickel were found to be free of any segregation, while sulphur segregation was unambiguously detected at asymmetrical facets of this boundary [534, 535]. Similarly, the symmetrical {113} grain boundary does not exhibit

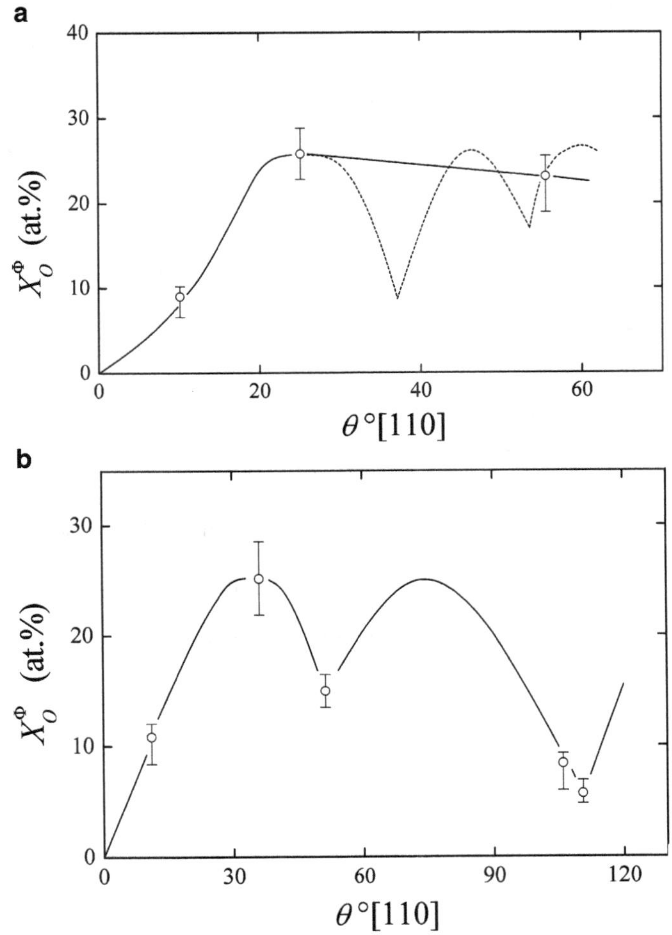


**Fig. 5.11** Orientation dependence of oxygen grain boundary segregation. (**a**) [100] grain boundaries; (**b**) [110] grain boundaries [528]. The *dotted line* in (**a**) represents an alternative possible course of the dependence [20]

segregation, while the asymmetrical grain boundaries oriented closely (up to 2°) to (112)/(117), (112)/(115), (225)/(114), (113)/(114) and (116)/(112) were found enriched by sulphur [536]. In contrast to the {113} grain boundary, the other $\Sigma = 11$, {233} symmetrical tilt grain boundary in nickel was found to be very unstable and no bicrystal with this boundary could be produced. During the growth of the bicrystal with this boundary, the boundary curving occurred with the asymmetrical parts highly segregated by sulphur [537, 538]. In $YBa_2Cu_3O_{7-\delta}$, oxygen depletion is observed at the (230)/(209) grain boundary while no change of chemical composition exists at the (010)/(001) grain boundary [520].

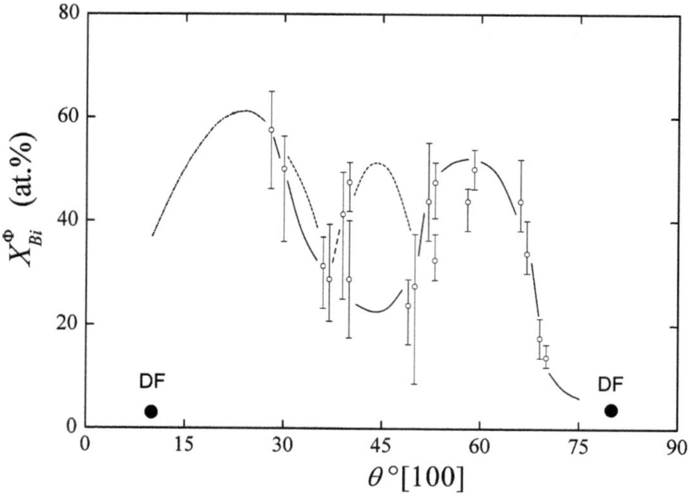


**Fig. 5.12** Orientation dependence of grain boundary concentration of bismuth at [100] symmetrical tilt grain boundaries in copper at 773 K. The points DF represent ductile fracture indicating no segregation [531]. The *dotted line* suggests another interpretation of the data [20]

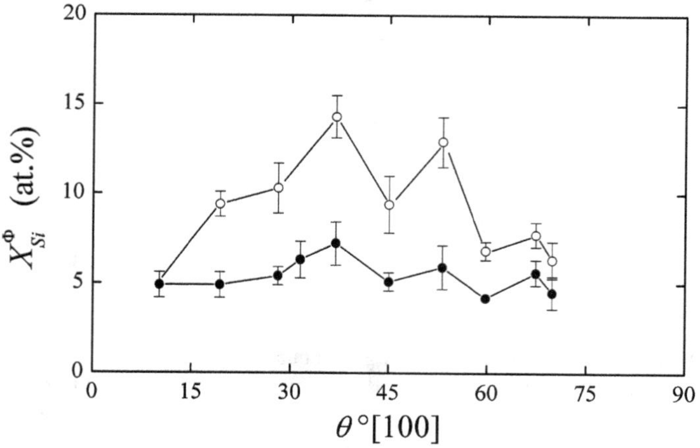


**Fig. 5.13** Orientation dependence of grain boundary concentration of silicon at [100] symmetrical tilt grain boundaries in 17Cr–13Ni austenitic stainless steel containing 0.3 mass.%Si (*solid circles*) and 0.8 mass.%Si (*empty circles*) at 923 K [533]

Suzuki et al. [539] studied phosphorus segregation at grain boundaries of known orientation in polycrystalline bcc iron. In this case, phosphorus enrichment was found to be predominantly dependent on crystallographic orientation of the boundary *plane* albeit not on *misorientation angle* (Fig. 5.14). As it is indicated in Fig. 5.14, the degree of segregation was found to be large on a high-index boundary plane and low on a low-index plane, independently of the orientation of the other

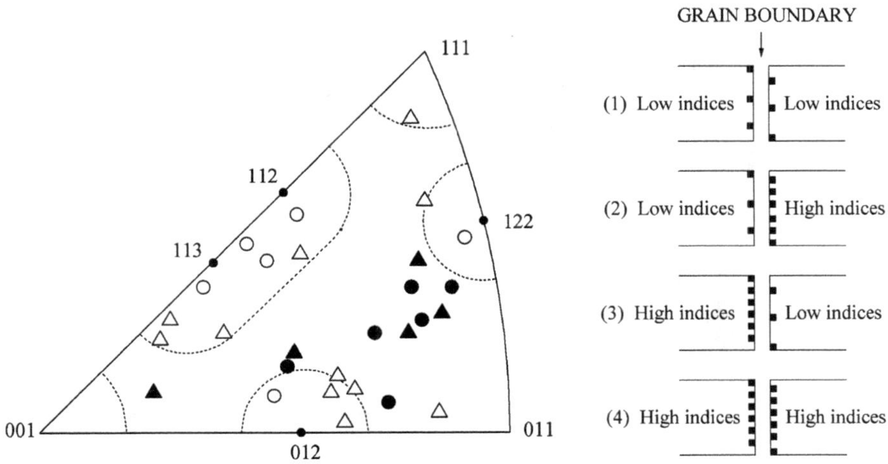


**Fig. 5.14** Phosphorus segregation on grain boundaries of bcc iron [539]. (**a**) Relation between Auger peak-to-peak ratios $r_P$ and the crystallographic orientation of grain boundary planes (*empty circles*: $r_P < 0.4$, *empty triangles*: $0.4 < r_P < 0.5$, *solid circles*: $0.5 < r_P < 0.6$, *solid triangles*: $r_P > 0.6$); (**b**) combinations of grain boundary planes with low and high index (*dark marks* schematically depict amount of segregated phosphorus)

matching boundary plane. In fact, this model is in agreement with the hypothesis of occurrence of solute segregation in different grain boundary positions [540].

Pang and Wynblatt [541] found that the strongest segregation of niobium in tetragonal rutile ($TiO_2$) occurs at the grain boundary planes laying along the (001)–(011)–(010) edge of the stereographic triangle, while the weakest segregation is observed for grain boundary planes close to the (110)–(010) edge.

The best way how to represent the anisotropy of grain boundary segregation precisely and independently, is an establishment of orientation dependence of its thermodynamic parameters – standard enthalpy and entropy of grain boundary segregation. Orientation dependence of the standard enthalpy of segregation of silicon, phosphorus and carbon in bcc iron can serve for an example [542]. As is seen in Fig. 5.15, this dependence for [100] symmetrical tilt grain boundaries is qualitatively similar for all three elements and is characterised by pronounced minima of $-\Delta H_I^0$ at the 22.6°[100], {015}, 36.9°[100], {013} and 53.1°[100], {012} grain boundaries. This result is in good agreement with classification of grain boundaries by means of the CSL model, because the above-mentioned grain boundaries are all characterised by low values of $\Sigma$ (13, 5 and 5, respectively). However, the CSL characterisation fails when the asymmetrical grain boundaries are considered, as was already mentioned in Chap. 2. For example, the $\Sigma = 5$, 36.9°[100] tilt orientation relationship covers all types of the grain boundaries – special, vicinal and general – according to their inclination from the {013} symmetrical orientation (Fig. 5.16) [86]. The lowest tendency to segregation represented by the lowest absolute value of $\Delta H_I^0$, was found – besides both symmetrical {013} and {012} grain boundaries – also for the asymmetrical (011)/(017) grain boundary. All these interfaces

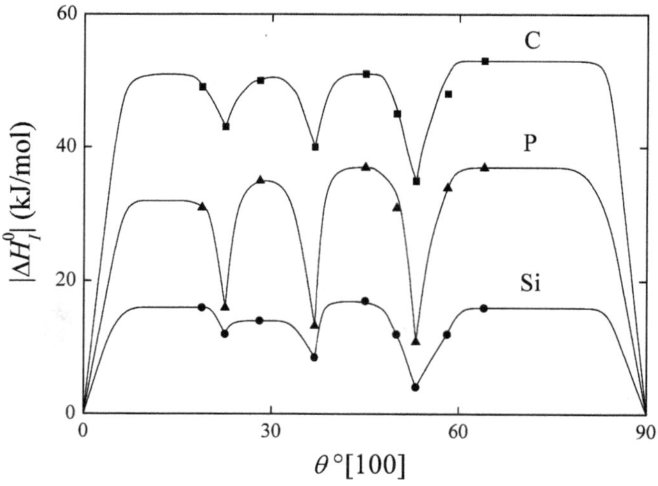


**Fig. 5.15** Dependence of absolute value of the standard enthalpy of segregation of silicon (*circles*), phosphorus (*triangles*) and carbon (*squares*) on misorientation angle of [100] symmetrical tilt grain boundaries in bcc iron [542]

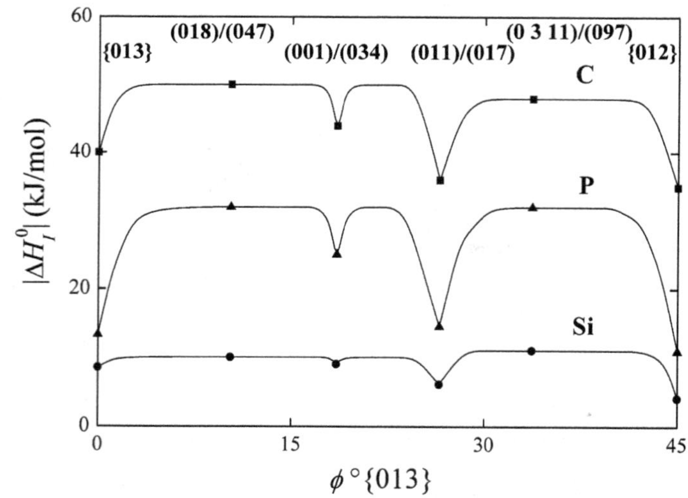


**Fig. 5.16** Dependence of the absolute value of the standard enthalpy of segregation of silicon (*circles*), phosphorus (*triangles*) and carbon (*squares*) of 36.9°[100] tilt grain boundaries on inclination angle $\phi$ from symmetrical orientation {013} in bcc iron [86]

are considered as special. The (001)/(034) grain boundary exhibits vicinal-like behaviour, and the (018)/(047) and (0 3 11)/(097) grain boundaries are general (Fig. 5.16) [86, 94].

The differences in segregation behaviour were also found in case of two tilt grain boundaries characterised by the same non-coincidence ($\Sigma \rightarrow \infty$) 45°[100]

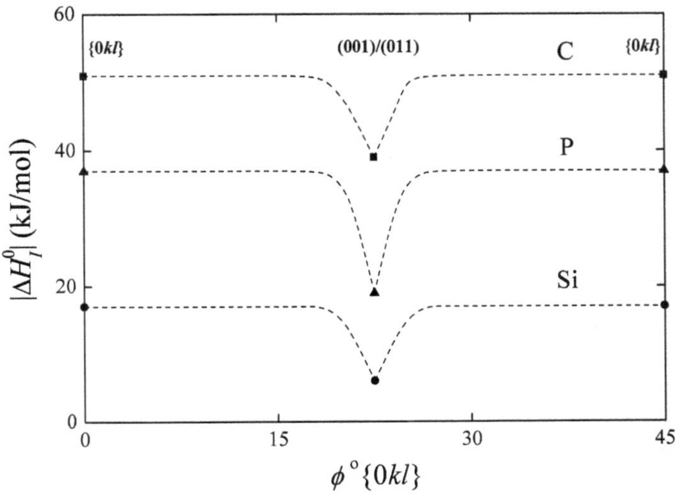


**Fig. 5.17** Dependence of the absolute value of the standard enthalpy of segregation of silicon (*circles*), phosphorus (*triangles*) and carbon (*squares*) of 45° [100] tilt grain boundaries on deviation angle $\phi$ from symmetrical orientation $\{0kl\}$ in bcc iron [94]

orientation relationship – symmetrical $\{0kl\}$[1] and asymmetrical (001)/(011) ones. While $\{0kl\}$ behaves as typical general grain boundary, the asymmetrical grain boundary exhibits very low absolute values of the segregation enthalpy comparable with those of the {013} and {012} grain boundaries (Fig. 5.17) and, therefore, can be classified as special [87,94,506].

An interesting anisotropy of grain boundary segregation was also observed for asymmetrical tilt grain boundaries formed by the (001) grain boundary plane (Fig. 5.18) and the (011) grain boundary plane (Fig. 5.19). In the former case, a low tendency to segregation was indicated for the (001)/(011) and the (001)/(013) grain boundaries, while the other interfaces of this kind exhibit "transitive" values of segregation enthalpy between the level of general grain boundaries and special grain boundaries. From this point of view, they might be considered as "vicinal" [94,98,100]. In case of the latter interfaces, all high-angle grain boundaries possess low-absolute values of segregation enthalpy and are special [94,98,100]. Obtained results are in agreement with the statement of Pang and Wynblatt [541] that a boundary can exhibit strong segregation only if its matching halves also exhibit strong segregation. The results obtained for impurity segregation in bcc iron were summarised in classification of individual grain boundaries as shown in Fig. 2.9.

[1] The 45° [100] grain boundary is incommensurate and the rate $k/l$ is irrational. Therefore, no integer Miller symbols can be found for the symmetrical interface.

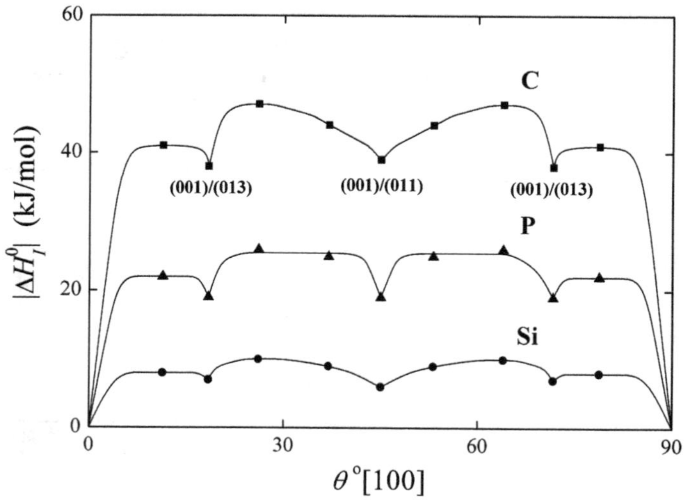


**Fig. 5.18** Dependence of absolute value of the standard enthalpy of segregation of silicon (*circles*), phosphorus (*triangles*) and carbon (*squares*) of (001) asymmetrical tilt grain boundaries on misorientation angle $\theta$ of both adjoining grains in bcc iron [94]

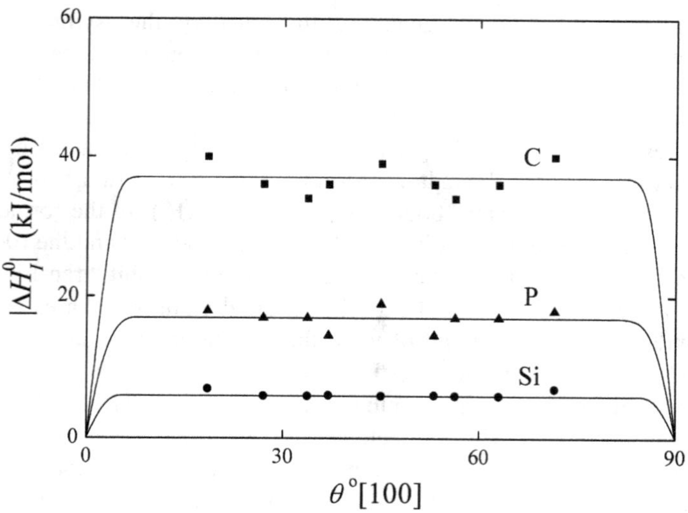


**Fig. 5.19** Dependence of absolute value of the standard enthalpy of segregation of silicon (*circles*), phosphorus (*triangles*) and carbon (*squares*) of (011) asymmetrical tilt grain boundaries on misorientation angle $\theta$ of both adjoining grains in bcc iron [94]

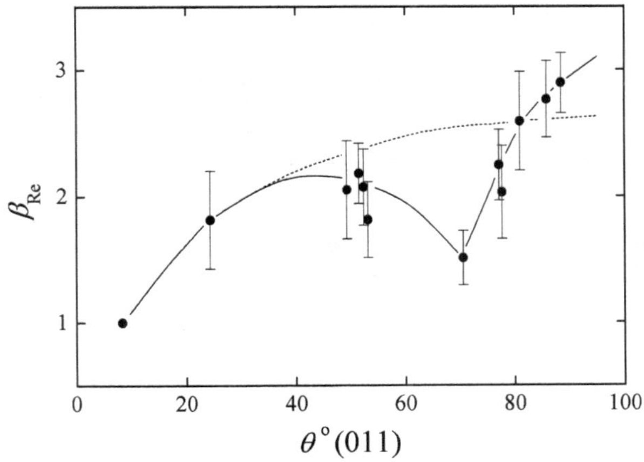


**Fig. 5.20** Dependence of the enrichment factor $\beta_{Re}$ for (011) twist grain boundaries on misorientation angle $\theta$ of both adjoining grains in a W–25at.%Re alloy [309]

The studies of anisotropy of solute segregation at high-angle *twist* grain boundaries are rather limited but the differences in segregation behaviour were detected as well. A minimum of rhenium segregation in tungsten at the 70.5°(011) twin grain boundary ($\Sigma = 3$) was only found in the whole misorientation range [309, 543] (Fig. 5.20) albeit not for the interface corresponding to the $\Sigma = 11$ relationship, where minims of segregation were detected for some tilt grain boundaries [309,544]. This disproportion can arise from probably larger ability of twist interfaces to grain boundary segregation as compared to tilt ones [543].

Anisotropy of solute segregation was also detected in computer simulations of the grain boundary structure and energy in binary systems. One of the first calculations of the grain boundary segregation was performed at 53.1°[100]{012} and 61.9°[100]{035} symmetrical tilt grain boundaries in Cu–Bi, Cu–Ag and Au–Ag dilute binary systems using molecular statics method [545]. The variations of the segregation energy $E_{seg}$ (which is nearly equal to the $\Delta H_I$) exist for different boundary sites (Table 5.1). Although both, positive and negative values of $E_{seg}$ were found, solute segregation can only occur at those positions characterised by negative values of $E_{seg}$ that are associated with hydrostatic tension. This suggests that bismuth segregation in copper is governed by the size effect (the atomic volume of bismuth is three times larger than that of copper). These results comply with the early model predictions of McLean [19].

The sequence of segregation of solute atoms at individual sites at a grain boundary can be documented, for example of bismuth segregation at 36.9°[100]{013} symmetrical tilt grain boundary in copper [546]. Individual grain boundary sites are shown in Fig. 5.21 and the values of $E_{seg}$ are listed in Table 5.1. The first bismuth atom segregates at the site of type "2" (Fig. 5.21) due to the highest negative value of $E_{seg}$. After filling all these positions at the whole boundary, bismuth segregation proceeds at the site of type "5". After all sites "2" and "5" are occupied by bismuth,

**Table 5.1** Energy (in kJ/mol) of segregation of bismuth and silver in copper, and of silver in gold at individual sites of 53.1°[100]{012} and 61.9°[100]{035} symmetrical tilt grain boundaries [545] and of segregation of bismuth in copper at 36.9°[100]{013} grain boundary [546]

| System | Boundary | Boundary site | | | | | |
|---|---|---|---|---|---|---|---|
| | | 1 | 2 | 3 | 4 | 5 | 6 |
| Cu(Ag) | {035} | +61 | +82 | +55 | +33 | +56 | |
| | {012} | +17 | +27 | | | | |
| Cu(Bi) | {035} | −33 | +208 | +43 | −17 | +222 | |
| | {012} | −106 | +147 | | | | |
| | {013} | +78 | −195 | +34 | +68 | −195 | −176 |
| Au(Ag) | {035} | −41 | −40 | −63 | −35 | −43 | |
| | {012} | −51 | −65 | | | | |

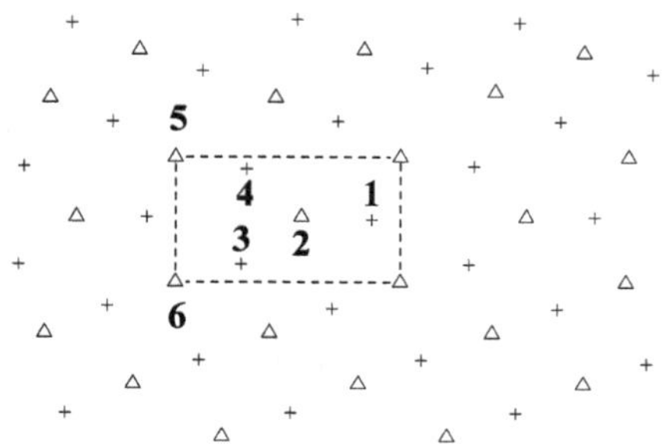


**Fig. 5.21** Individual sites at the 36.9°[100]{013} grain boundary of copper [546]

a 2-D ordered structure can be distinguished in the boundary layer, where each bismuth atom is surrounded by copper atom and vice versa [546]. Similar behaviour was detected at other grain boundaries [547, 548].

A thorough theoretical study of phosphorus and boron segregation at 36.9°[100]{013} and 38.9°[110]{114} symmetrical tilt grain boundaries in bcc iron were performed by molecular dynamics calculations [281, 549, 550]. The most advantageous sites for segregation of both elements were found in the central positions of corresponding structural units – capped trigonal prism and pentagonal bipyramid, respectively. These sites are not occupied by iron atoms in the structure of pure grain boundaries and represent *interstitial* grain boundary positions. The local atomic arrangement at the respective boundaries is close to $Fe_9P$ and $Fe_7P$ and structurally very similar to the $Fe_3P$ compound [551]. The values of the segregation energy [281] show high site sensitivity; however, its highest absolute value (217 kJ/mol) (Table 5.2) suggesting the most probable segregation positions for phosphorus at these grain boundaries are by about one order of magnitude higher than those of the segregation enthalpy determined experimentally not only for special {013} grain boundary (13 kJ/mol) [350] but also for general interfaces in polycrystals (21–38 kJ/mol) [328, 339, 343, 346, 552].

A rather simple method – tight-binding type electronic theory of s-, p- and d-basis orbitals – was used to study segregation of sp-valence impurities Mg, Al, Si, P and

**Table 5.2** Energy (in kJ/mol) of segregation of boron and phosphorus in bcc iron at individual sites of 36.9°[100]{013} and 38.9°[110]{114} symmetrical tilt grain boundaries [281, 549, 550]

| System | Boundary | Boundary site | | | |
|---|---|---|---|---|---|
| | | 1 | 2 | 3 | 4 |
| Fe(P) | {013} | −208 | −217 | +2 | −78 |
| | {114} | −207 | −190 | +93 | −3 |
| Fe(B) | {013} | −236 | +54 | −102 | −100 |
| | {114} | −195 | +20 | −116 | −116 |

Cl at the {013} grain boundary in bcc iron [270]. Calculated segregation energies for the sites at the exact boundary plane seem to depend on filling the sp-band: Maximum values of $E_{seg}$, suggesting the weakest segregation tendency, were found for silicon with the half-filled sp-bands. In contrast to the extremely high absolute values of the segregation enthalpy found above, the values of $E_{seg}$ for phosphorus (−14.6 kJ/mol) and silicon (−9.8 kJ/mol) calculated in this simple way are in an excellent agreement with the values of segregation enthalpy measured experimentally for the same elements at this boundary [350] (−13.3 kJ/mol and −8.5 kJ/mol, respectively).

Similar level of the grain boundary segregation of platinum in gold represented by $\Gamma_{Pt}$ was observed at corresponding $\Sigma = 5$ interfaces, 53.13°[100] {012} symmetrical tilt grain boundary and 53.13°(100) twist grain boundary at 850 K. On the other hand, five times higher segregation of platinum was surprisingly found at the above tilt grain boundary compared to the twist one in the system Ni–Pt [294].

The variations of the degree of copper segregation at various (001) twist grain boundaries of nickel using Monte Carlo simulations at 800 K showed surprisingly the strongest segregation tendency for the low-$\Sigma$ misorientation in the series $\Sigma = 5$, 13 and 61 orientation relationships. The grain boundary with $\Sigma = 61$ misorientation exhibited the weakest segregation among all these boundaries [70, 553]. However, variations of chemical composition are extended in more planes parallel to the grain boundary. In case of equiatomic alloy, an oscillatory depth distribution of copper occurs, which is similar to that found for free surfaces [289, 554]. This result was confirmed using a simulation method based upon a point approximation for the configuration entropy, Einstein model for vibrational contributions to the free energy with respect to atomic co-ordinates and composition of each site [295, 555]. In addition, the segregation was found to increase with increasing (001) twist rotation angle [296].

The value of the segregation enthalpy of gold segregation at (001) twist grain boundaries of platinum were calculated as decreasing from −1.1 kJ/mol for 5°(001) grain boundary to −11.7 kJ/mol for 36.9°(001) grain boundary [70, 284]. The average enrichment ratio $\beta_{Au}$ was found to depend on both temperature and misorientation angle as

$$\beta_{Au} = 1 + m(T)\sin(\theta/2) \tag{5.8}$$

with the slope $m(T)$ depending only on temperature. These Monte Carlo simulations indicate that segregation primarily occurs in cores of the grain boundary

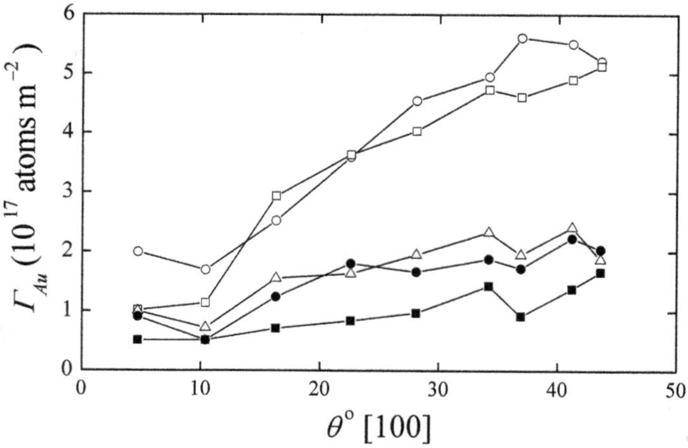


**Fig. 5.22** Orientation dependence of the grain boundary concentration of gold, $\Gamma_{Au}$, at (001) twist grain boundaries on misorientation angle $\theta$ of both adjoining grains in Pt–1at.%Au alloy at 900 K (*empty circles*), 1,000 K (*empty squares*), 1,300 K (*empty triangles*), 1,500 K (*solid circles*) and 1,900 K (*solid squares*) (according to [284])

dislocations that comprise the (001) twist grain boundaries. The density of the grain boundary dislocations increases with increasing misorientation angle up to the value of about 35°: if such misorientation is exceeded, saturation establishes [284]. However, a question arises why the temperature dependence of gold segregation was theoretically found as non-monotonous in some cases (Fig. 5.22). Solute atom enrichment was also determined by Monte Carlo simulations for (001) low-angle twist grain boundaries in the Ni–Pt system. A saturation of the grain boundaries was detected at about 22°(001) grain boundary: The enrichment of the nickel boundaries by platinum is about twice higher than the nickel enrichment of platinum interfaces [285]. In contrast to this system, only gold segregation was determined by the same method at (001) low-angle twist grain boundaries in Au–Pt system on both sides of the concentration range [286].

Monte Carlo and molecular statics simulations were used to determine detail anisotropy of palladium segregation at different boundary positions of numerous special and general [110] symmetrical tilt grain boundaries in nickel [341]. The values of the segregation energy and entropy of individual sites at 23 grain boundaries were calculated. The orientation dependence of minimum average grain boundary concentration of palladium, $\Gamma_{Pd}$, is shown in Fig. 5.23. It is apparent that no segregation was calculated for the 109.47°[110] {111} symmetrical tilt grain boundary, which represents the coherent twin grain boundary. This corresponds very well with the zero-value of the energy of this grain boundary [341]. The orientation dependence of palladium grain boundary concentration copies well that of grain boundary energy except the range 50.48°–109.47°, where the increase of the grain boundary energy is not followed by appropriate increase of the palladium grain boundary concentration. Although the values of the free energy of segregation at

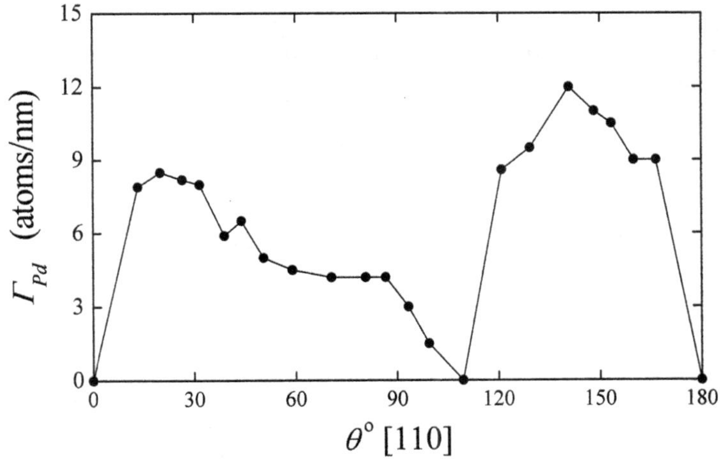


**Fig. 5.23** Grain boundary concentration of palladium, $\Gamma_{Pd}$, at numerous [110] symmetrical tilt grain boundaries in Ni–4at.%Pd alloy at 800 K (according to [341])

enhanced temperature were calculated for individual grain boundary sites for the first time, no clear correlation between their values and grain boundary structure were found [341].

To elucidate the relationship between the structure and solute segregation to particular grain boundaries, high-angle grain boundaries were modelled as planar defects characterised by the thickness and the atomic density [556]. Model calculations showed that the grain boundary enrichment strongly depends on the atomic density of grain boundaries albeit not on grain boundary thickness. Therefore, a special grain boundary with the near-bulk density in simple metal will show low electronic binding energy and an important elastic binding energy. It suggests that segregation may occur at such boundaries although at low levels. Additionally, grain boundaries with small extra volume may show high adsorptive capacity. Although a change of the grain boundary density represents only one of the possible contributions to segregation, the grain boundary atomic density is considered as the most important physical parameter for segregation at periodic high-angle grain boundaries [557].

## 5.5 Nature of Segregating Element

### *5.5.1 Truncated BET Isotherm*

The grain boundary energy and consequently, the Gibbs energy of segregation are also affected by the nature of the solute and the matrix element. This results in pronounced difference in segregation extent. For example, the grain boundary enrichment ratio (4.19) of silicon in α-iron at temperatures 700–900 K was found to range between $10^1$ and $10^0$, while the grain boundaries of copper may accumulate $10^6$ times higher amount of bismuth as compared to grain interior [307].

Seah and Hondros [307] formulated the truncated BET theory (cf. Chap. 4) for interfacial segregation in a dilute binary system as

$$\frac{X_I^{\Phi}}{X^0 - X_I^{\Phi}} = \frac{X_I}{X_I^*}\exp\left(-\frac{\Delta G_I'}{RT}\right) \tag{5.9}$$

with

$$\Delta G_I' = \Delta G_I^0 - \Delta G_I^{\text{sol}}, \tag{5.10}$$

where $X_I^*$ is the solid solubility limit of solute $I$ in bulk matrix $M$, and $\Delta G_I^{\text{sol}}$ is the Gibbs energy of solution. As it is apparent from (5.9) and (5.10) where the excess Gibbs energy term is missing, that this model assumes a non-interactive segregation in an ideal solid solution.

The truncated BET isotherm offers an interesting consequence. For low interfacial enrichment ($X_I^{\Phi} \ll X^{0\Phi}$), it transforms into

$$\beta_I^{\Phi} = \frac{X_I^{\Phi}}{X^{0\Phi}}\frac{1}{X_I^*} = \frac{\exp(-\Delta G'/RT)}{X_I^*}. \tag{5.11}$$

The analysis of numerous experimental data on grain boundary segregation in various binary and pseudobinary systems showed that $\Delta G'$ possesses relatively low but similar values ranging between −20 and 0 kJ/mol for all systems. Equation (5.11) can thus be simplified as

$$\beta_I^{\Phi} = \frac{K}{X_I^*} \tag{5.12}$$

with $K = \exp(-\Delta G'/RT)$ ranging from 1.8 to 10.8, i.e. within one order of magnitude for a wide variety of systems [13, 20, 307]. Equation (5.12) can thus be used to predict the grain boundary enrichment ratio of a solute in a matrix knowing only its bulk solid solubility (Fig. 5.24).

There are objections against the close relationship between the grain boundary segregation and the solid solubility as expressed by (5.12) and represented in Fig. 5.24. One of the most serious arguments for this disagreement results from comparison of experimental data on phosphorus and antimony segregation in bcc iron and can be summarised as follows (a) the amount of antimony segregation is much less than that of phosphorus and (b) the segregation of antimony increases with increasing bulk concentration up to the solubility limit and remains constant after reaching the solid solubility limit by bulk concentration while phosphorus enrichment of the boundary reaches its maximum for bulk concentrations below the solid solubility limit and then remains constant when the majority of primary segregation sites are occupied [336]. Another argument contra this relationship is the relatively low tendency of antimony to segregate to grain boundaries ($\Delta H_{\text{Sb}}^0 = -19$ kJ/mol) in comparison with phosphorus ($\Delta H_{\text{P}}^0 = -34$ kJ/mol) or tin ($\Delta H_{\text{Sn}}^0 = -23$ kJ/mol)

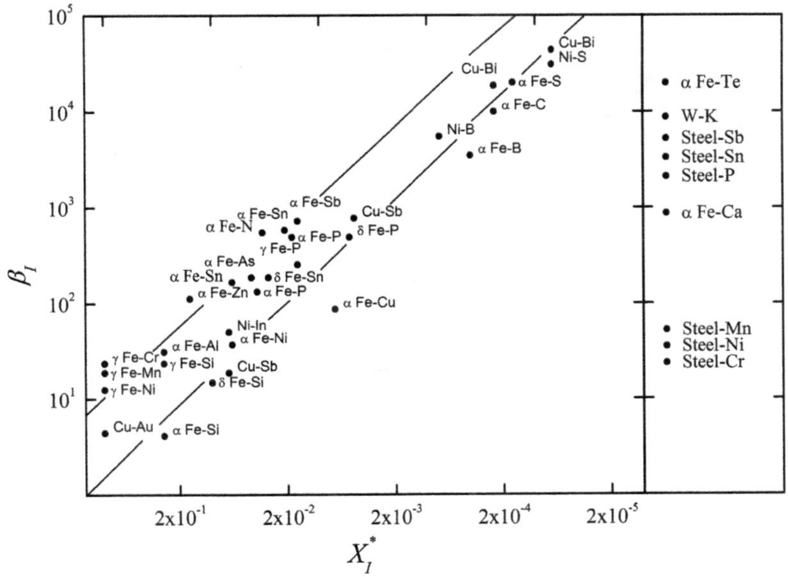


**Fig. 5.24** Dependence between the experimental values of the grain boundary enrichment ratio $\beta_I$ and the solid solubility $X_I^*$ for numerous binary systems. The solid solubility of the solutes in the system given in the right column of the figure is unknown [20]

although both these elements exhibit similar solid solubility [124]. This argument is based on incorrect interpretation of the data for antimony segregation [338] as discussed in detail in Chap. 4. It is clear that (5.12) provides us with a rough estimate of the grain boundary segregation only, which does reflect neither the temperature of the grain boundary segregation nor the grain boundary structure and energy of individual sites at the grain boundary. These effects are covered by relatively large limits of the above model (within one order of magnitude as marked by solid lines in Fig. 5.24) [338]. As was shown above, the value of $\Delta H^0_{\mathrm{Sb}} = -19$ kJ/mol showing "the low tendency to segregation" was determined under the simplified assumption $X^0 = 1$, which does not correlate the experimental data well (cf. Fig. 4.6a). When the limited saturation level is considered, a more reliable value $\Delta H^0_{\mathrm{Sb}} = -23$ kJ/mol is obtained [338], which is identical with the segregation enthalpy of tin.

BET isotherm was also applied to correlate experimental data on sulphur and antimony segregation in iron alloys, while the correlation according to both the site competition and the Guttmann model failed [503].

### 5.5.2 *Grain Boundary Segregation Diagram*

The main drawback of the truncated BET approach is that it does not take into account the dependence of the grain boundary segregation on two main factors,

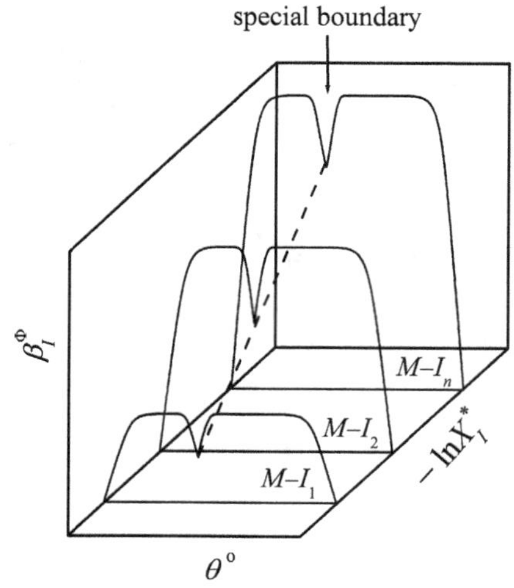


**Fig. 5.25** Schematic depiction of a grain boundary segregation diagram proposed by Watanabe et al. (according to [526])

interface structure and temperature, which can alter the grain boundary concentration [20]. All these changes are supposed to be included in the scatter of $K$ in (5.12).

Watanabe et al. [526] tried to overcome this drawback and suggested to extend the dependence between $\beta_I^\Phi$ and the solid solubility by new dimension – the grain boundary orientation. In this sense, they proposed the construction of so-called *grain boundary segregation diagrams* (Fig. 5.25).

Thermodynamic analysis of the dependence between the standard Gibbs energy of interfacial segregation, $\Delta G_I^0$, and the solid solubility of an element in a chosen matrix, $X_I^*$, was done with respect to different structure of the grain boundary, $\Phi$ [17, 337, 375]. Chemical potential, $\mu_I^*$, of the solute $I$ in saturated bulk solid solution is

$$\mu_I^* = \mu_I^0 + RT \ln a_I^*, \tag{5.13}$$

where $a_I^*$ is its activity in the system $M$–$I$ at the solubility limit $X_I^*(T)$ [337]. We can then express the segregation free energy of the solute $I$, $\Delta G_I^*$, as

$$\Delta G_I^* = \left(\mu_I^{0,\Phi} - \mu_I^*\right) - \left(\mu_M^{0,\Phi} - \mu_M^0\right) = \Delta G_I^0 - RT \ln a_I^*. \tag{5.14}$$

It can be simply shown that

$$\Delta S_I^* = \Delta S_I^0 - R \left(\frac{\partial [T \ln a_I^*]}{\partial T}\right)_{P,X_i}. \tag{5.15}$$

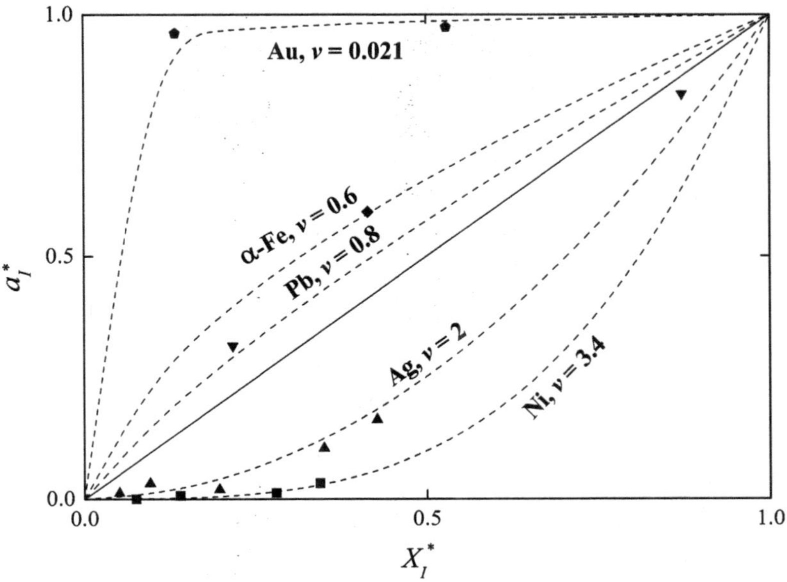


**Fig. 5.26** Plot of the values of activity vs. atomic concentration of a solute at its solubility limit in α-iron [337] (data from [558])

The plot of numerous pairs of the values of activities and corresponding concentrations at the solid solubility level in various systems found in [558] revealed a simple power relationship between these two values for different systems and temperatures (Fig. 5.26),

$$a_I^* = \left(X_I^*\right)^\nu, \tag{5.16}$$

where the parameter $\nu$ depends on matrix element $M$ but not on nature of the solute element $I$ [337]. Using condition (5.15), (5.16) may be written as

$$\Delta S_I^* = \Delta S_I^0 - \nu R \left( \frac{\partial [T \ln X_I^*]}{\partial T} \right)_{P,X_i}. \tag{5.17}$$

The product on the right-hand side of (5.17), $T \ln X_I^* = \Delta G_I^{\mathrm{sol}}/R$, was proved to be nearly independent of temperature for various systems [337, 559] (Fig. 5.27). Thus, the term in brackets in (5.17) is equal to zero and consequently, we can write (5.14) as

$$\Delta H_I^* = \Delta H_I^0 - RT \ln a_I^*. \tag{5.18}$$

As a result, the grain boundary segregation diagram can be represented by a dependence of $\Delta H_I^0$ on both the grain boundary orientation and the product of temperature and logarithm of bulk solid solubility, i.e.

$$\Delta H_I^0(\Phi, X_I^*) = \Delta H^*(\Phi, X^* = 1) + \nu R[T \ln X_I^*(T)], \tag{5.19}$$

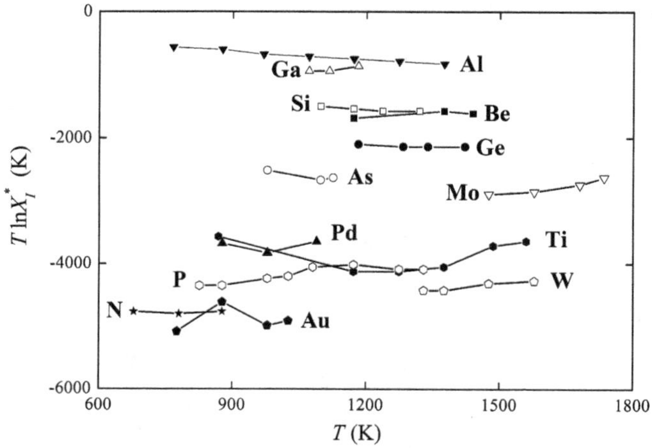


**Fig. 5.27** Temperature dependence of the product of temperature and logarithm of solid solubility of various solutes in α-iron [337] (data from [560])

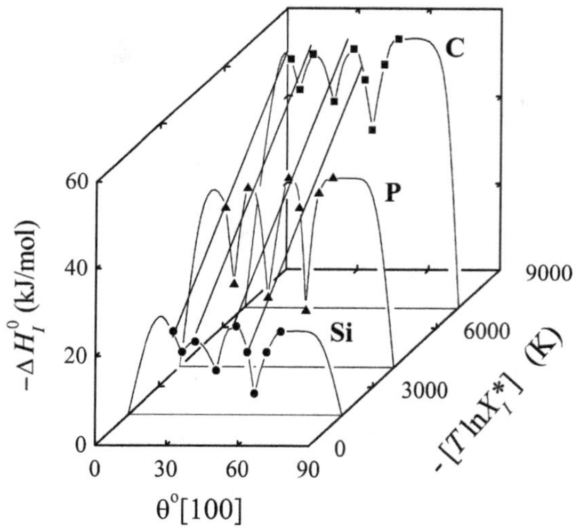


**Fig. 5.28** Grain boundary segregation diagram for [100] symmetrical tilt grain boundaries in α-iron [337]

where $\Delta H^*(\Phi, X^* = 1)$ is the enthalpy of segregation of a completely soluble element at grain boundary Φ. The first experimental grain boundary segregation diagram was constructed in 1991 from the well-defined data also listed in Appendix A [337, 559].

The grain boundary segregation diagram is shown in Fig. 5.28. It is apparent that individual connecting lines (the dependence of $\Delta H_I^0$ on the solubility term for the same grain boundary) are parallel suggesting that the two terms of the right-hand side of (5.19) are mutually independent and the slope of the dependence is the same

**Table 5.3** Parameters of grain boundary segregation diagrams (5.19) for α-iron [337, 375]

| Grain boundary type | $\Delta H^*$ (kJ/mol) | $\nu$ |
|---|---|---|
| General | −8 to −4 | 0.77 |
| Vicinal | −2 to +2 | 0.77 |
| Special | +5 to +8 | 0.77 |

for all grain boundaries. In fact, (5.19) represents an extension of the model of Seah and Hondros [307] by considering (a) the anisotropy of grain boundary segregation ($\Delta H_I^0(\Phi) \neq$ const.) and (b) non-ideal behaviour of the solid solutions at the solubility limit ($\nu \neq 1$) [337]. The characteristic values of $\Delta H^*$ for special, vicinal and general grain boundaries and the values of $\nu$ (5.19) in α-iron-based binary alloys [94] are listed in Table 5.3.

There are some interesting consequences of the grain boundary segregation diagrams. First, for segregation of highly soluble elements, positive values of $\Delta H_I^0$ may appear at special grain boundaries according to the prediction. Such values have no physical meaning and therefore, it is supposed that $\Delta H_I^0 = 0$ describes such segregation most realistically [337]. Second, let us notice that the values of $\Delta H^*$ range between −8 and +8 kJ/mol for all grain boundaries (Table 5.3). This implies maximum scatter of the values of $\Delta H_I^0$ due to segregation anisotropy to be of ±8 kJ/mol at all grain boundaries despite the character of the segregated element. It is clear because this scatter represents structural variations of the grain boundaries in the chosen matrix element.

### 5.5.3 *Enthalpy–Entropy Compensation Effect*

The *enthalpy–entropy compensation effect* is the linear dependence between the characteristic enthalpy, $\Delta H^{\text{ch}}$, and entropy, $\Delta S^{\text{ch}}$, of a process or equilibrium state,

$$\Delta S^{\text{ch}} = a\Delta H^{\text{ch}} + b, \tag{5.20}$$

where $a$ and $b$ are constants. Frequently, the compensation effect is also expressed as the linear dependence between the characteristic enthalpy and the logarithm of the pre-exponential factor of an Arrhenius relationship. The compensation effect is a general phenomenon detected for many processes and states in chemistry, physics, material science, biology and other fields (e.g. [12, 48, 561, 562]). Nevertheless, it is one of the concepts that cause considerable confusion and divide the scientific community to "enthusiasts" and "sceptics." The enthusiasts accept the linear enthalpy–entropy correlation and apply it in order to generalise a particular behaviour to groups of systems and propose models to explain these effects for selected phenomena mainly on basis of atomic binding (c.f. [561]). The sceptics consider it as an empirical relationship generated by statistical treatment of data [563] or as a purely mathematical consequence of the formulae employed [564] and point out the sensitivity of the obtained data to procedure variables. Obviously, the problem in understanding arises from inappropriate application of incorrect enthalpy–entropy pairs employed in construction of the compensation graph.

The principles of the compensation effect can be elucidated by a thermodynamic analysis [48,561]. Let us assume that a "process" (chemical reaction, diffusion, ...) or an equilibrium "state" (interfacial segregation, solubility...) is controlled by a change of its characteristic Gibbs energy, $\Delta G^{ch}$. In case of the process, $\Delta G^{ch}$ is an activation Gibbs energy, in case of the equilibrium state, $\Delta G^{ch}$ is a non-zero part of the total Gibbs energy ($\Delta G = 0$), that controls the equilibrium state. In general, $\Delta G^{ch}$ depends on $N$ intensive variables (potentials), $\Psi_j$, such as electric and magnetic fields, specific defect energy, solubility and bond strength. The total differential of $\Delta G^{ch}$ in respect to the variables $\Psi_j$ at constant temperature and pressure can be, thus, expressed as [561]

$$\mathrm{d}\Delta G^{ch} = \sum_{j=1}^{N} \left( \frac{\partial \Delta G^{ch}}{\partial \Psi_j} \right)_{T,P,\Psi_i \neq \Psi_j} \mathrm{d}\Psi_j . \tag{5.21}$$

Analogously,

$$\mathrm{d}\Delta H^{ch} = \sum_{j=1}^{N} \left( \frac{\partial \Delta H^{ch}}{\partial \Psi_j} \right)_{T,P,\Psi_i \neq \Psi_j} \mathrm{d}\Psi_j , \quad \mathrm{d}\Delta S^{ch} = \sum_{j=1}^{N} \left( \frac{\partial \Delta S^{ch}}{\partial \Psi_j} \right)_{T,P,\Psi_i \neq \Psi_j} \mathrm{d}\Psi_j . \tag{5.22}$$

Evidently, $\mathrm{d}\Delta H^{ch}(\Psi_j)$ and $\mathrm{d}\Delta S^{ch}(\Psi_j)$ are general, non-zero real numbers and therefore, a range of the changes of the variables $\Psi_j$ must exist for that the constant temperature $T_{CE}$ is defined as [561]

$$T_{CE} = \frac{\mathrm{d}\Delta H^{ch}}{\mathrm{d}\Delta S^{ch}} = \frac{\sum_{j=1}^{N} \left( \frac{\partial \Delta H^{ch}}{\partial \Psi_j} \right)_{T,P,\Psi_i \neq \Psi_j} \mathrm{d}\Psi_j}{\sum_{j=1}^{N} \left( \frac{\partial \Delta S^{ch}}{\partial \Psi_j} \right)_{T,P,\Psi_i \neq \Psi_j} \mathrm{d}\Psi_j} . \tag{5.23}$$

It follows from (5.21) and (5.23) with (4.42) that

$$\mathrm{d}\Delta G^{ch}(T_{CE}) = \sum_{j=1}^{N} \left( \frac{\partial \Delta G^{ch}(T_{CE})}{\partial \Psi_j} \right)_{T=T_{CE},P,\Psi_i \neq \Psi_j} \mathrm{d}\Psi_j = 0. \tag{5.24}$$

This provides us with a very important consequence: *at temperature $T_{CE}$, $\Delta G^{ch}$ does not change with changing variables $\Psi_i$ in their specific ranges although the values of $\Delta H^{ch}$ and $\Delta S^{ch}$ may vary significantly*. Notice that at $T_{CE}$, $\Delta G^{ch}$ may possess any value and is not a priori equal to zero.

Integration of (5.23) results in $\Delta H^{ch} = T_{CE}(\Delta S^{ch} + \Delta S')$ with the integration constant $\Delta S'$, and thus,

$$\Delta S^{\mathrm{ch}}(\Psi_j) = \frac{\Delta H^{\mathrm{ch}}(\Psi_j)}{T_{\mathrm{CE}}} - \frac{\Delta G^{\mathrm{ch}}(\Psi_j, T_{\mathrm{CE}})}{T_{\mathrm{CE}}}. \tag{5.25}$$

Equation (5.25), which is in its form identical with empirical (5.20), represents the mathematical form of the *compensation effect*. It suggests that the change of the characteristic enthalpy $\mathrm{d}\Delta H^{\mathrm{ch}}$ of the process caused by the change of the variable(s) $\Psi_j$ is *compensated* by the corresponding change of the characteristic entropy $\mathrm{d}\Delta S^{\mathrm{ch}}$. $T_{\mathrm{CE}}$ is the compensation temperature.

As we did not specify the process or state until now, the compensation effect is general and applicable to both the dynamic processes and the equilibrium states. Therefore, it is not surprising that it is observed in many fields of science as mentioned above [561]. The presented thermodynamic treatment concerning the compensation effect is also general and does not need any assumption about its mechanism, bonding, etc. Let us stress out that the true compensation effect can exist exclusively for a well-defined single process or state and is related to unambiguously defined state conditions. In this case, the characteristic enthalpy, $\Delta H^{\mathrm{ch}}$, and entropy, $\Delta S^{\mathrm{ch}}$, must be well-defined and have clear physical meaning. In interfacial segregation, the only functions applicable for the compensation effect are $\Delta H_I^0$ and $\Delta S_I^0$. Here, the intensive variables $\Psi_j$ provide constancy of the segregation mechanism. For example, the interfacial energy varies with the interface orientation but also with the nature of the segregating element. The integration constant $\Delta S' = -\Delta G_I^0(T_{\mathrm{CE}})/T_{\mathrm{CE}}$ is related to the configurational entropy of the system at $T_{\mathrm{CE}}$ [322, 561]. Any overlapping process such as species interactions have to be avoided because it may provide an additional complex contribution to the values of $\Delta H^{\mathrm{ch}}$ and $\Delta S^{\mathrm{ch}}$ and therefore, it may result in a mechanism differing from that for which the compensation effect is considered.

This can be documented in Fig. 5.29 by comparing the plots of the values of $\Delta S_{\mathrm{C}}^0$ vs. $\Delta H_{\mathrm{C}}^0$ of carbon segregation at symmetrical tilt grain boundaries of α-iron [94] with the values of the segregation enthalpy and entropy, $\Delta S_I$ vs. $\Delta H_I$ [113]. Since the values of both $\Delta S_I^0$ and $\Delta H_I^0$ are principally independent of temperature (and concentration), there is no indication of temperature. It is apparent that this plot exhibits a pronounced linear dependence between $\Delta S_{\mathrm{C}}^0$ and $\Delta H_{\mathrm{C}}^0$ (Fig. 5.29a) albeit not between $\Delta S_I$ and $\Delta H_I$ for carbon at 923 K (Fig. 5.29b). This is because the latter thermodynamic functions involve two contributions to the actual segregation of carbon (a) the tendency of carbon to segregate at grain boundaries of iron in infinitesimally diluted (ideal) solid solution and (b) the interaction of carbon with other segregating elements, phosphorus and silicon. Figure 5.29 clearly demonstrates the above mentioned necessary condition of the linear dependence between characteristic enthalpy and entropy – its exclusivity for a single, well-defined mechanism of the process or state [561].

The range of existence of the compensation effect (i.e. $T_{\mathrm{CE}} = \mathrm{const}$) for grain boundary segregation in α-iron, represented by correlated pairs of $\Delta H_I^0$ and $\Delta S_I^0$, is quite large covering very different grain boundaries and also different segregating elements (Fig. 5.30) [561]. As is clearly seen in Fig. 5.30, the linear dependence between $\Delta H_I^0$ and $\Delta S_I^0$ is very well fulfilled not only for individual

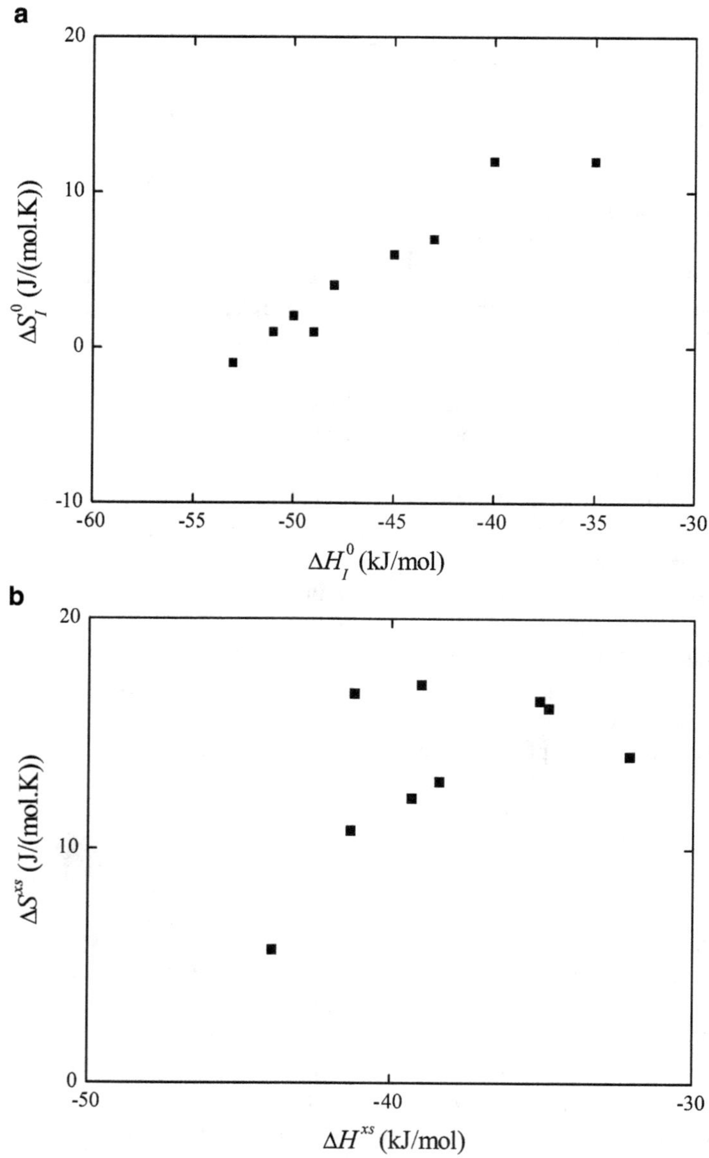


**Fig. 5.29** Interdependence of characteristic entropy and enthalpy of segregation of carbon at well-characterised symmetrical tilt grain boundaries in α-iron. (**a**) Standard entropy and enthalpy of carbon segregation, $\Delta S_C^0$ and $\Delta H_C^0$, and (**b**) entropy and enthalpy of carbon segregation, $\Delta S_C$ and $\Delta H_C$, at 923 K (according to [113])

grain boundaries (sites) in case of carbon, phosphorus and silicon segregation in α-iron (solid symbols) [98, 565] but also for various other solutes segregating preferentially at general interfaces in polycrystals [561]. The compensation effect for interfacial segregation splits into two branches reflecting, thus, two different atomic

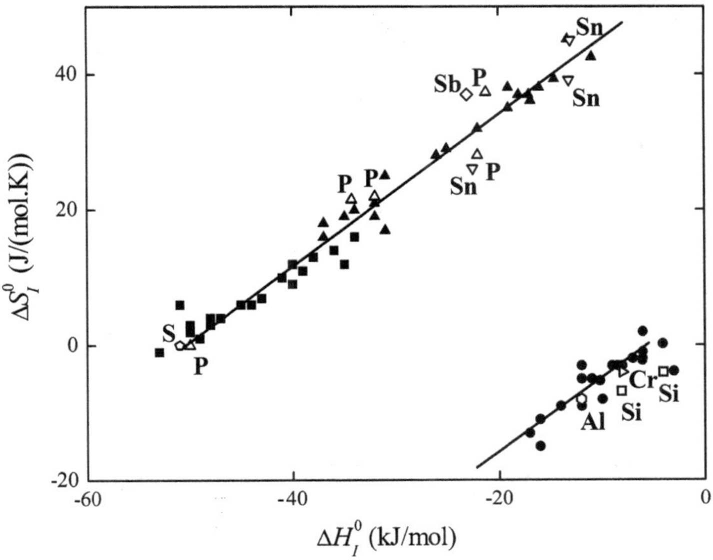


**Fig. 5.30** Dependence of the standard molar entropy, $\Delta S_I^0$, and the standard molar enthalpy, $\Delta H_I^0$, of grain boundary segregation in α-iron. *Full symbols* depict segregation of carbon (*squares*), phosphorus (*triangles*) and silicon (*circles*) at individual grain boundaries, the *empty symbols* represent the data measured in polycrystalline iron ("average" grain boundaries) that were found in literature. The upper and lower branches of the dependence represent the best fits of the data on solute segregation at interstitial sites and in substitutional positions, respectively [97, 565]

mechanisms of grain boundary segregation – interstitial (upper branch) and substitutional (lower branch) [98]. This proves the sensitivity of the compensation effect to the character of the single process for which the compensation effect can exist. The value of $T_{CE} = 930$ K remains, however, identical for both branches of the compensation effect suggesting that the compensation temperature is a characteristic of the matrix element ($\Delta G_I^0(T_{CE})/T_{CE} = -56$ J/(mol K) and $-5$ J/(mol K) were deduced for interstitial segregation and for substitutional segregation, respectively). Due to wide existence of the compensation effect it can also be used to predict the values of $\Delta H_I^0$ and $\Delta S_I^0$ for many grain boundaries and many segregating elements (see Sect. 5.7.1) [375].

Although the nature and physical meaning of the compensation temperature is still open for discussion at present, its existence is well established and has a very interesting and important consequence. According to (5.24), $\Delta G_I^0(T_{CE}) = \text{const}$ for all interfaces (and also for different solutes segregating at the interfaces supposing identical mechanism of segregation – substitutional or interstitial). It means that there exists a joint cross-section of $\Delta G_I^0$ on its temperature dependence for different grain boundaries as indicated in Fig. 5.31. It means that for two interfaces (sites), A and B, for which ${}^{A}\Delta G_I^0(T_1) < {}^{B}\Delta G_I^0(T_1)$ at $T_1 < T_{CE}$, the reverse relationship is valid at $T_2 > T_{CE}$, ${}^{A}\Delta G_I^0(T_2) < {}^{B}\Delta G_I^0(T_2)$. This is obvious from

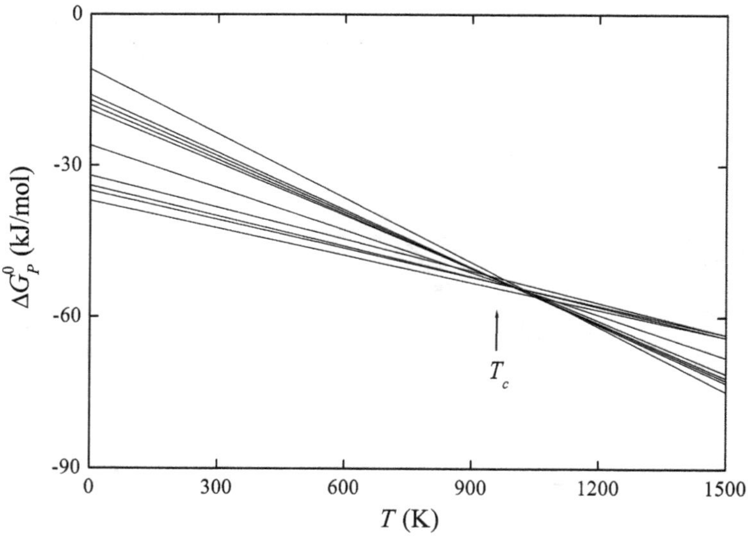


**Fig. 5.31** Temperature dependence of the standard molar Gibbs energy of phosphorus segregation, $\Delta G_P^0$, in α-iron at various grain boundaries (data from [94])

Fig. 5.32 for phosphorus segregation at grain boundaries in an Fe–Si–P–C alloy [94]. This is also supported by much higher silicon segregation at {013}, {012} and {023} special grain boundaries in austenitic stainless steel as compared to general ones [533] (Fig. 5.13). We may well deduce that these measurements were performed at $T > T_{CE}$. Existence of the compensation effect confirms the importance of the segregation entropy in interfacial segregation that cannot be neglected in any case.

A clear enthalpy–entropy compensation effect was also detected for palladium segregation at different sites in different grain boundaries of nickel obtained from computer simulations [341]. However, the type of their enthalpy and entropy terms is not completely clear: Because it is assumed in [341] that both functions are independent of temperature, one should suppose that these data represent the effective $\Delta H_{seg}^{eff}$ and $\Delta S_{seg}^{eff}$. Only when $\Delta G_I(T)$ deviates negligibly from a linear dependence on temperature these values could be considered as equivalent to $\Delta H_I^0$ and $\Delta S_I^0$. However, this information is not known. The results suggest that in the above case of Ni–Pt, (a) the averaging over a temperature range does not play important role and (b) the segregation mechanism is considered identical for all grain boundaries (grain boundary sites). Therefore, the necessary conditions of the compensation effect are fulfilled. The compensation effect was also reported for gold segregation at nickel surface [567].

It should be emphasised that special care is necessary in selecting appropriate data to verify the compensation effect. All fundamentals have to be well understood before making general conclusions, particularly in constructing arguments against the importance of the compensation effect.

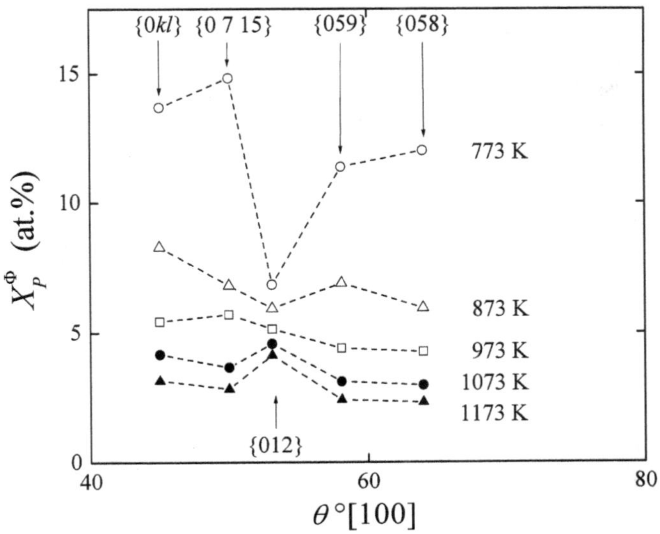


**Fig. 5.32** Dependence of phosphorus concentration at [100] symmetrical tilt grain boundaries in an Fe–3.55at.%Si–0.0089at.%P–0.014at.%C alloy on misorientation angle at various temperatures [566]

## 5.6 Grain Size

The size of individual grains in a polycrystalline material does not belong among intensive thermodynamic parameters affecting interfacial segregation of solutes. In a wide range of the coarse grain size, it also does not affect the interfacial segregation. It is because the total number of the interfacial positions available for segregation in coarse-grained materials is negligible compared to the amount of the solute atoms in grain volume [568]. However, the grain size becomes important for the segregation effects if it reaches the nanometer range. In this case, the fraction of the grain boundary sites substantially increases and consequently, the grain boundaries become strongly curved. As shown by Siegel [569], the volume fraction of the grain boundaries is approximately 25% for the material with the grain size of 10 nm and approximately 50% for the grain size of 5 nm supposing the thickness of the grain boundary being 1 nm. Thus the amount of the grain boundary positions available for segregation is comparable with the volume sites. As a consequence, segregation of the solute atoms at the grain boundaries evokes reduction of the bulk concentration to one-half. This implies that the solute segregation is strongly suppressed in nanocrystalline materials. It also means that the solubility of a solute in nanocrystalline materials will be enhanced comparing to the classical fine-grained polycrystals [569, 570]. This was confirmed by the finite element simulations of the kinetics of sulphur segregation at grain boundaries of nickel. This simulation showed that the saturation level of more than 80 at.% of sulphur is reached in nickel with bulk concentration of sulphur of 20 ppm at 800 K if the grain size is 1 mm or

0.1 mm. However, if the grain size is lowered to 10 μm, maximum sulphur segregation reaches 30 at.% and in case of the grain size of 1 μm, the sulphur segregation reaches only less than 5 at.% [571].

The total concentration of the solute atoms in a polycrystalline material, $X_I^{\mathrm{p}}$, is given as [568]

$$X_I^{\mathrm{p}} = X_I f + X_I^{\Phi}(1 - f), \tag{5.26}$$

where $f$ is the volume fraction of the grain boundaries. In conventional materials (bicrystals, general grain boundaries in polycrystals, see Chap. 4), the number of available positions for segregation is negligible comparing to that of the amount of the solute atoms and therefore, the bulk concentration $X_I$ does not change during the segregation process so that $X_I^{\mathrm{p}} \cong X_I$. Rigorously, the volume fraction of the grain boundaries can be expressed as [568]

$$f = \frac{\delta}{2} \times \frac{4\pi \bar{r}^2}{\frac{4}{3}\pi \bar{r}^3} = \frac{3\delta}{2\bar{r}}, \tag{5.27}$$

where $\delta$ is the grain boundary thickness and $\bar{r}$ is the average grain radius. Introducing (5.27) into the Langmuir–McLean segregation isotherm, (4.61), we obtain

$$\frac{X_I^{\Phi}}{1 - X_I^{\Phi}} = \frac{X_I^{\mathrm{p}} - \frac{3\delta}{2\bar{r}} X_I}{1 - X_I^{\mathrm{p}} - \frac{3\delta}{2\bar{r}}(1 - X_I)} \exp\left(-\frac{\Delta G_I}{RT}\right). \tag{5.28}$$

In conventional materials, $3\delta/2\bar{r} \ll 1$, (5.28) is identical to (4.38). Using this model, Ishida [568] showed that the extent of variations in the grain boundary segregation is negligible for the grains larger than 300 μm. However, the critical value will be different for particular segregation systems. On the other hand, a clear size dependence of calcium segregation was found at grain boundaries of synthesised rutile below the critical size of 150–350 nm. In case of larger grain size, the grain boundaries were found to be saturated by calcium at the level of about one-half of a monolayer [570]. Theoretical calculations suggest that there is no practical change in grain boundary segregation of sulphur in nickel polycrystals with the grain size larger than 100 μm; however, it is about three times lower when the grain size is reduced to 10 μm and about 20 times lower in case of 1 μm grain size [182, 572].

Solute diffusion of silver in nanocrystalline γ-Fe–40mass%Ni alloy revealed its strong segregation characterised by $\Delta H_{\mathrm{Ag}} = -47$ kJ/mol. This value is comparable with that found for coarse-grained polycrystals [573].

At lower temperatures and small grain sizes, there exists a critical concentration, $X_I^{\mathrm{c}}$,

$$X_I^{\mathrm{c}} = X_I^{\mathrm{p}} \frac{2\bar{r}}{3\delta} \tag{5.29}$$

representing the state when all solute atoms segregate at the grain boundaries. Corresponding relationships can be derived for multi-component systems [568].

Cerium, calcium and strontium effectively segregate at the grain boundaries of AZ91 magnesium-based alloy and suppress the grain growth in this material.

Alloying with pure cerium causes reduction of the grain size from 107 to 36 μm in as-cast material. After a suitable mechanical and thermal treatment, the grain size was further reduced to 20 and 30 μm using combination of cerium with calcium and with strontium, respectively [574].

Thermodynamic analysis performed by Weissmüller [575] revealed a very important feature in the field of *nanocrystalline* materials – an existence of a state where the alloy is stable with respect to the variation of the total grain boundary area. For a nanocrystalline material, the Langmuir–McLean segregation isotherm was extended to yield the total Gibbs energy of an alloy nanocrystal, $G^{nx}$, as function of pressure $P$, temperature $T$, concentrations of solvent $M$ and solute $I$, and of the total grain boundary area $A$. Supposing random substitutional solid solution with crystal volume and the grain boundaries with fixed numbers of sites $N = N_I + N_M$ and $N^\Phi = N_I^\Phi + N_M^\Phi$, the Gibbs energy of the system is given as

$$G^{nx} = \left(N_I + N_I^\Phi\right)\mu_I^0 + \left(N_M + N_M^\Phi\right)\mu_M^0 + \sigma^0 A + N_I \Delta H_{I\,\mathrm{in}\,M}^{\mathrm{sol}} + N_I^\Phi \Delta H_{I\,\mathrm{in}\,\Phi}^{\mathrm{sol}} + RT\left[N_I \ln\left(\frac{N_I}{N}\right) + N_M \ln\left(\frac{N_M}{N}\right) + N_I^\Phi \ln\left(\frac{N_I^\Phi}{N^\Phi}\right) + N_M^\Phi \ln\left(\frac{N_M^\Phi}{N^\Phi}\right)\right], \tag{5.30}$$

where $\sigma^0$ is the grain boundary energy in pure solvent, both at chosen temperature and pressure. Stirling formula was applied to the logarithms of $N_i$-factorials.

Supposing, for simplicity, spherical grains in a polycrystal, the grain boundary area $A$ can be related to the crystal volume $V$ as $A \approx 3V/2\bar{r}$. The detail treatment of the problem shows that $G^{nx}$ reduces with decreasing $\bar{r}$, because the increase of the grain boundary energy connected with the decrease of the grain size is less than the decrease of the Gibbs energy connected with the solute segregation. At very low grain sizes, however, $G^{nx}$ increases again because the majority of the solute atoms are already present at the grain boundary and the grain boundary area further increases. Thus the boundary energy then prevails over the gain in reduction of the Gibbs energy due to segregation. As a result, there exists a minimum of $G^{nx}$ for a grain size $\bar{r}$ in nanometer range [575].

The existence of a minimum of the Gibbs energy suggests the segregation-induced stability of nanosized materials, for example stopping the grain growth when this "equilibrium" size is reached [576]. This inhibition and, thus, stabilisation of the grain size in the nanometer range is of high importance for advantageous applicability of nanocrystalline materials. According to (4.15) the grain boundary adsorption is related to the change of the grain boundary energy with the change of the chemical potential. Supposing an ideal solid solution and Henry law for the solvent, we can rewrite (4.15) as [577]

$$\left(\frac{\partial \sigma}{\partial X_I}\right)_{T,P} = -\frac{\Gamma_{I,M}^\Phi RT}{X_I}. \tag{5.31}$$

Integration of (5.31) results in

$$\sigma - \sigma_0 = -RT\delta\rho X^{0\Phi}\left[1+\frac{1}{\exp(-\Delta G_I^0/RT)-1}\right]\times\ln\left[1-X_I+X_I\exp\left(-\frac{\Delta G_I^0}{RT}\right)\right], \tag{5.32}$$

where $\rho$ is the density and $\delta$ is the thickness of the grain boundary. In (5.32) the Langmuir–McLean segregation isotherm (4.61) was also applied. Assuming for simplicity $|\Delta G_I^0| \gg RT$, (5.32) may be written as

$$\sigma = \sigma_0 - \Gamma^{0\Phi}\left[RT\ln X_I - \Delta G_I^0\right]. \tag{5.33}$$

In (5.33) $\Gamma^{0\Phi} = X^{0\Phi}\delta\rho$ is the Gibbs adsorption in saturation. The main message of (5.32) is that $\sigma$ decreases from the value of $\sigma_0$, because the Gibbs energy of grain boundary segregation, $\Delta G_I^0$, is essentially negative. In a real system, a segregation isotherm should be used accounting for interaction in the system, e.g. the Fowler isotherm ((4.32) and (4.79) for a binary system).

To relate $X_I$ to the grain boundary area $A$ or to the grain size, we can again suppose for simplicity the spherical grains with an average radius $\bar{r}$. Then

$$X_I = X_I^{\mathrm{p}} - \frac{3V\Gamma_{I,M}^{\Phi}}{2\bar{r}} \tag{5.34}$$

with $V$ being the molar volume of the alloy [577]. According to (5.34), the changes in grain size represented by $\bar{r}$ evoke changes in volume concentration of solute $I$ as already proposed above. According to (5.33) and (5.34), the grain size $2\bar{r}^*$ corresponding to the metastable state is

$$\bar{r}^* = \frac{3V\Gamma_{I,M}^{\Phi}}{2\left[X_I - \exp\left(\frac{\sigma_0+\Gamma^{0\Phi}\Delta G_I}{RT\Gamma^{0\Phi}}\right)\right]}. \tag{5.35}$$

Under the conditions of (5.35), the grain boundary energy is equal to zero [577]. Such a boundary is then intact and thus rather stable. Indeed, any change of external conditions (temperature, pressure, magnetic field...) causes a disturbance of this state and further process leading either to reaching another metastable state or the equilibrium one [578]. Knowledge on the relationship between the level of grain boundary segregation, grain size and thermal condition may be successfully used to predict the metastable grain size of a nanocrystal [567, 579, 580].

Stabilization of nanocrystalline grain sizes up to the temperatures close to the melting points was reported for Pd–Zr [581], Fe–Zr [582] and Cu–Nb [583] alloys. In case of the grain size stabilization in Pd–Zr and Fe–Zr alloys, the thermodynamic reasons for the stabilization seem to be important [584]. The behaviour of the Cu–10 at.% Nb alloy is more complex because of presence of niobium second phase particles in metastable solid solution of niobium in copper. Nevertheless, the 50 nm size of the grains did not change even after annealing at 1,000 °C, i.e. at

$0.94T_m$ [583] proving that nanocrystalline grain structures can be stabilised up to the temperatures near the melting point [584].

Shvindlerman and Gottstein considered the relationship between the grain size and the grain boundary volume [585]. The Gibbs–Duhem expressions for an interface can be written at constant temperature as

$$d\sigma = -\sum_{i}^{k} \Gamma_i^{\Phi} d\mu_i, \tag{5.36}$$

i.e.

$$d\sigma = -\Gamma_0^{\Phi} \Omega dP, \tag{5.37}$$

where $\Gamma_0^{\mathrm{F}}$ is the auto-adsorption in one-component system and $\Omega$ is the atomic volume. Evidentially,

$$\frac{d\sigma}{dP} = -\Gamma_0^{\Phi} \Omega. \tag{5.38}$$

The grain boundary volume, $\Delta V^{\Phi}$, can be written as

$$\Delta V^{\Phi} = \frac{3}{2\bar{r}} \Gamma_0^{\Phi} \Omega. \tag{5.39}$$

The force $F_{\Delta V^{\Phi}}$ on the grain boundary is

$$F_{\Delta V^{\Phi}} = \frac{3}{2\bar{r}} P \Gamma_0^{\Phi} \Omega. \tag{5.40}$$

Analogously to (5.26) the Gibbs energy of the polycrystal is

$$\Delta G = (V_0 - A_t\delta)\Delta G_{\mathrm{sol}}(X_I) + \sigma(X_I)A_t, \tag{5.41}$$

where $\Delta G_{\mathrm{sol}}(X_I)$ is the formation Gibbs energy of the system with the concentration $X_I$, $V_0$ is the sample volume, $A_t$ is the total grain boundary area and $\delta$ is the grain boundary thickness. Actual impurity concentration of $I$ in the grain is

$$X_I = \frac{V_0 X_I^{\mathrm{p}} - X_I^{\Phi} A_t \delta}{V_0 - A_t \delta}, \tag{5.42}$$

where $X_I^P$ is the concentration of I in the sample. When $A_t$ increases due to reduction of $\bar{r}$, $V_0$ becomes comparable to the product $A_t\delta$ and the term of $\Delta G_{\mathrm{sol}}(X_I)$ approaches to zero [585]. Then

$$\frac{d\Delta G}{dA_t} \approx \sigma(X_I). \tag{5.43}$$

If the grain boundary segregation could substantially reduce the grain boundary energy so that $\sigma(X_I) \approx 0$, $\Delta G$ reaches minimum in respect to $A_t$ and corresponding grain size is equilibrium. This case could be considered as *grain size stabilization by segregation*. However, this situation has not been observed till now and the driving force for grain growth always exists.

Recently, an analytical model was proposed by Trelewicz and Schuh [586] which is based on statistical mechanical approach to a regular solution model for a binary polycrystalline system supposing grain size is a state variable. According to this model, the energetics of the system is strongly affected by solute segregation in such an extent that it should control the equilibrium grain size. Enhanced bulk concentration of the solute results in increased segregation, which should further reduce the grain size of the polycrystal [586].

In connection with the nanocrystalline materials, one has to think about the structure of grain boundaries connecting small grains. Such interfaces are in no means planar and will cover wide ranges of orientations [569]. Many experimental studies disclosed that atomic structures of these grain boundaries may be considered as random and do not possess any order normally found in grain boundaries of conventional coarse-grained polycrystals [569, 584].

## 5.7 Prediction of Grain Boundary Segregation

Knowledge on grain boundary segregation in a variety of systems including multicomponent alloys is important for many applications, for example to disclose detrimental effects of rarely used alloying elements on brittleness, corrosion resistance or electric conductance in technological materials. Therefore, appropriate measurements of grain boundary segregation are necessary to elucidate these aspects. In cases when the measurement of grain boundary concentration may become impossible from technical reasons (cf. Chap. 3), any estimate of the grain boundary segregation is welcome.

### *5.7.1 Model of Lejček and Hofmann*

Obviously, the values of the enthalpy and entropy for each involved solute and for each grain boundary (or boundary site) are necessary to completely describe the grain boundary segregation. In the diluted (ideal) limit, the values of $\Delta H_I^0$ and $\Delta S_I^0$ are satisfactory. Their values for segregation of any solute at any grain boundary can be predicted on basis of the above-mentioned grain boundary segregation diagrams $(\Delta H_I^0)$ and using the compensation effect $(\Delta S_I^0)$.

The prediction procedure is very simple: From a phase diagram (e.g. [560]), the solubility of a solute in the matrix element can be read at any temperature. Knowing the values of $v$ and $\Delta H^*$ (for α-iron these values are given in Table 5.3), the values of

$\Delta H_I^0$ can be completely determined according to (5.19). Knowing the value of $T_{CE}$ and of $\Delta G_I^0(T_{CE})/T_{CE}$ depending on the type of solute segregation (interstitial or substitutional), the values of $\Delta S_I^0$ can be obtained using (5.25) (for above-mentioned case of α-iron, $T_{CE} = 930$ K and $\Delta G_I^0(T_{CE})/T_{CE} = -56$ J/(mol K) for interstitial segregation and $\Delta G_I^0(T_{CE})/T_{CE} = -5$ J/(mol K) for substitutional segregation, cf. Sect. 5.5.3). According to (4.42) and (4.66) we can then estimate the values of the concentration of this solute at chosen grain boundaries for any temperature and bulk concentration (within the range of existence of the solid solution) [375].

The values of $\Delta H_I^0$ and $\Delta S_I^0$ for segregation of numerous solutes at chosen grain boundaries in α-iron are listed in Appendix B.

For more concentrated (real) solid solutions, a correction to mutual interaction has to be made resulting in the values of $\Delta H_I$ and $\Delta S_I$.

The predicted interfacial segregation can be compared with the experimental data using the values of $\Delta G_I^0$, $\Delta H_I^0$ and $\Delta S_I^0$, or with $X_I^\Phi$ of some solutes in α-iron-based binary systems published in literature. Usually, such data were most often obtained by AES measurements on polycrystalline samples. Because the segregation enthalpy and entropy data listed in some publications are not well specified and obviously possess non-standard character, it is necessary to correct for interactions or site limitations in order to obtain values that really represent the standard Gibbs energy or enthalpy and entropy of segregation required for a meaningful comparison. The predicted values are not compared to computer simulations as these results frequently represent the values for $\Delta G_I$, $\Delta H_I$ and $\Delta S_I$.

As regards the solutes segregated in substitutional sites, Table 5.4 shows an excellent agreement between prediction and experiment for *aluminium* and *silicon*, and fairly good agreement for *chromium* and *molybdenum*. The differences between the predicted and the experimental values of $\Delta H_{Cr}^0$ and $\Delta S_{Cr}^0$ are within the experimental error (±5 kJ/mol and ±5 J/(mol K)), the predicted and experimental values of $\Delta G_{Cr}^0$ differ only by 1 kJ/mol. The value $\Delta G_{Mo}^0 = -20$ kJ/mol differs substantially from the prediction; however, this value was determined from the measurements of the grain boundary composition in a (Mo,P)-low alloy steel with rather weak temperature dependence that may introduce a large error. From this point of view, the agreement of these data within the above scatters is remarkable. The experimental and predicted values for *nickel* differ substantially. In this case too, the value of $\Delta G_{Ni}^0$ was determined from measurements of the composition of the grain boundaries in a complex ternary Fe–Ni–Sb alloy, which exhibits strong attractive interaction between nickel and antimony and thus, it is questionable whether the experimental value really represents $\Delta G_{Ni}^0$.

Among the interstitially segregating elements, an excellent agreement between the predicted and experimental values was often achieved namely for *boron, phosphorus, sulphur* and *tin*. The value of $\Delta H_P^0 = -50$ kJ/mol determined assuming $\Delta S_P^0 = 0$ also represents de facto the value of $\Delta G_P^0$. Some values of the enthalpy and entropy of phosphorus segregation given in Table 5.4, which differ substantially from the prediction, were determined from AES measurements in polycrystalline α-iron and ferritic steels and – as seen from Fig. 5.30 – they fit very well with the linear dependence between $\Delta H_P^0$ and $\Delta S_P^0$. It is highly probable that the sets of

**Table 5.4** Predicted and experimental characteristics of grain boundary segregation in various α-iron-based binary systems

| $I$ | Prediction | | | | | Experiment | | | | | Ref. |
|---|---|---|---|---|---|---|---|---|---|---|---|
| | $\Delta H_I^0$ | $\Delta S_I^0$ | $T$ | $\Delta G_I^0$ | $X_I$ | $X_I^\Phi$ | $X_I^\Phi$ | $\Delta G_I^0$ | $\Delta H_I^0$ | $\Delta S_I^0$ | |
| Al | **−13** | **−9** | 1,073 | **−3.3** | | | | **−3.4** | **−12** | **−8** | [587] |
| B | −76 | −26 | 673 | −58.5 | 0.04 | **5.4** | **4** | | | | [227] |
| C | **−50** | +2 | | | | | | | **(−80** | **0)** | [588] |
| | | | | | | | | | **(−37.7** | **+43.2)** | [354] |
| | | | | | | | | | **(−57** | **+21.5)** | [589] |
| | | | | | | | | | **(−79** | **−13)** | [396] |
| Cr | **−13** | **−9** | 773 | **−6.0** | | | | **−4.9** | **−8** | **−4** | [587] |
| Mo | −28 | −25 | 773 | **−8.7** | | | | **−12 to −8.8** | | | [497] |
| | | | 853 | **−6.7** | | | | **−20.0** | | | [590] |
| Ni | −20 | −16 | 823 | **−6.8** | | | | **−14.4** | | | [418] |
| P | **−32** | **+22** | | | | | | | **−34.3** | **+21.5** | [328] |
| | | | | | | | | | **−32** | **+22** | [552] |
| | | | | | | | | | **(−50** | **0)** | [588] |
| | | | | | | | | | **−21.2** | **+37.3** | [355] |
| | | | | | | | | | **−22** | **+28** | [329] |
| | | | | | | | | | **(−38** | **0)** | [343] |
| | | | 773 | **−48.8** | | | | **−44 to −33.5** | | | [497] |
| | | | 853 | **−50.6** | | | | **−50** | | | [590] |
| S | −68 | −17 | 823 | **−54.0** | 0.0035 | **8.6** | **6.1** | **−51.5** | | | [381] |
| Sb | **−30** | **+24** | | | | | | | **(−19** | **+28)** | [335] |
| | | | | | | | | | **−23** | **+37** | [338] |
| | | | | | | | | | **(−13** | **−)** | [396] |
| | | | 823 | **−49.8** | | | | **−32.7** | | | [397] |
| Si | −16 | −12 | 823 | **−7.0** | 3 | **7.0** | **10.3** | **−9.0** | | | [395] |
| | | | 873 | **−8.2** | 4 | **8.0** | **8.0** | **−5.3** | | | [591] |
| Sn | **−24** | **+30** | 823 | **−48.8** | | | | **−44.0** | **−13.0** | **+45** | [592] |
| | | | | | | | | **−50.0** | **−22.5** | **+26.1** | [593] |
| | | | | | | | | **−45.3** | **−13.1** | **+39.1** | [593] |

Concentrations in at.%, temperature in K. $\Delta G_I^0$ and $\Delta H_I^0$ in kJ/mol, $\Delta S_I^0$ in J/(mol K). The data in bold are to be compared [375]. The data in bracket evidently represent the values of $\Delta G_I$, $\Delta H_I$ and $\Delta S_I$ and cannot be directly compared with the values of $\Delta G_I^0$, $\Delta H_I^0$ and $\Delta S_I^0$

thermodynamic functions [329,355] can be attributed to "less general" grain boundaries characterised by lower absolute values of the segregation enthalpy and higher entropy values. Only the values $\Delta H_P = -38$ kJ/mol and $\Delta S_P = 0$ [343] do not fit with the linear dependence although the value of $\Delta H_P$ is quite close to the predicted one. This is probably due to a specific way of determination of $\Delta H_P^0$ (the authors used only the maximum values of $X_I^\Phi$ at each temperature for determination of the thermodynamic parameters of segregation [343]) that can be misleading, as discussed elsewhere [342]. The predicted and experimental values of $\Delta H_{Sb}^0$ and $\Delta S_{Sb}^0$ fit well with the compensation effect although a larger difference is apparent between these sets of the data. Similarly to phosphorus, this difference can be explained by measuring "less general" grain boundaries on the fracture surface of polycrystals. The value $\Delta G^0_{\mathrm{Sb}}(823\,\mathrm{K}) = -32.7\,\mathrm{kJ/mol}$ was determined from the measurements in a ternary Fe–Ni–Sb system by correcting for the excess contribution caused by attractive interaction between nickel and antimony according to the Guttmann procedure [416]. It is remarkable that this value is rather close to the predicted segregation enthalpy, $\Delta H^0_{\mathrm{Sb}} = -30\,\mathrm{kJ/mol}$.

Comparison of the predicted and experimental data marginally differs in case of *carbon*. As mentioned in Chap. 3, the determination of the thermodynamic functions from AES measurements of carbon grain boundary concentration is very complicated for several reasons. First, due to the ductilising effect of carbon on grain boundary brittle fracture, its interfacial segregation is often studied in a multi-component system containing an embrittling element and – similarly to the Fe–Ni–Sb system – the interaction effects between carbon and other solutes should be then considered. Second, carbide precipitation may occur in the bulk causing a substantial reduction of the concentrations of the solved carbon and other elements in α-iron solid solution. Last but not least, progressing contamination of the fracture surface by carbon from the residual gas atmosphere in the apparatus during the study may increase the carbon peak height measured by AES [594]. Probably, a combined effect of all these sources affects the quantification of AES spectra and is the origin of the discrepancies between predicted and experimental data for carbon (Table 5.4). Let us stress one important fact: There are clear differences of more than 40 kJ/mol in the experimental values of the enthalpy and of more than 50 J/(mol K) in the entropy reported by individual sources [354, 395, 588, 589] (Table 5.4). We can well assume that the analysed in situ fractured grain boundaries were general. In this case, very similar experimental values of $\Delta H^0_{\mathrm{C}}$ and $\Delta S^0_{\mathrm{C}}$ should be expected [94], as evident from the measurements on individual grain boundaries of different character, i.e. special, vicinal and general in model bicrystals [94]: The respective differences are 20 kJ/mol in enthalpy and 17 J/(mol K) in entropy for all grain boundaries while only 6 kJ/mol and 7 J/(mol K), respectively, for the general grain boundaries (cf. Fig. 5.30 and Appendix A). This suggests that the values of the thermodynamic functions [354, 395, 588, 589] represent a kind of $\Delta H^{\mathrm{eff}}_{\mathrm{seg}}$ and $\Delta S^{\mathrm{eff}}_{\mathrm{seg}}$ albeit not $\Delta H^0_{\mathrm{C}}$ and $\Delta S^0_{\mathrm{C}}$.

In majority of cases, there is a very good agreement within ±5 kJ/mol and ±5 J/(mol K) for $\Delta H^0_I$ and $\Delta S^0_I$, respectively: these limits are comparable with experimental uncertainties of measured values of segregation of any solute at general grain boundaries in α-iron and ferritic steels. Some discrepancies can be well explained on basis of complexity of the measurements of grain boundary segregation and of the systems studied. The above-mentioned remarkable agreement strongly supports the proposed prediction method, which is easy and straightforward. It was also shown that this method can predict chemical composition of low-alloy steels [375] as well as cast irons [17, 595] with fairly high precision. Let us mention that the equilibrium composition of ferrite solid solution has to be considered in case of these complex iron systems instead the nominal composition of the alloy: The solutes bound in precipitates are not allowable for segregation. This is very important mainly in case of carbon, whose concentration in solid solution is substantially reduced by precipitation [17, 375, 595].

### 5.7.2 Model of Wynblatt and Shi

Another approach to predict grain boundary segregation was proposed by Wynblatt and Shi [596]. Their model is based on the regular solution formalism of interfacial segregation [313, 324, 325]. The energy of the two adjoining crystals is described in terms of the nearest neighbour bonds. Each crystal consists of atoms located on the lattice points of (*hkl*) plane which terminates at the grain boundary so that no structural relaxations are permitted. In general, different (*hkl*) planes can meet at the grain boundary. They can be distinguished as $(hkl)_1$ and $(hkl)_2$ in respect to the two grains. The grains will be mutually rotated by a twist angle, $\phi$, about the grain boundary normal. The five grain boundary DOFs are then used to define each of the two terminating (*hkl*) planes and the twist angle. Let us choose the description in which the indices of the terminating planes are $h \geq k \geq l$. The (*hkl*) planes in each crystal are denoted by index $i$, where $i = 1$ for the grain boundary plane. The second index, $j$, identifies the distance of the plane from plane $i$. The maximum value of $j$ is denoted as $J_{max}$, which represents the farthest plane containing nearest neighbours of the atoms in the $i$th plane. In fcc crystals $J_{max} = \delta(h + k)$, where $\delta = 1/2$ if all $h$, $k$, $l$, are odd, and $\delta = 1$ for mixed $h$, $k$, $l$. The arrangement and the indices $i$ and $j$ are schematically shown in Fig. 5.33. Let us assume a fcc binary substitution solid solution A–B in which the solute is the component B. The composition of the $i$th atomic plane on one side of the grain boundary is given by (4.38). If the value of the segregation entropy is neglected, the Gibbs energy of segregation, $\Delta G_B^i$, is replaced by the enthalpy of segregation, $\Delta H_B^i$ [596]. $\Delta H_B^i$ includes both the nearest neighbour bond as well as the elastic strain energy terms. In context of a nearest neighbour bond model, the relative locations of atoms across a grain boundary are

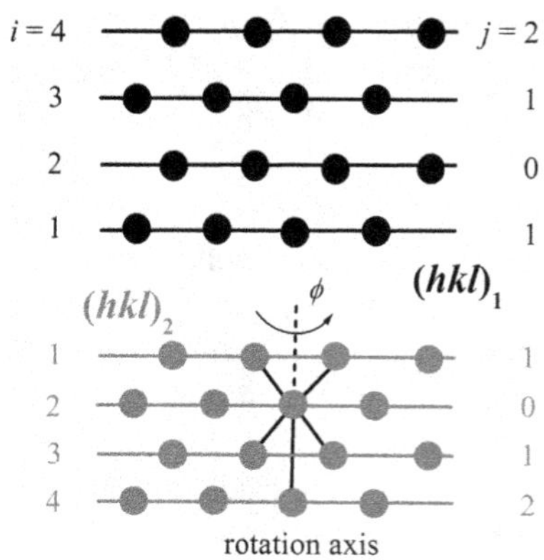


**Fig. 5.33** Grain boundary between two crystals terminated by crystallographic planes $(hkl)_1$ and $(hkl)_2$, which are twisted by an angle $\phi$ about the grain boundary normal. The indices $i$ number the planes away from the grain boundary plane. The indices $j$ are illustrated here for the planes $i = 2$. A dangling bond from an atom in plane $i = 2$ is shown as a *dashed line*. Dangling bonds are reconnected to the other side of the grain boundary with probability $P$ (see text) (according to [596])

not compatible with the nearest neighbour distances. Certain fraction of the dangling bonds of the atoms on one side of the grain boundary will be reconnected to the atoms on its other side. The fraction of the bonds reconnected to the other side of the grain boundary is given by a parameter, $P$, which depends on the Miller indices of the both boundary planes and on the twist angle $\phi$ [596].

The segregation enthalpy of the $i$th atom plane ($i \leq J_{\max}$) is given by

$$\Delta H_{\mathrm{B}}^{i} = 2\omega \begin{bmatrix} ZX - Z^{i}X^{i} - \sum_{j=1}^{J_{\max}} Z^{j}X^{i+j} - \sum_{j=1}^{i-1} Z^{j}X^{i-j} \\ -P\sum_{j=1}^{J_{\max}} Z^{j}X' - \frac{1}{2}(1-P)\sum_{j=1}^{J_{\max}} Z^{j} \end{bmatrix} - \frac{1}{2}(1-P)(\varepsilon_{\mathrm{BB}} - \varepsilon_{\mathrm{AA}})\sum_{j=1}^{J_{\max}} Z^{j} - \Delta E_{\mathrm{el}}^{i} \quad (5.44)$$

and

$$\Delta H_{\mathrm{B}}^{i} = 2\omega \left[ ZX - Z^{i}X^{i} - \sum_{j=1}^{J_{\max}} Z^{j}(X^{i+j} + X^{i-j}) \right] - \Delta E_{\mathrm{el}}^{i} \quad (5.45)$$

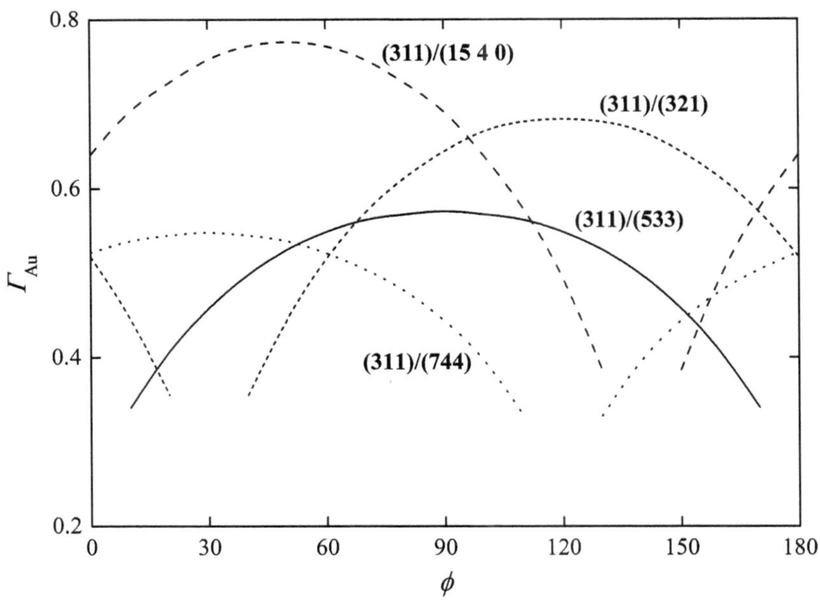


**Fig. 5.34** Prediction of gold concentration at selected twist grain boundaries, terminated by a (311) plane on one side, as a function of the twist angle $\phi$ for a Pt–1at%Au alloy at 1,000 K (according to [597])

$X' = \sum_{i,j=1}^{J_{\max}} Z^j X^i / \sum_{i,j=1}^{J_{\max}} Z^j$. The equation for the determination of the composition of a plane $i$ according to (5.44) or (5.45) in combination with (4.38) must be solved by iterative numerical methods [597].

Model of Wynblatt and Shi was applied to predict the composition of individual grain boundaries formed by the (311) plane on one side and by different (*hkl*) planes on the other side in a Pt–1at.%Au system at 1,000 K. In addition, the twist rotation by angle $\phi$ about the grain boundary normal was considered (Fig. 5.34). The results exhibit anisotropy of the gold segregation at the grain boundaries that is characterised by variations of the gold concentration approximately by factor three. A more pronounced anisotropy could be expected at lower temperatures or by considering a wider spectrum of the grain boundaries. The model also predicts the composition profiles across the interface. In an ideal solution, the composition on one side of the grain boundary depends only on its orientation and is independent of the orientation of the other side at the interface because no interaction occurs. In a real system, the degree of interaction will increase with increasing concentration and therefore, the interaction across the grain boundary is more effective [596, 597].

# Chapter 6
# Principles of Non-equilibrium Segregation

Although the basic goal of this book is to review fundamental aspects of equilibrium grain boundary segregation, it can also be reasonable to give brief information about the principles of non-equilibrium segregation, which occurs in many practical applications and plays an important role in mechanical properties of materials. In principle, this problem can be divided into two parts (a) kinetics of reaching equilibrium grain boundary segregation and (b) non-equilibrium segregation.

## 6.1 Kinetics of Grain Boundary Segregation

By annealing the sample containing grain boundaries, the diffusion processes run in the material thus redistributing the components to reduce the Gibbs energy of the system. This redistribution is time dependent and represents kinetics of reaching the final state – equilibrium redistribution of the solutes, that is equilibrium segregation. In principle, kinetics of grain boundary segregation can be considered from the point of view of semi-infinite solution of Fick equation applied to the system containing a grain boundary [19, 598]. In the following, we will use the description given by duPlessis and van Wyk [311].

### *6.1.1 Semi-infinite Solution of Fick Equation*

Let us suppose an ideal binary system containing a grain boundary between two semi-infinite bulk crystals in which the solute atoms diffuse towards the grain boundary. Such diffusion of solute $I$ represented by the changes of its concentration $X_I$ can be described by the second Fick law [311]

$$\left(\frac{\partial X_I}{\partial t}\right) = D_I\left(\frac{\partial^2 X_I}{\partial x^2}\right), \tag{6.1}$$

where $D_I$ is the diffusion coefficient of solute $I$. The driving force of the diffusion is the concentration gradient, that is the flux of the atoms into the grain boundary region ($x = 0$),

$$D_I \left( \frac{\partial X_I}{\partial x} \right)_{x=0} = \delta^{\Phi} \left( \frac{\partial X_I^{\Phi}}{\partial t} \right)_{x=0}. \tag{6.2}$$

In (6.2), $\delta^{\Phi}$ is the thickness of the segregated layer. The grain boundary concentration $X_I^{\Phi}$ is time-dependent and reaches the values $X_I^{\Phi,\infty}$ at $t \to \infty$ while $X_I^{\Phi} = X_I^{\Phi,0}$ at $t = 0$. According to McLean [19],

$$X_I^{\Phi,0} = \frac{X_I^{\Phi}}{\beta_I^{\Phi}}, \tag{6.3}$$

where $\beta_I^{\Phi}$ is the grain boundary enrichment factor (cf. (4.19)). The initial conditions for solution of (6.1) are $X_I(x > 0, t = 0) = X_I^{\Phi,0}$. The boundary condition is given by (6.3).

The diffusion equation (6.1) can be solved taking Laplace transform as [311]

$$\bar{X}_I = M \exp(-qx) + \frac{X_I^{\Phi,0}}{p}, \tag{6.4}$$

where $q^2 = p/D$. $M$ is the constant, which can be determined as

$$M = \frac{\left(1 - \beta_I^{\Phi}\right) X_I^{\Phi,0} \delta^{\Phi}}{D_I q \left(\beta_I^{\Phi} q \delta^{\Phi} + 1\right)} \tag{6.5}$$

assuming the boundary condition

$$D_I \left( \frac{\partial \bar{X}_I}{\partial x} \right)_{x=0} = \beta_I^{\Phi} \delta^{\Phi} \left( p \bar{X}_I - \frac{X_I^{\Phi,0}}{p} \right). \tag{6.6}$$

Finally, we obtain an error function solution

$$X_I(x) = X_I^{\Phi,0} \left\{ 1 - \left( 1 - \frac{1}{\beta_I^{\Phi}} \right) \exp\left( \frac{x}{\beta_I^{\Phi} \delta} + \frac{D_I t}{\beta_I^{\Phi 2} \delta^{\Phi 2}} \right) \operatorname{erfc}\left[ \frac{x}{2\sqrt{D_I t}} + \frac{\sqrt{D_I t}}{\beta_I^{\Phi} \delta^{\Phi}} \right] \right\} \tag{6.7}$$

which transforms for $\beta_I^{\Phi} \gg 1$ to

$$X_I^{\Phi} = \beta_I^{\Phi} X_I \left[ 1 - \exp\left( \frac{D_I t}{\beta_I^{\Phi 2} \delta^{\Phi 2}} \right) \operatorname{erfc}\left( \frac{\sqrt{D_I t}}{\beta_I^{\Phi} \delta^{\Phi}} \right) \right] \tag{6.8}$$

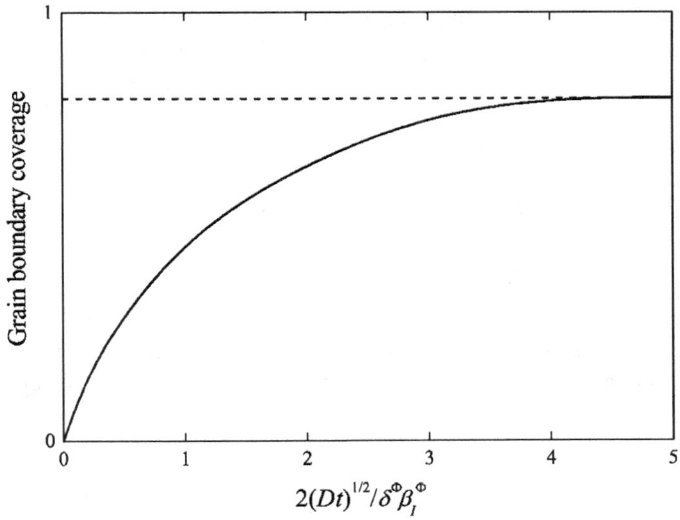


**Fig. 6.1** Time dependence of the kinetics of grain boundary segregation determined according to (6.8). According to [47]

or

$$\frac{X_{I,t}^{\Phi} - X_{I,t=0}^{\Phi}}{X_{I,t\to\infty}^{\Phi} - X_{I,t=0}^{\Phi}} = 1 - \exp\left(\frac{D_I t}{\beta_I^{\Phi 2}\delta^{\Phi 2}}\right)\operatorname{erfc}\left(\frac{\sqrt{D_I t}}{\beta_I^{\Phi}\delta^{\Phi}}\right). \tag{6.9}$$

For short time $t$, a parabolic law is obeyed [311] (Fig. 6.1). It is supposed that the bulk concentration remains unchanged during the segregation process: As was shown above, this is valid for bicrystals and for polycrystalline materials of a wide spectrum of grain sizes albeit not for nanosized materials. Analytical solution of (6.9) assumes that the parameter $\beta_I^{\Phi}$ does not change with the grain boundary concentration. This assumption is well fulfilled for low values of the grain boundary concentration.

The kinetics of the grain boundary segregation described above using the Langmuir–McLean segregation isotherm may be modified by application of more complex models of equilibrium grain boundary segregation such as Guttmann models (cf. Chap. 4). In this way, Seah [599] predicted the grain boundary segregation of impurities in steels for different times of annealing at various temperatures supposing mutual interaction between nickel and phosphorus. Theory of segregation kinetics in ternary systems including site competition and mutual interactions is shown in detail in [600] as well as other variations such as non-homogeneous initial distribution of solutes [601] or existence of surface phase transformations [415,602].

A rather complex and unexpected behaviour was observed in case of sulphur segregation at (100) surfaces in an Fe–6at%Si alloy showing the influence of the sample history on the kinetics of the process [603]. These results evoked to modify the kinetic equations for accounting sulphide formation and pipe diffusion along dislocations, and possible surface phase transitions for the case of higher coverage effects. Considering all these effects, the interfacial concentration of the segregated element represents the result of three contributions (a) bulk diffusion, (b) pipe diffusion along extended defects in the material (dislocations, grain boundaries) and (c) dislocation enrichment and precipitation. Thus,

$$X^{\Phi}_{I,t} = X^{\Phi}_{I,t=0} + X_{I,t} + X^{\mathrm{d}}_{I,t} + X^{pv}_{I,t}, \tag{6.10}$$

where $X^{\Phi}_{I,t=0}$ is the interfacial concentration of $I$ at beginning of the process, $X_{I,t}$ is the contribution following from bulk diffusion, $X^{\mathrm{d}}_{I,t}$ is the contribution of pipe diffusion and $X^{pv}_{I,t}$ is the additional contribution to interfacial segregation due to the solute enrichment of dislocations. The complete kinetics of surface segregation is then expressed as

$$X^{\Phi}_{I,t} - X^{\Phi}_{I,t=0} = \frac{2X_{I,t=0}}{\delta^{\Phi} X^{\Phi}_{I,t\to\infty}}\sqrt{\frac{D_I t}{\pi}} + \frac{2d_{\mathrm{d}} X^{\mathrm{d}}_{I,t=0}}{\delta^{\Phi} X^{\Phi}_{I,t\to\infty}}\sqrt{\frac{D^{\mathrm{d}}_I t}{\pi}}$$
$$+ \frac{4d_{\mathrm{d}}\pi\left(X_{I,t=0} - X^*_I\right)}{3\delta^{\Phi} X^{\Phi}_{I,t\to\infty}} \frac{D^{\mathrm{d}}_I t}{a^2 \ln(L/a)}\sqrt{\frac{D^{\mathrm{d}}_I t}{\pi}}, \tag{6.11}$$

where $d_{\mathrm{d}}$ is the dislocation density related to the total amount of the atomic sites, $2L$ is the average distance between precipitates at the dislocations, $2a$ is the inner width of the dislocation core, and $X^*_I$ is the solubility limit of the solute. Despite some simplification, this model was successfully used to interpret the kinetics of surface segregation in the above mentioned Fe–6at%Si alloy [603].

### 6.1.2 *Layer-by-Layer Model (Model of Hofmann and Erlewein)*

Hofmann and Erlewein [604] supposed that the course of the potential (Gibbs) energy in direction perpendicular to the interface is periodic with the barrier $\Delta G_{\mathrm{D}}$ in the bulk. In the interface layer 1, the minimum is lower by $\Delta G(\Phi)$ (Fig. 6.2). During diffusion, the atoms jump from layer to layer and overcome the barriers. Supposing the fluxes between the layers, $J_{i,j}$, with $j = i \pm 1$, we can solve the Fick laws. The first Fick law

$$J_x = -D_{I,x}\frac{\partial X_I}{\partial x} \tag{6.12}$$

describes the flux through the plane in distance $x$ from the interface ($x = 0$) in one-dimensional case. For small changes in concentration, we can approximate for cubic structure $\partial x = a$, where $a$ is the lattice parameter and $\partial X_I = X_{I(i+1)} - X_{I(i)}$ is the difference in concentrations of two neighbour atomic layers,

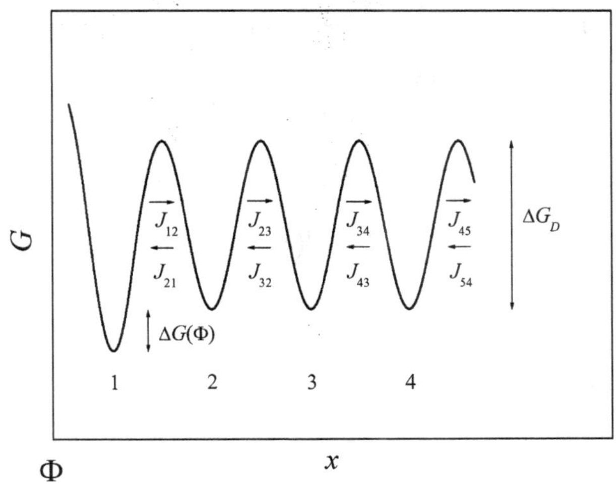


**Fig. 6.2** Schematic depiction of the layer-by-layer course of the potential Gibbs energy in direction $x$ perpendicular to the interface $\Phi$. According to [604]

$$J_{i+1,i} = -D_{I,i+1,i}\frac{X_{I+1} - X_{I,i}}{a}. \tag{6.13}$$

Supposing

$$D_I = D_{0,I}\exp\left(-\frac{Q}{RT}\right) = a^2\nu\exp\left(-\frac{Q}{RT}\right), \tag{6.14}$$

the evaluation of (6.13) for individual layers provides us with

$$J_{12} = \frac{X_{I,1}W_{12}\nu_1}{a^2}\exp\left(-\frac{\Delta G_{\mathrm{D}} + \Delta G(\Phi)}{RT}\right), \tag{6.15}$$

$$J_{21} = \frac{X_{I,2}W_{21}\nu_2}{a^2}\exp\left(-\frac{\Delta G_{\mathrm{D}}}{RT}\right), \tag{6.16}$$

and

$$J_{i,i+1} = \frac{X_{I,i}W_{i,i+1}\nu_i}{a^2}\exp\left(-\frac{\Delta G_{\mathrm{D}}}{RT}\right), \tag{6.17}$$

etc., where $\nu_i$ is the frequency of atomic jumps in layer $i$ and $W_{i,i+1}$ is the jump probability. In the bulk, the probability of the jumps is

$$W_{i,i+1} = 1 - X_{I,i\pm 1}, \tag{6.18}$$

while in vicinity of the interface layer

$$W_{21} = \left(1 - \frac{X_{I,1}}{X_I^{\Phi}}\right)^{X_I^{\Phi}}, \tag{6.19}$$

where $X_I^{\Phi}$ is the equilibrium interfacial concentration at this temperature.

The rate of the concentration changes can be expressed using the second Fick law

$$\frac{\mathrm{d}X_{I,1}}{\mathrm{d}t} = a^2(J_{21} - J_{12}). \tag{6.20}$$

Thus [604]

$$\frac{\mathrm{d}X_{I,1}}{\mathrm{d}t} = \frac{D_I}{a^2}\left[W_{21}X_{I,2} - W_{12}X_{I,1}\exp\left(-\frac{\Delta G(\Phi)}{RT}\right)\right] \tag{6.21}$$

and

$$\begin{aligned}\frac{\mathrm{d}X_{I,i}}{\mathrm{d}t} &= a^2(J_{i-1,i} + J_{i+1,i} - J_{i,i-1} - J_{i,i+1}) \\ &= \frac{D_I}{a^2}\left[W_{i-1,i}X_{I,i-1} + W_{i+1,i}X_{i+1} - W_{i,i-1}X_{I,i} - W_{i,i+1}X_{I,i}\right]. \end{aligned} \tag{6.22}$$

Numerical solution of the above equations for high values of $\Delta G(\Phi)$ results in parabolic dependence [604]. Model of Hofmann and Erlewein was extended for ternary alloys by Boudjemaa and Moser [605].

### 6.1.3 Model of Limited Reaction Rates

In cases when the bulk diffusion is faster than that at interface (e.g. in case of diffusion of carbon in tungsten [606]) the flow of solutes to the interface is constant and the rate of segregation is controlled by concentration difference at the interface between the equilibrium and actual concentrations, $X^{\Phi}_{I,\mathrm{eq}}$ and $X^{\Phi}_I(t)$, respectively,

$$\frac{\mathrm{d}X_I}{\mathrm{d}t} = k\left(X^{\Phi}_{I,\mathrm{eq}} - X^{\Phi}_I(t)\right), \tag{6.23}$$

where $k$ is the reaction rate constant. Integration of (6.23) results in

$$X^{\Phi}_I(t) = X^{\Phi}_{I,\mathrm{eq}}\left[1 - \exp(-kt)\right]. \tag{6.24}$$

## 6.2 Non-equilibrium Segregation

During some processes in which the non-equilibrium character prevails, additional (non-equilibrium) point defects may be introduced into the materials containing the grain boundaries. As a result of the interaction of the solute atoms with these point defects the chemical composition of grain boundaries may be substantially changed as compared to equilibrium segregation. This effect is called *non-equilibrium segregation*. The mechanism of the non-equilibrium segregation

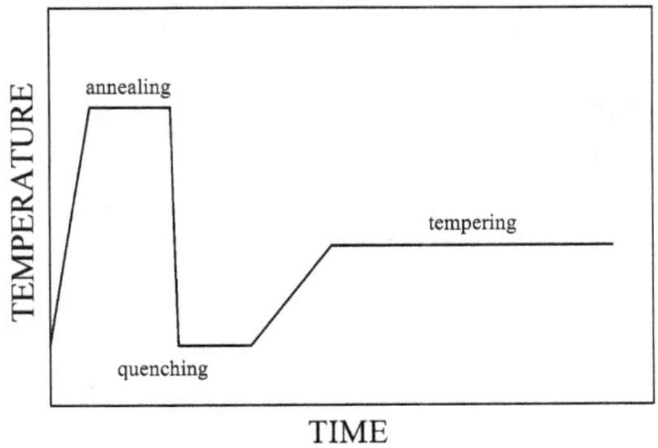


**Fig. 6.3** Schematic depiction of thermal treatment needed to produce non-equilibrium segregation. According to [18]

consists in formation of impurity–vacancy complexes in the volume [18, 47]. There are three basic types of non-equilibrium segregation (a) thermally induced segregation, (b) radiation-induced segregation and (c) stress-induced segregation [47]. Additionally, we may also consider the segregation at moving grain boundaries as non-equilibrium segregation.

### 6.2.1 *Thermally Induced Segregation*

One of the processes producing additional point defects is quenching after a heat treatment. In this way, the vacancy supersaturation occurs. During subsequent heating at moderate temperatures, (Fig. 6.3) the concentration of non-equilibrium vacancies changes and redistributes region-to-region in the microstructure. Because the interfaces are efficient sinks for vacancies, their concentration in vicinity of the grain boundaries is quickly reduced while their supersaturation in the matrix preserves. Consequently, a concentration gradient of vacancies establishes between the crystal volume and the grain boundaries and the vacancies should move towards the grain boundaries. However, the vacancies in matrix – mainly in case when a substantial misfit between solute and solvent atoms exists – form the impurity–vacancy complexes, which drag this movement but help to bring the impurity atoms close to the boundary. Consequently, an enrichment of the regions in vicinity of the grain boundaries occurs on the scale of several nanometres. This is not an equilibrium state. Indeed, a prolonged tempering can erase the effect of non-equilibrium segregation [18, 47]. Such processes were firstly observed about 50 years ago [607, 608].

According to Bercovichi et al. [609], the number of complexes $N_c$ reaching the grain boundary during quenching is

$$N_c = \int_0^{t_f} x(t)\left(\mathrm{d}X_c - \mathrm{d}X_c^{\Phi}\right), \tag{6.25}$$

where $x(t)$ is the distance of impurity diffusion in time $t$, $X_c^{\Phi}$ and $X_c$ are the concentrations of the complexes at the grain boundary and in the grains, respectively, and $t_f$ is the time of quenching. This model is physically realistic but is limited to numerical solution of the integral.

In the model of Doig and Flewitt [610], the error function solution of the Fick equation was used to analyse the vacancy concentration profile in the grain boundary region for a series of small time intervals of quenching, $\Delta t$, as

$$\frac{X_{V,x} - X^{\Phi}_{V,\Delta t}}{X^{\Phi}_{V,t=0} - X^{\Phi}_{V,\Delta t}} = \operatorname{erfc}\left(\frac{x}{2\sqrt{D_V \Delta t}}\right), \tag{6.26}$$

where $X_{V,x}$ is the vacancy concentration at a distance $x$ from the boundary, $X^{\Phi}_{V,\Delta t}$ is the vacancy concentration at the grain boundary at the quenching temperature for the quenching time $\Delta t$, $X^{\Phi}_{V,\Delta t=0}$ is the vacancy concentration at the grain boundary at the quenching temperature at zero time (= volume concentration of vacancies) and $D_V$ is the vacancy self-diffusivity in matrix [18]. The temperature $T_n$ reached during the cooling over a period $\Delta t$, can be estimated according to

$$T_n = T_s \exp(-\varphi_n \Delta t) \tag{6.27}$$

with $T_s$ and $\phi_n$ being the starting temperature and the number of the time step. For each temperature and time interval, a series of vacancy concentration profiles can be obtained. The spatial extent of the vacancy concentration profile, $x_n$, is expressed as

$$x_n = 2\sqrt{D_{v(n)} \Delta t}. \tag{6.28}$$

The final situation is then represented by an envelope of the curves [18, 47, 610].

A general and simple model of non-equilibrium segregation was also developed by Faulkner [611]. The non-equilibrium grain boundary enrichment ratio $\beta_I^{\Phi}$ (4.19) is predicted as

$$\beta_I^{\Phi} = \frac{X_I^{\Phi}}{X_I} = \exp\left(\frac{E_c^b - E_V^f}{kT_i} - \frac{E_c^b - E_V^f}{kT_{0.5T_m}}\right)\frac{E_c^b}{E_V^f}. \tag{6.29}$$

In (6.29), $E_c^b$ and $E_V^f$ are the binding energy of the complex and the vacancy formation energy, respectively, and $T_i$ and $T_m$ are the absolute starting temperature and absolute melting temperature of the material. For simplicity, all calculation is done at $T_i$. An effective time of quenching is then

$$t = \frac{K^* k T_i^2}{\varphi E_a}, \tag{6.30}$$

where $E_a$ is the effective activation energy for bulk self-diffusion and impurity diffusion and $K^* \approx 0.01$ is a constant. This model additionally admits existence

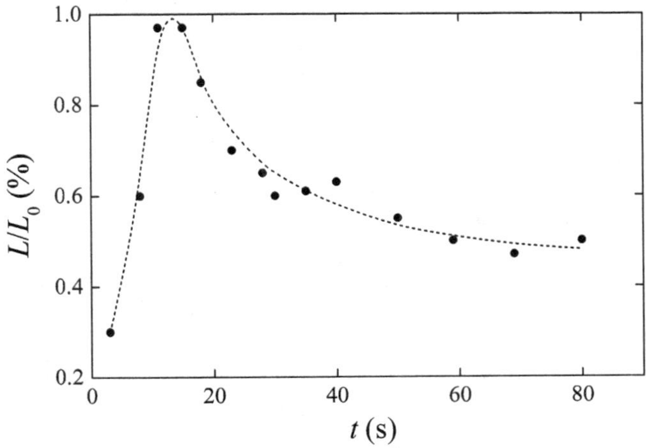


**Fig. 6.4** Time dependence of the ratio $L/L_0$ of an Fe–30%Ni(B) alloy hold at 1,050°C after isothermal annealing at 1,250°C for 0.5 h followed by quenching to ice-water. $L_0$ is the total length of the grain boundaries in a unit cross-section area and $L$ is the length of the grain boundaries in the same area. According to [613]

of the critical time $t_c$ of slow quenching or fast heating after that the equilibrium segregation establishes. The critical time is given by

$$t_c = \frac{d^2 \ln(D_c/D_I)}{4\delta(D_c - D_I)}, \tag{6.31}$$

where $d$ is the grain size, $D_c$ and $D_I$ are the complex and the impurity bulk diffusivities, respectively, and $\delta \approx 0.05$ is a constant [18, 47, 611, 612]. Existence of $t_c$ was first experimentally confirmed in case of boron segregation in an Fe–30%Ni base alloy [613] (Fig. 6.4) and further observed for boron segregation in a Mn–Mo–B steel [614] and in Fe–40%Al intermetallics [128]. Due to the temperature dependence of the diffusion coefficients, the rate of grain boundary segregation increases with increasing temperature and the critical time is shorter [615]. As a result, a temperature should exist at which the non-equilibrium segregation reaches maximum for an alloy quenched from a chosen annealing temperature (Fig. 6.5) [374, 615, 616].

Maier and Faulkner [617] point out that the final segregation level at the grain boundary may also be affected by the history of the sample. They document this statement, for example of the weld chemistry in phosphorus-containing Mn–C steel. The final segregation amount is the sum of each single step. Phosphorus segregation, which is primarily controlled by equilibrium segregation, depends on concentration of alloying elements and on temperature, but is independent of the microstructure and the thermal history. In fact, its segregation is governed by the final annealing conditions. On the other hand, manganese segregation is dependent

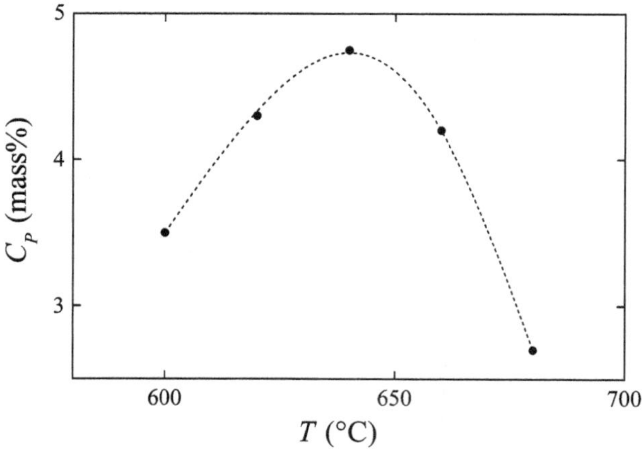


**Fig. 6.5** Grain boundary concentrations of phosphorus in steel aged at 600, 620, 640, 660, and 680°C for 70 h after quenching from a temperature of 1,050°C. According to [615]

on the microstructure (grain size) and the welding temperature, but independent of presence of the additional elements and only slightly dependent on the service temperature.

## *6.2.2 Radiation-Induced Segregation*

Another source of production of point defects is an irradiation of the sample, usually by neutrons. The major difference between this source and production of non-equilibrium point defects by thermal shocks is that interstitials also affect the segregation process. In the irradiated materials, a stronger binding occurs between interstitials and impurity atoms than with vacancies [18,47]. Besides the interstitial–impurity binding energy, the other main factors controlling the magnitude of the radiation-induced segregation are the relative diffusion rates of the free impurities and the impurities forming complexes in the matrix. Let us note that strong interaction between the interstitial and impurity atoms only occurs in case of a negative misfit, that is when the impurity atom is smaller than the matrix atom. This contrasts with interaction of vacancies with impurity atoms that form complexes despite of the character of the misfit [18, 47]. Typical concentration profiles of individual mechanisms of non-equilibrium segregation are schematically depicted in Fig. 6.6.

To describe the radiation-induced segregation, the rate theory has also been applied. Generally, the total concentrations of vacancies, $X_V$, interstitials, $X_{int}$, and solute atoms, $X_I$, are given as

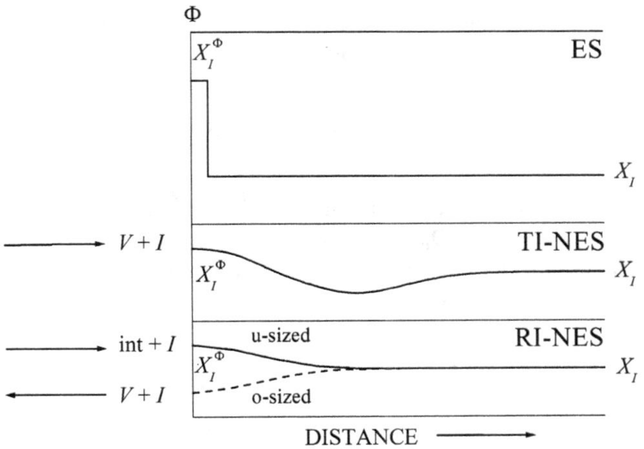


**Fig. 6.6** Schematic representation of concentration profiles of a solute in cases of thermally and radiation-induced non-equilibrium segregation. For comparison, the same representation for the equilibrium segregation is also shown. *ES* equilibrium segregation, *TI-NES* thermally induced non-equilibrium segregation, *RI-NES* radiation-induced non-equilibrium segregation, *u-sized* undersized, *o-sized* oversized solutes *I*, *V* and *int* vacancies and interstitials, respectively

$$\left(\frac{\partial X_{\mathrm{V}}}{\partial t}\right) = K - \alpha X_{\mathrm{int}} X_{\mathrm{V}} - L_{\mathrm{V}} - \nabla I_{\mathrm{V}}, \tag{6.32}$$

$$\left(\frac{\partial X_{\mathrm{int}}}{\partial t}\right) = K - \alpha X_{\mathrm{int}} X_{\mathrm{V}} - L_{\mathrm{int}} - \nabla I_{\mathrm{int}}, \tag{6.33}$$

$$\left(\frac{\partial X_I}{\partial t}\right) = -\nabla I_I, \tag{6.34}$$

where $I_{\mathrm{V}}$, $I_{\mathrm{int}}$ and $I_I$ are the fluxes of vacancies, interstitials and solute atoms, respectively, $K$ is the production rate for vacancies and interstitials, $\nabla$ is the recombination coefficient and $L_{\mathrm{V}}$ and $L_{\mathrm{int}}$ are the respective dose rates for vacancies and interstitials to dislocations and/or grain boundary sinks. Taking into account that some of the point defects can disappear within the material, the boundary condition $I_{\mathrm{V}} = I_{\mathrm{int}} = I_I$ should be fulfilled at the grain boundary, the thickness of the boundary is spread to about ten atomic spacings, the fluxes of the point defects (pd $= V$, int) from the surface to the material are given as

$$I_{\mathrm{pd}} = D_{\mathrm{pd}}^{\mathrm{eff}} \left(X_{\mathrm{pd}} - X_{\mathrm{pd}}^{\mathrm{s}}\right) / \mu_{\mathrm{pd}} + N_{\mathrm{pd}} v, \tag{6.35}$$

where $X_{\mathrm{pd}}^{\mathrm{s}}$ is the equilibrium concentration of pd at the surface, $\mu_{\mathrm{pd}}$ is the rate limiting parameter for the absorption of point defects pd and $v$ is the velocity of the receding surface due to the sputtering.

In another approach, the variables such as the diffusivities of the solute and the interstitial–solute complex, grain size, dislocation density, neutron dose, dose rate, interstitial formation, binding energies of the guest atoms with either the solute or the impurity atoms and temperature are considered. The maximum grain boundary segregation, $X^{\Phi}_{I,\max}$, is given as [47,618–620]

$$X^{\Phi}_{I,\max} = X_I \frac{E^{\mathrm{b}}_I}{E^{\mathrm{f}}_I}\left[1 + \frac{B_{\mathrm{R}}G^*}{KD_{\mathrm{int}}k_{\mathrm{d}}^2}\exp\left(\frac{E^{\mathrm{f}}_I}{kT}\right)\right], \quad (6.36)$$

where $E^{\mathrm{b}}_I$ and $E^{\mathrm{f}}_I$ are the solute–interstitial binding energy and the interstitial formation energy, respectively, $D_{\mathrm{int}}$ is the interstitial diffusivity, $K$ is a geometrical constant, $G^*$ is the production rate of the point defects that is proportional to the neutron dose rate, $B_{\mathrm{R}}$ is the dose rate correction factor and $k_{\mathrm{d}}$ is the sink strength of the grain volume for interstitials,

$$k_{\mathrm{d}} = \sqrt{K_Z\rho_{\mathrm{disl}}} \times \sqrt{6\Big/d + \sqrt{K_Z\rho_{\mathrm{disl}}}}. \quad (6.37)$$

In (6.37), $\rho_{\mathrm{disl}}$ is the dislocation density, $K_Z$ is a bias parameter defining preferred interaction between interstitials and dislocations compared with vacancies and dislocations and $d$ is the grain size [47]. Similarly to the thermally induced segregation (6.31), there exists a critical time $t_{\mathrm{c}}$ of the slow quenching or fast heating after that the equilibrium segregation establishes. The critical time is given by [47]

$$t_{\mathrm{c}} = \frac{\delta d^2 \ln(D_{\mathrm{int}}/D_I)}{4(D_{\mathrm{int}} - D_I)}. \quad (6.38)$$

Frequently, the radiation-induced segregation is explained on basis of the inverse Kirkendall mechanism (e.g. [621–623]). Similarly to equilibrium segregation, anisotropy of non-equilibrium segregation is also observed. For example, the $\Sigma = 3$ twin grain boundaries in 304 grade stainless steel were found to be resistant to solute segregation, while the level of chromium segregation increases with increasing value of $\Sigma$ [622]. Generally, the oversized solute atoms suppress the radiation-induced segregation at grain boundaries in austenitic stainless steels. The mechanisms of this effect may be different. Platinum reduces the diffusion rates of the point defects and influences subsequent defects aggregation, while hafnium increases the recombination rate of point defects possibly through formation of additive-vacancy complexes [624]. The flow of the undersized nickel solute toward the grain boundary may by facilitated by the grain boundary migration [625]. In nickel, however, the segregation of oversized atoms – manganese, palladium and niobium – is enhanced by interaction with interstitials [626].

A model of radiation-induced segregation and migration based on diffusion and reaction rate equations was proposed by Sakaguchi et al. [627]. The model considers the rearrangement of the grain boundary plane during its migration induced by radiation. The non-equilibrium point defects induced by this process, their recombination, annihilation and rearrangement are taken into account. The calculations indicate a progressive expansion of the chromium-depleted zone along the grain boundary during its migration. The results of the calculations are in good agreement with experimental data on the behaviour of an austenitic stainless steel obtained by the same authors. It was also shown that presence of third elements and their segregation suppresses the chromium segregation level and width [628].

A possible mechanism of the chromium segregation depends on whether it is oversized or undersized relatively to other solutes and in dependence on its concentration and electronic and/or magnetic effects. If it is undersized, the complexes chromium–self-interstitial iron may be preferentially formed and chromium atoms will be solute dragged towards grain boundaries. The oversized chromium atoms will drift away from grain boundaries due to a negative binding solute-drag mechanism [629]. Chromium segregates, therefore, during radiation of Fe–13Cr–1Si, HT–9, 12CrMoVNb, HCM12A and T91 steels, while it is depleted from the boundaries in high chromium ferritic/martensitic and austentic steels F82H, E911, 13Cr2MoVNbB, 13Cr2Mo+TiO, Fe–5Cr, Fe–13Cr–1Ti [623]. Similar effect in suppressing intergranular stress corrosion cracking have zirconium and – in some extent – also hafnium [630], platinum and titanium [631]. Irradiation of 304 austenitic stainless steel evokes the grain boundary segregation of nickel and the depletion of the grain boundaries by chromium and manganese. This results in reduction of intergranular cohesion of this material [632].

### *6.2.3 Stress-Induced Segregation*

Plastic deformation induces formation of non-equilibrium vacancies in alloys. The supersaturated vacancies can interact with solute atoms and create supersaturated vacancy–solute complexes thus inducing non-equilibrium grain boundary segregation of a solute via complex diffusion to the grain boundary [633]. However, the number of studies of grain boundary segregation during plastic deformation has been rare till now.

Annihilation of vacancies at grain boundaries results in reduction of their concentration and in its approaching to the equilibrium value. As the vacancy concentration in the grain volume remains non-equilibrium since there are no vacancy sinks, the vacancy concentration gradient is formed between the grain interior and the boundary. As a consequence, there also exists the concentration gradient of the vacancy–solute complexes, and these complexes diffuse to the grain boundary.

It is assumed that during high temperature plastic deformation, the annihilation of vacancies within the grains is diffusion controlled and takes place predominantly at dislocations that are the sinks of vacancies. The concentration of non-equilibrium vacancies in the steady-state, $X_{se}$, at a given strain rate, $\dot{\varepsilon}$, can be expressed as [620]

$$X_{se} = \frac{\kappa^2\mu^2\Omega_0}{D_V} A^{1/n} \exp\left(-\frac{Q_{def}}{nkT}\right) \times \left[\frac{\chi b^2}{Q_f^V}\dot{\varepsilon}^{(1-1/n)} + \frac{\xi X_j}{4b} A^{1/n} \exp\left(-\frac{Q_{def}}{nkT}\right)\dot{\varepsilon}^{(1-2/n)}\right], \quad (6.39)$$

where $\kappa$ is the structural parameter describing the distribution of dislocations, (for homogeneous distribution, $\kappa = 1$), $\mu$ is the shear modulus, $\Omega_0$ is the atomic volume,

$D_V$ is the vacancy diffusivity, $Q_f^V$ is the energy of vacancy formation, $A$ and $n$ are constants, $Q_{def}$ is the apparent activation energy for deformation, $b$ is the Burgers vector of the dislocations, $X_j$ is the concentration of thermal jogs, $\chi$ is the constant associated with the mechanical production of jogs, and $\xi$ is the parameter depicting the neutralisation effect produced by the presence of vacancy emitting and vacancy absorbing jogs. $X_j$ can be expressed as

$$X_j = \exp\left(-\frac{E_j}{kT}\right), \tag{6.40}$$

where $E_j$ is the energy of jog formation. The vacancy diffusivity in pure metals is given by

$$D_V = D_{V0} \exp\left(-\frac{Q_m^V}{kT}\right) \tag{6.41}$$

where $Q_m^V$ and $D_{V0}$ are the vacancy migration energy and the pre-exponential factor for vacancy diffusion, respectively. In case of alloys the vacancy may migrate by exchanging the position either with the host atom or with the solute atom. Therefore, (6.41) only represents an estimate which is reasonable for dilute alloys where the vacancy migration preferentially realises via exchange with the host atoms [634].

The concentration of vacancies in the matrix in a steady state, $X_{Vs}$, is equal to the sum of the concentrations of non-equilibrium vacancies, $X_{se}$, and the equilibrium vacancy concentration, $X_{th}$, which can be expressed for a pure metal as

$$X_{th} = B \exp\left(-\frac{Q_f^V}{kT}\right), \tag{6.42}$$

where $B$ is a constant. In case of an alloy, (6.42) is again only an estimate, although well acceptable for dilute alloys as discussed above for (6.41).

The concentration of vacancy–solute complexes, $X_{IV}$, is given by [633, 634]

$$X_{IV} = m X_I X_V \exp\left(\frac{Q_{IV}}{kT}\right), \tag{6.43}$$

where $m$ is a constant, $X_V$ is the vacancy concentration, and $Q_{IV}$ is the binding energy between a vacancy and a solute atom. In the regions far away from the grain boundary, $X_V = X_{VI}$. Assuming that $X_I$ is invariable, $X_V = X_{th}$ in vicinity of the grain boundaries, maximum grain boundary concentration of solute $I$, $X_{I,\max}^{\Phi}$, is [634]

$$X_{I,\max}^{\Phi} = X_I^{\Phi} + X_I \frac{Q_{IV}}{Q_f^V} \left\{ 1 + \frac{\kappa^2 \mu^2 \Omega_0 A^{1/n}}{D_V B} \exp\left(-\frac{Q_{def}}{nkT}\right) \right.$$
$$\left. \times \left[ \frac{\chi b^2}{Q_f^V} \dot{\varepsilon}^{(1-1/n)} + \frac{\xi X_j A^{1/n}}{4b} \exp\left(-\frac{Q_{def}}{nkT}\right) \dot{\varepsilon}^{(1-2/n)} \right] \times \exp\left(\frac{Q_f^V}{nkT}\right) \right\}. \tag{6.44}$$

$X^{\Phi}_{I,\max}$ can be obtained from the expressions for the kinetics of segregation (6.38) and (6.39).

A significant amount of vacancies can also be produced during creeping a material at temperatures above $0.4T_{\mathrm{m}}$. In such processes, the dimension of the material is changed to accommodate the stresses by the emission of vacancies from grain boundaries in a transverse direction to the direction of the stress. Simultaneously, there can move the over- or undersized solute or impurity atoms. As considered previously, vacancy–solute complexes may occur and if their binding energy is advantageous, they may drag the solute atoms moving to grain boundaries [47]. The effect of applied stress can be divided into two mechanisms (a) the reduction of grain boundary segregation due to modified diffusion rate of solute atoms in grain volume, and (b) the reduction of the absorption ability of grain boundaries for the solute atoms due to change of the grain boundary free energy [620]. Another model proposes production of non-equilibrium vacancies by non-conservative motion of jogs on screw dislocations at low temperatures [635]. Solute segregation is closely connected with growth of the cracks under stress and many models in this field were developed to describe the cracking of the material [47]. This consideration is out of scope of this book.

### 6.2.4 Grain Boundary Segregation and Migration

The dragging effect of segregated solutes on the grain boundary migration is notoriously known (cf. [48]). The atmosphere of the solute atoms tends to move with migrating grain boundary, however, the velocity of its diffusion is substantially lower than the rate of the grain boundary migration as the latter process is diffusionless. The interaction of the slowly diffusing solutes with the boundary results in a dragging force. The rate $v$ of the boundary motion is slowed down

$$v = m(P - P_v(v)), \tag{6.45}$$

where $P$ and $P_v(v)$ are the driving force of the grain boundary migration and the dragging term, respectively. Increasing bulk concentration of the solute has a more pronounced effect to the general grain boundaries as compared to the special ones [636]. The strongest effect of the solute segregation to the anisotropy of grain boundary migration is probably observed in case of very pure materials. In absolutely pure metals, it is supposed that special grain boundaries will migrate hardly because there is a high activation barrier to change their orientation by curving, which is necessary for the boundary migration. On the other hand, general boundaries should move rather simply in pure metals. Due to increasing concentration of impurities, the general grain boundaries are more strongly segregated than the special grain boundaries, and therefore, the rate of their motion decreases. At a suitable bulk concentration of impurities, the rate of migration of all boundaries is nearly the same. If the bulk concentration of impurities further increases, a reversed effect appears: Special grain boundaries can migrate more quickly than general ones. After saturation of the grain boundaries by the segregated impurities, the migration of all boundaries is again equalised [48].

It is obvious that during the grain boundary migration, the concentration distribution of the solute atoms becomes asymmetrical. Supposing migration of a planar grain boundary in $x$-direction with constant velocity $v$, the grain boundary concentration of solute $I$, $X_I^{\Phi}$, can be described as a unidirectional case [48]

$$\frac{\partial X_I(x,t)}{\partial t} = D_I \frac{\partial^2 X_I(x,t)}{\partial x^2} + \frac{\partial}{\partial x}\left(\frac{D_I X_I(x,t)}{kT}\frac{\mathrm{d}U}{\mathrm{d}x}\right), \tag{6.46}$$

where $D_I$ is the volume diffusivity and $X_I(x,t) = X_I$ for $x \to \pm\infty$. The co-ordinate system is supposed to have its origin located in the boundary plane, so moving with the grain boundary. The solute distribution across the migrating grain boundary can reach a steady state so that $\partial X_I(x,t)/\partial t = 0$. To simplify the problem, Gottstein and Shvindlerman [48] assume

$$U(x) = 0 \quad \text{for} \quad |x| \geq a \quad \text{and} \quad U(x) - H_0 \quad \text{for} \quad |x| \leq a. \tag{6.47}$$

Therefore, $\mathrm{d}U/\mathrm{d}x = 0$ except the discontinuities at $x = \pm a$. Supposing the Galilei transformation $x = x' - vt$, the diffusion equation in the moving co-ordinate system reduces to

$$\frac{\partial X_I(x,t)}{\partial t} = D_I \frac{\partial^2 X_I(x,t)}{\partial x^2} + v\frac{\partial X_I(x,t)}{\partial x} \tag{6.48}$$

for $x \neq \pm a$. The solution of this equation provides

$$X_I(x,t) = X_I + X_I(a)\exp\left(-\frac{vx}{D_I}\right). \tag{6.49}$$

The concentration behind the grain boundary is $X_I(x,t) = X_I$ for $x \to -\infty$. The concentration of the solute in front of the boundary is

$$(X_I)_{\mathrm{F}} = X_I + (X_I(a) - X_I)\exp\left(-\frac{vx}{D_I}\right). \tag{6.50}$$

Let us repeat that the potential $U$ is discontinuous for $x = \pm a$. Therefore, the constant $X_I^{\Phi}$ as well as the concentration distribution cannot be determined in closed form [48]. Supposing the steady state is reached during the migration, the diffusion flux is constant at any $x$, that is $J_X = vX_I$. Assuming the thickness of the grain boundary is comparable to a single atomic layer in the bulk, the flux through the interface plane between the grain boundary $\Phi$ and the bulk in front of the grain boundary ($x = +a$) is

$$X_I^{\Phi}\frac{D_I}{b}\exp\left(-\frac{H_0}{kT}\right) - X_I(a)\frac{D_I}{b} - vX_I(a) = -vX_I, \tag{6.51}$$

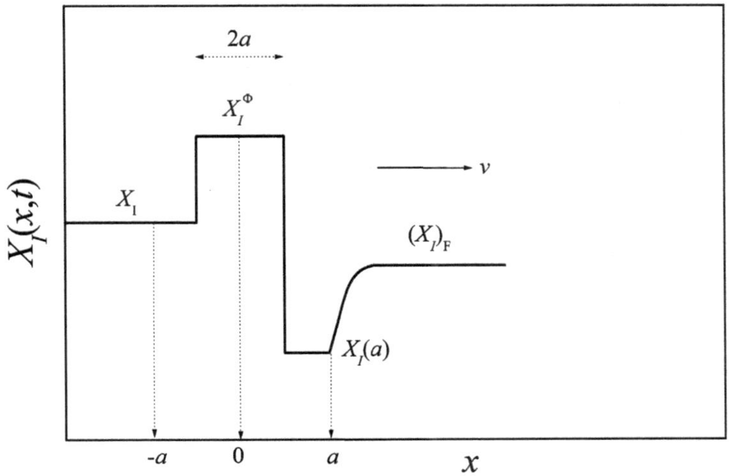


**Fig. 6.7** Model of a concentration distribution of the solutes in the vicinity of a grain boundary migrating in a steady state. According to [48]

where $b$ is the atom diameter. The flux through the interface plane between the boundary $\Phi$ and the bulk ($x = -a$) is (Fig. 6.7)

$$X_I \frac{D_I}{b} - X_I^\Phi \frac{D_I}{b} \exp\left(-\frac{H_0}{kT}\right) - vX_I^\Phi = -vX_I. \tag{6.52}$$

Since the first two terms of (6.51) describe the diffusion fluxes through the mentioned interface plane, we can write them as

$$X_I^\Phi v \exp\left(-\frac{H_0 + H_\mathrm{D}}{kT}\right) - X_I(a)v \exp\left(-\frac{H_\mathrm{D}}{kT}\right) \tag{6.53}$$

with $H_\mathrm{D}$ representing the activation enthalpy of volume diffusion. Equation (6.52) can be treated similarly. Then

$$X_I^\Phi = X_I \frac{D_I + bv}{D_I \exp(-H_0/kT) + bv} \geq X_I \tag{6.54}$$

and

$$X_I = 1 - \frac{D_I bv(1 - \exp(-H_0/kT))}{(D_I + bv)(D_I \exp(H_0/kT) + bv)}. \tag{6.55}$$

Corresponding concentration profile is shown in Fig. 6.7.

Similar result was obtained by the phase field model of Cha et al. [637]. They constructed a movable abstract boundary between two grains with rapid diffusion within it while the diffusion in the grain is neglected. Considering two variables, the solute concentration and the phase field, $f$ ($f = 1$ for one grain and $f = 0$ in the

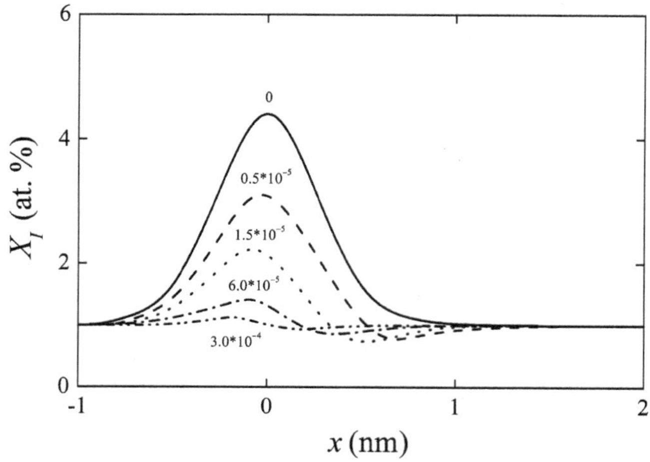


**Fig. 6.8** Concentration profiles for various velocities: equilibrium segregation (no migration, $v = 0$): *full line*, low velocity: *dashed line*, velocity of the maximum drag: *dotted line*, the inflection in the drag–velocity curve: *dashed-dotted line*, high velocity: *dashed-dotted-dotted line*. According to [637]

other grain). The interfacial region is a mixture of the grain and the grain boundary with different concentrations but with the same chemical potential. The concentration profile across the grain boundary in dependence on its migration velocity is shown in Fig. 6.8. Similar concentration distribution of the solute around the migrating grain boundary was found in simulations of kinetics of grain boundary migration and prediction of solute-drag-effect induced abnormal grain growth [638].

Simulations performed on basis of the phase field model showed that particularly in strongly segregating systems the concentration of a solute at the moving grain boundary can increase with increasing the migration velocity to such an extent that the level of segregation can be higher than that in equilibrium. The drag force of the solute segregation also causes deviations of the relationship between the migration rate and the curvature from linearity [639].

Similar results were obtained by means of a self-consistent continuum model of grain boundary segregation and segregation transition based on gradient thermodynamics and its relations to the discrete lattice model. This model stresses out the role of some distinctive terms that were ignored in previous models such as concentration gradient, spatial variation of the gradient-energy coefficient and concentration dependence of solute–grain boundary interactions and can predict the segregation transition (i.e. the transition from low to high segregation) that takes place with changing temperature, bulk composition and/or grain boundary velocity. This transition, which is first-order with a hysteresis, is responsible for the observed sharp transition of grain boundary mobility with temperature and its dependence on bulk composition observed experimentally [640].

In contrast to the solute drag of the migrating boundary discussed above, only little is known about the reverse effect – the non-equilibrium segregation of solutes at the moving grain boundaries, for example during recrystallisation and grain growth. This is despite the fact that the solute enrichment can be stronger than that of equilibrium segregation. For example, 1.6 times higher concentration of boron than its bulk concentration was detected on moving grain boundaries of an Fe–3mass.%Si alloy while no boron was found at static grain boundary [641]. Boron segregation is intensified with continuing recrystallization. If the velocity of the grain boundary migration decreases, boron grain boundary segregation also decreases and eventually disappears. The cause of these changes in the segregation behaviour is obviously the change of the width of the grain boundary during its migration. The broadening of the boundary results from its interaction with volume dislocations disappearing during grain boundary migration. The width $\delta$ of the moving grain boundary can be expressed as

$$\delta = \delta_0 + v\tau\delta_\tau \Delta\rho, \tag{6.56}$$

where $\delta_0$ is the width of the equilibrium grain boundary, $v$ is the boundary velocity, $\tau$ is the relaxation time for dislocation annihilation by moving grain boundary, $\delta_\tau$ is the increment of the average width of unit area of the boundary during $\tau$ and $\Delta\rho$ is the difference of the dislocation density in deformed and new grains. The enrichment ratio of the solute at moving grain boundaries is then

$$\beta_I^\Phi = 1 + \frac{\delta}{v} \exp\left(\frac{U_0}{RT} - 1\right). \tag{6.57}$$

In (6.57), $U_0$ is the potential of the solute atoms near the grain boundary. According to (6.56) and (6.57), the value $\beta_\mathrm{B}^\Phi = 1.7$ was determined for boron segregation at moving grain boundaries in an Fe–3mass.%Si alloy during its recrystallization at 1,273 K after 20% deformation [642].

# Chapter 7
# Grain Boundary Segregation and Related Phenomena

Grain boundary segregation affects various physical and chemical properties of the materials, which further control material behaviour. It is mainly the consequence of the close relationship between the grain boundary segregation on the one hand and the grain boundary energy and bonding state on the other hand. The chart listing some metallurgical phenomena affected by grain boundary segregation is shown in Fig. 7.1.

In this chapter, we will present some examples of the effect of grain boundary segregation on material behaviour.

## 7.1 Grain Boundary Cohesion

Historically, the widely manifested consequence of the grain boundary segregation is its effects on grain boundary cohesion as expressed in a number of well-known forms of intergranular fragility. This continues to provide much of the focus for work in this field. The central problem is the role of solute atoms on atomic cohesion at the interface. As shown in Fig. 7.1, the embrittlement can be of different nature, for example temper embrittlement, hydrogen embrittlement and liquid–metal embrittlement.

### *7.1.1 Grain Boundary Cohesion and Temper Embrittlement*

The majority of the work on grain boundary cohesion has been connected with the effect of additions in iron because of the technological importance of this material [643]. It is now well established that the elements such as copper, zinc, silicon, germanium, tin, phosphorus, arsenic, antimony, bismuth, sulphur, selenium, tellurium, oxygen and manganese all cause intergranular weakness in iron. Similar solutes also weaken copper and nickel. We now consider the two questions (1) why do certain elements weaken the grain boundaries? and (2) what is the relative embrittling potency of these elements? [13].

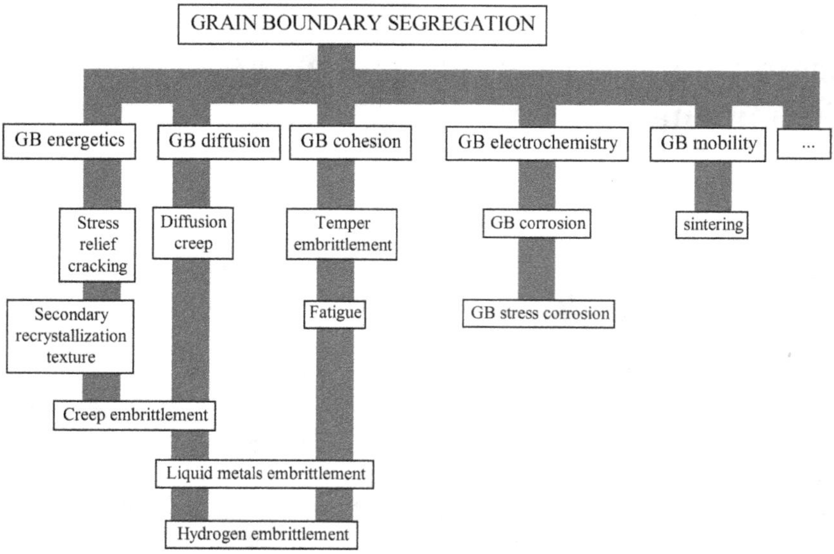


**Fig. 7.1** Some materials phenomena influenced by grain boundary segregation (according to [13])

A number of theories have been proposed to account for the role of the solute atoms on grain boundary cohesion. All have a common approach but differ in their interpretation of the way to carry out the numerical calculations. As first, McLean [19] proposed in 1957 that the total work of fracture in low temperature intergranular fracture is the sum of the ideal work of fracture, $\gamma$, and the irreversible plastic work of deformation, $\gamma_p$. $\gamma$ is the energy to open the bonds across the interface, and $\gamma_p$ remains absorbed in the crystal as dislocations and phonons over a region well away from the boundary. $\gamma_p$ is principally much larger than $\gamma$, however, if $\gamma$ increases, the forces at a crack tip must also be enhanced to propagate the fracture and therefore, $\gamma_p$ will increase. For a given grain boundary, $\gamma$ and $\gamma_p$ are thus directly related. The precise relation will depend on the micro-structure and also on the orientations of the two grains and the grain boundary. The analyses showed that $\gamma_p \propto \gamma^n$ where $n =$ 2–5 [644, 645]. To be able to answer the above questions the effects of segregated solutes on the ideal work of the grain boundary fracture must be understood.

The ideal work of fracture of a non-segregated grain boundary with the energy $\sigma^{\Phi,0}$ is [19]

$$\gamma^0 = 2\sigma^{s,0} - \sigma^{\Phi,0}, \tag{7.1}$$

where $\sigma^{s,0}$ is the surface energy of the pure material because by opening the grain boundary two free surfaces are created. An analysis of the values of $\sigma^{s,0}$ and $\sigma^{\Phi,0}$ for many metals shows that typically, $\sigma^{\Phi,0} \approx \sigma^{s,0}/3$ [371]. Accordingly, the ideal work of fracture of the pure grain boundaries is only slightly lower than that of an average plane in the crystal (approximately 5/6 of it). It does not mean automatically that the grain boundaries will always fail in preference to cleavage. The fracture manner will also depend on the orientation of the slip planes and the surface energy of the

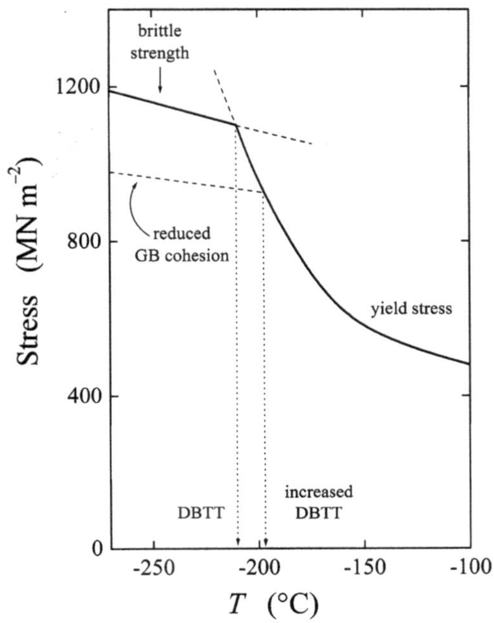


**Fig. 7.2** Schematic representation of the increase of the DBTT as a result of reduced grain boundary cohesion due to solute segregation in a 0.2% carbon steel (according to [13, 647])

low-index cleavage planes makes cleavage generally more likely than intergranular failure in clean bcc metals [646].

Let us comment now the balance between brittle and ductile failure. For example, the fracture behaviour of a 0.2% carbon steel is shown schematically in Fig. 7.2 in dependence on the test temperature [13, 647]. At high temperatures, the dislocations move easily and the yield stress is low. Therefore, the steel fails in a ductile manner. At lower temperatures, the dislocation movement becomes more difficult and the material eventually fails in a brittle manner. The highest temperature at which this occurs is the ductile–brittle transition temperature (DBTT). Sometimes, DBTT is also called as the fracture appearance transition temperature (FATT) or the onset of the upper shelf transition (OUST) [47]. The temperature dependence of the fracture energy is schematically depicted in Fig. 7.3. The DBTT corresponds to the inflection point of this dependence.

Solute segregation at the grain boundaries may lead to a sufficient reduction of the grain boundary cohesion and therefore, DBTT is increased so that the otherwise ductile material becomes brittle.

A thorough thermodynamic analysis of the ideal work of fracture of segregated grain boundaries, $\gamma$, has been made by Hirth and Rice [648]. According to their theory, the reduction of the ideal work of fracture if there is no redistribution of the

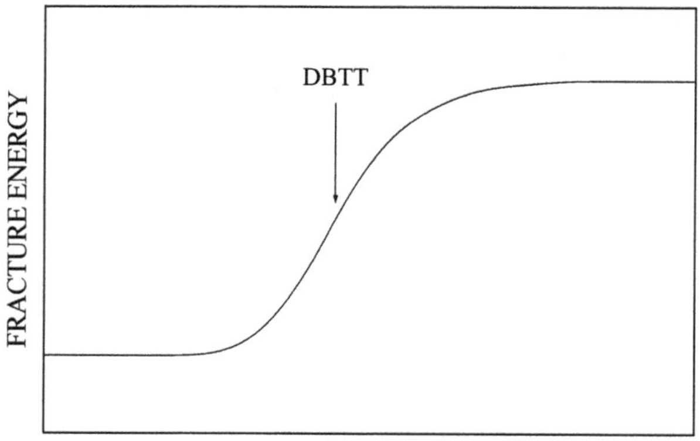


**Fig. 7.3** Schematic course of the temperature dependence of fracture energy defining the DBTT

segregated species is given by

$$\gamma = \gamma^0 - \int_0^{\Gamma_I^\Phi} \left[\mu_I^\Phi(\Gamma) - \mu_I^s(\Gamma/2)\right] \mathrm{d}\Gamma, \tag{7.2}$$

where $\mu_I^\Phi$ and $\mu_\mathrm{I}^s$ are the grain boundary and surface chemical potentials of solute $I$ in equilibrium with a level of the grain boundary adsorption, $\Gamma_I^\Phi$. This expression can be written as

$$\gamma = \gamma^0 + \Gamma_I^\Phi \left(\Delta G_I^s - \Delta G_I^\Phi\right) + RT \int_0^{\Gamma_I^\Phi} \ln \frac{\Gamma_{I,\max}^\Phi - \Gamma}{\Gamma_{I,\max}^s - \Gamma} \mathrm{d}\Gamma, \tag{7.3}$$

where $\Gamma_{I,\max}^\Phi$ and $\Gamma_{I,\max}^s$ are the maximum adsorption levels at the boundary and surface, respectively. $\Delta G_I^s$ and $\Delta G_I^\Phi$ are the Gibbs energies of segregation of solute $I$ at the surface and at the grain boundary, respectively [649]. In the dilute limit, the last term on the right-hand side of (7.3) equals to $-RT$ ln2 and usually, it can be neglected ($RT$ ln2 $= 5.76$ kJ/mol at 1,000 K). The ideal work of fracture thus decreases linearly with the level of segregation, and those elements that segregate more strongly to free surfaces than to grain boundaries reduce the ideal work of fracture and vice versa. Equation (7.3) also suggests the change of the DBTT with segregation, $\partial\mathrm{DBTT}/\partial\Gamma_I^\Phi \propto \left(\Delta G_I^s - \Delta G_I^\Phi\right)$ [366]. Measurements of $\Delta H_I^s$, $\Delta H_I^\Phi$ and $\Gamma_I^\Phi$ in $Ni_3Al$ bicrystals containing the artificial cavities at the grain boundary performed by Otterbein et al. showed the decrease of the grain boundary cohesion caused by sulphur segregation [650]. Similarly, the reduction of the cohesion energy of the {013} symmetrical tilt grain boundary by 56.4 mJ/m$^2$ (i.e. by 1.5 rel.%) was caused by complex segregation of silicon, phosphorus and sulphur in an Fe–Si alloy [651].

An alternative approach of Seah [652] giving the same result in the quasi-chemical pair bonding approximation is based on evaluation of the actual bond energies across the boundary before and after fracture,

$$\gamma = \gamma^0 + \Gamma_I^{\Phi} \left( \frac{Z_g}{Z \Gamma_I^{\Phi,0}} \right) \left( H_I^{\text{sub},*} - H_M^{\text{sub},*} - \frac{Z\Omega}{a_I^2} \right), \tag{7.4}$$

where $\Omega = H_{\text{mix}}/[ZX_I(1-X_c)]$ with the mixing enthalpy, $H_{\text{mix}}$. $Z_g$ is co-ordination number at the grain boundary, $Z$ is the bulk co-ordination number and $H_{I,M}^{\text{sub},*}$ are the sublimation enthalpies of solute $I$ and matrix element $M$. $\Gamma_I^{\Phi,0}$ is the value of $\Gamma_I^{\Phi}$ at one monolayer and $a_I$ is the atomic diameter of the segregated solute atom. In the regular solution approximation when $\Omega$ is neglected, the lower the value of $H_I^{\text{sub},*}$, comparing to that of the matrix element, the more will a segregated solute embrittle its grain boundaries. Figure 7.4 summarises the values of $H_I^{\text{sub},*}$ for numerous elements and clearly indicates that the solutes such as antimony, tin, sulphur, phosphorus, manganese, silicon and copper, with the values of $H_I^{\text{sub},*} < H_{\text{Fe}}^{\text{sub},*}$ below the limit represented by the dashed horizontal line, will all embrittle iron. On the other hand, the solutes with higher values of $H_I^{\text{sub},*}$ such as molybdenum and carbon, will improve the cohesion as was also confirmed experimentally [13].

There is numerous experimental evidence that the chart in Fig. 7.4 is generally valid for other materials than iron. For example, carbon increases the grain boundary cohesion of molybdenum and tungsten, but the effect of oxygen is quite opposite [653, 654], segregations of silicon and thorium reduce the grain boundary cohesion of iridium [389], lithium reduces the cohesion of aluminium [655] but boron segregation strengthens grain boundaries of $Ni_3Al$ [656] and copper [657].

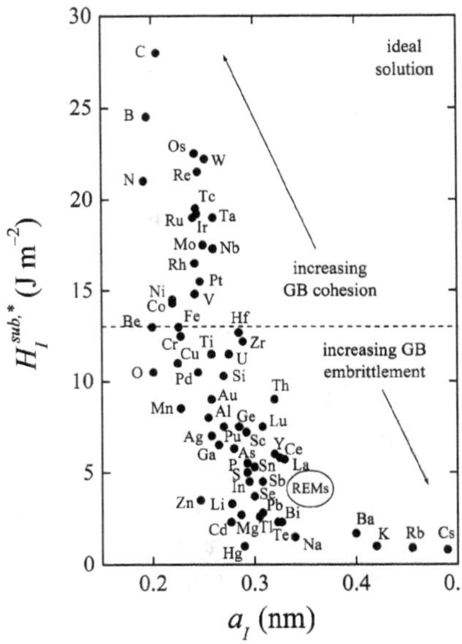


**Fig. 7.4** The plot of the sublimation enthalpy, $H_I^{\text{sub},*}$, vs. lattice parameter of the corresponding solute, $a_I$. The elements having the value of $H_I^{\text{sub},*}$ higher than any chosen matrix element (for example iron as suggested by the dashed horizontal line) increase its ductility while those with lower value of $H_I^{\text{sub},*}$ embrittle the matrix (according to [19, 652])

However, exclusive consideration of the effect of segregated solutes on the ideal work of fracture is not sufficient. One must keep in mind that there is a choice between dislocation emission and bond breaking at a crack tip that decides whether ductile cleavage or brittle fracture occurs. It can be shown that the segregated solutes causing the embrittlement increase the ease of dislocation emission at the crack tip, so offsetting the total embrittling effect discussed above. The correction is small but allows the correct prediction of the absolute embrittlement of cooper by bismuth [658].

A thorough study of the structure and solute effects on grain boundary fracture was performed on molybdenum [659]. It was shown that low angle and twist grain boundaries are strong but the majority of the interfaces are weak. It indicates an intrinsic brittleness of the grain boundaries due to the interatomic bonding of molybdenum. The weak grain boundaries can be strengthened by segregation of interstitial impurities such as carbon or nitrogen. In the latter case, a weak strengthening effect is due to precipitation of grain boundary nitride [659].

The aspects of chemical bonding in grain boundary segregation can be elucidated by ab initio calculations of chemical bonding in clusters with the same atomic arrangements as these in the structural units of a grain boundary (polyhedral structures). Application of the molecular orbital method on tetrahedral $Me_4I$ clusters to segregation of sulphur and boron at grain boundaries of nickel and iron as well as of segregation of carbon in iron revealed a redistribution of electron charge density in the grain boundary [503, 660, 661]. In case of electronegative sulphur, the charge is drawn from the surrounding metal atoms to that of sulphur and – consequently – the adjacent metal–metal bonds are weakened [660, 662]. The bonding parallel to the boundary is stronger than perpendicular one, and therefore, the cohesion across the boundary is reduced and grain boundary embrittlement results [663]. Qualitatively similar charge transfer was calculated for the case of phosphorus in bcc iron [664]. Strong bonding orbitals are formed between Fe(3d) and P(3p) orbitals in $Fe_xP$ clusters at the grain boundaries, and thus, the metallic bonds between $Fe_xP$ clusters and surrounding Fe atoms are weakened, which give rise to embrittlement [258]. The average d-band energy is higher with phosphorus (and also sulphur) than in pure metal. Then, more antibonding states are filled and cohesion is reduced [665]. In case of the less electronegative atoms of boron and carbon, relatively strong covalent-like bonds form with metal in direction perpendicular to the boundary, whereas much weaker bonding occurs in parallel direction. This kind of bonding enhances grain boundary cohesion [257, 258, 260, 662, 664]. In contrast, boron atoms occupy positions in the centre of compact polyhedra of iron atoms with little change in the structure of iron grain boundary. Therefore, the stress field is not changed by boron segregation [546], but weaker bonds in the iron grain boundary are strengthened due to the interaction of bulk dislocations with segregated atoms at the boundary [666]. Addition of boron to iron containing phosphorus evokes a site competition at the grain boundaries and phosphorus atoms are replaced by boron ones. Consequently, the material exhibits improved grain boundary ductility due to this "cleansing" effect of site competition [276]. Boron and carbon reduce the average d-band energy for the neighbouring iron atoms, fewer antibonding states are thus filled and cohesion increases [665]. Other elements such as nitrogen exhibit the bonds with some ionic character. This may be an important reason of nitrogen behaving differently from phosphorus [667]. In case of a more complex system with the grain boundary segregation of titanium, boron and oxygen, co-segregation of titanium and boron, both of which enhance grain boundary cohesion of iron, increases the cohesion. If titanium co-segregates with oxygen, the embrittling effect of the latter solute is completely reduced [668].

Similar situation occurs in Ni–S system [661]. As sulphur segregates at the grain boundaries of nickel, both the cohesive strength in vicinity of the segregated atom and the shear strength are reduced. Overcoming a critical grain boundary concentration, the network of S–S bonds within a grain boundary is formed. As a consequence, the shear stress begins to increase but the cohesive strength continues to reduce so that the conditions for brittle fracture are fulfilled. If a different solute such as boron segregates, no apparent decrease of the cohesive strength occurs [669]. Boron forms similarly strong bonds with nickel as sulphur, however, the valence orbitals of boron are compact because large ion core is absent in contrast to those of sulphur that are more extended. As a result, boron forms bonds at smaller hole sites with less intrinsic strain comparing to sulphur. Consequently, the metal–metal bond is weakened in vicinity of sulphur atoms in contrast to the case of boron, which needs less additional strain [668].

The effect of sulphur on reduced cohesion can also be explained by repulsive interaction between sulphur atoms in nickel resulting in prolongation of S–S bonds and to a consequent interfacial decohesion [670]. Another explanation for the sulphur-induced decohesion of nickel is based on consideration of directional change of chemical bonding [671]: The question of the true mechanism of the decohesion remains still open [672]. Similarly to sulphur, other metalloids are reported to reduce usually the grain boundary cohesion except beryllium, boron, carbon and silicon while aluminium and phosphorus are indicated as indifferent to nickel embrittlement [673]. This result seems not to agree completely with the results of Seah and Hondros [13, 652] (Fig. 7.4) according to which beryllium and silicon as well as phosphorus and aluminium should also be grain boundary embrittlers. This discrepancy may be affected by the fact that the calculations were performed for the $\Sigma = 5$, $\{012\}$ special grain boundary with principally low segregation energy.

The charge changes induced by solute segregation are probably the reason for grain boundary embrittlement of aluminium by gallium. First-principles pseudopotential calculations showed that gallium at the grain boundary in aluminium draws charge from the surrounding matrix atoms as a result of the difference in electronegativity between these two elements. As a result, the charge density is reduced between gallium and aluminium atoms as well as between aluminium atoms along the grain boundary [255].

Besides the case of sulphur embrittlement of nickel [670–672], a heavy discussion concerns the effect of bismuth on intergranular embrittlement of copper. A redistribution of the charge density due to grain boundary segregation of bismuth in copper was described to directly weaken the bonds between dissimilar atoms in $Cu_6Bi$ grain boundary clusters [674]. There exist three models for bismuth embrittlement of copper. Two models are based on changing electronic structure (a) the strengthening of the bonds of the surrounding metal by bismuth and thereby inhibiting the bond mobility that is a necessary condition for plasticity [675], and (b) the weakening of the bonds across the GB by bismuth [668]. These models assume that the weakening is an electronic effect, where the more electronegative segregated atom withdraws electrons from the metal d-bands thus reducing the bond strength. This interpretation is based on electron microscopic observation of the "white lines" in the inelastic electron scattering intensity observed at bismuth segregated grain boundaries of copper. Recently, the role of the electronic effect was supported when almost no expansion was found at the 36.9°[001], {310} symmetrical tilt grain boundary segregated by bismuth [674]. The embrittlement was attributed to an electronic effect: the Cu d-band retreats below the Fermi level, reducing sd hybridisation. However, it is not clear whether the grain boundary was embrittled either in the calculations or in the measurements. In addition, the grain boundary concentration of bismuth observed in this case 2.9 at/nm$^2$ is about 5 times lower than measured at grain boundaries [676]. The electronic effect is rather small in this case and cannot account for the reduction of fracture toughness. Let us add that sodium should also be a good embrittler of copper albeit silver acts as a modest cohesion enhancer [677].

The third model for embrittlement of copper due to the grain boundary segregation of bismuth is a size effect suggested by Sutton and Vitek [545]. This model is based on two necessary conditions for copper embrittlement due to grain boundary segregation: The solute is virtually insoluble and has much larger atomic radius than the matrix element. The solute then segregates to the boundary thus pushing apart copper atoms across the interface and weakening the interatomic bonding. The first-principles quantum-mechanical calculations [261] support the latter model. In addition, the size effect seems also to be responsible for copper embrittlement induced by lead and mercury segregation: The latter two solutes are practically indistinguishable in the properties contrary to the expectations following from the models of changing electronic structure [678].

The results of computer simulations proved that the grain boundary structure also plays an important role in the fracture properties of brittle materials [679]. For example, low energy fracture was observed at room temperature for the high-angle grain boundaries of molybdenum containing traces of oxygen and carbon. This behaviour was observed for the [100] symmetrical tilt grain boundaries with the misorientation angles 20°, 30°, 50°, 52° and 70°. On the other hand, high energy is necessary to fracture the boundaries misoriented by 5°,10° and 88° [680]. This behaviour very probably reflects the level of carbon and oxygen segregation at high angle grain boundaries [653]. Similar results were observed for tungsten [654].

The structure of the grain boundaries in conjunction with corresponding solute segregation also affects strongly the fracture characteristics. Recently, the role of the grain boundary plane orientation in the segregation-induced embrittlement was studied in case of bismuth segregation in copper [681]. Both experimental and theoretical studies of the asymmetrical tilt grain boundaries in the $\Sigma = 9$ [110] showed that highly segregated general grain boundaries exhibit brittle intergranular fracture while in case of the special ones containing low level of bismuth segregation, the transgranular ductile fracture is observed. It is interesting to note that the special grain boundaries are characterised by {111}, {112} and {114} planes on one side of the grain boundary [681].

Another approach to the problem of grain boundary embrittlement is based on direct consideration of impurity–impurity bonds [682]. Particularly, the effect of formation of covalent *I–M–I* bonds on material embrittlement depends on the type and valence of impurity $I$. For example, sulphur needs two electrons to complete its 3p-subshell so that it will form two bonds at a nickel boundary. One of these bonds is parallel and the other perpendicular to the boundary. The redistribution of the metal charge will then occur preferentially into bonds parallel to the grain boundary. The energy difference between the two types of the bonds then determines the embrittling potency of the segregated elements: in the case of sulphur an embrittlement occurs. On the other hand, boron with only one p-electron forms only bonds across the interface so that the charge flow exists in this direction: in this case, no embrittlement occurs.

### 7.1.2 Hydrogen Embrittlement

Another effect, which reduces the grain boundary cohesion, is hydrogen embrittlement. This effect has received a considerable interest during past decades, because hydrogen is present in the environment [683]. Absorbed hydrogen atoms preferentially segregate at the micro-crack surface along the grain boundaries thus increasing the stress ahead its tip. It is possible that some hydrogen atoms also recombine to molecules and the stress at the micro-crack tip is further enhanced although this effect is probably not so large. As a result of these stresses, micro-cracks grow and form micro-cavities. Segregated hydrogen can possibly cause two different situations, dislocation-screened tips and dislocation-assisted crack tips [638].

The microscopic model of hydrogen embrittlement starts from a hydrogenated elastic–plastic material with a crack embedded along the grain boundary. The crack tip is screened by dislocations. Under stressing this body, the crack is loaded in the direction perpendicular to the crack plane by a remote applied stress. The dislocation-screened crack preserves a local stress in the dislocation-free zone and produces a stress intensity. The crack tip can maintain a local equilibrium expressed as [638]

$$\varphi = \cos^{-1}\left(\frac{\gamma^{\Phi}}{2\gamma^{s}}\right), \tag{7.5}$$

where $\phi$ is the half crack-tip angle, $\gamma^{\Phi}$ and $\gamma^{s}$ are the grain boundary and the crack surface energy, respectively. At the grain boundaries, the equilibrium hydrogen segregation should then occur. Since this segregation at the crack surfaces and grain boundaries affects both $\gamma^{\Phi}$ and $\gamma^{s}$, the crack tip profile, the local stress and the ideal work change. Eventually, the relationship for theoretical stress-intensity factor, $K_{th}^{c}$, can be derived from the local energy balance,

$$K_{\text{th}}^{\text{c}} = K_0 \left(\frac{\gamma}{2\gamma_0}\right)^{[(n+1)/4n]} \left[\frac{\tan(\varphi_0/2)}{\tan(\varphi/2)}\right]^{[(1-n)/4n]}, \tag{7.6}$$

where $n$ is the work hardening coefficient, the index 0 indicates the quantities in the absence of hydrogen segregation. The dependence of $K_{\text{th}}^{\text{c}}$ on the bulk concentration of hydrogen is the function of the energy of the hydrogen binding at the crack surfaces and the grain boundaries as well as plastic deformation characterised by the parameter $n$ and the yield stress [638].

Latanision and Opperhauser [684] showed that the segregated solutes at the grain boundary reduce the recombination of hydrogen atoms and promote penetration of hydrogen along the grain boundaries. These solutes, such as phosphorus, sulphur, arsenic and antimony, which have a strong tendency to segregate to grain boundaries in steel, have also a stronger intrinsic binding to hydrogen than iron [685, 686]. Hydrogen atoms, which dissociated from molecules adsorbed from the vapour or from solution, bond strongly with the segregated atoms and reside at the grain boundary sites much longer than at the single ones. Consequently, there is a greater chance for hydrogen to be absorbed into the lattice. Therefore, the role of these segregated solutes consists in increasing the amount of hydrogen trapped in the grain boundary core. If the mechanism of hydrogen-embrittlement requires the concentration of hydrogen in the metal just ahead of the crack tip, the above solutes segregated at the grain boundary should therefore aid this embrittlement [13]. Kameda and McMahon detected the first intergranular micro-cracks induced by hydrogen in antimony doped Ni–Cr steel using acoustic emission and obtained corresponding threshold stress for its onset. This stress depends on the grain boundary segregation of antimony. Once the threshold stress is reached, the crack spreads and the decohesion occurs [687]. Similarly, grain boundary segregation of phosphorus and sulphur assists to hydrogen embrittlement of an X–750 nickel-based superalloy. Occurred micro-cracks can then facilitate stress corrosion cracking in water environment [688].

As expected, hydrogen serves at the grain boundaries of iron as electron acceptor and the H–Fe bonding is rather ion-like than covalent but strong. Presence of hydrogen thus leads to a similar reduction of the charge density in the region between iron atoms across the grain boundary as in case of other metalloid impurities. The charge transfer from Fe to H at both the grain boundary and free surface seems to be the key mechanism for hydrogen embrittlement [689]. In fact, hydrogen atoms are extremely mobile and their segregation is not expected to be too large. The main role of hydrogen in intergranular embrittlement consists in nucleation of the cracks rather than in their spreading [690] due to its dynamic effect occurring in course of loading the sample.

The reduced cohesion of the grain boundaries resulting from hydrogen segregation (hydrogen embrittlement) is sometimes used to facilitate intergranular fracture of less segregated or less brittle grain boundaries for studies of grain boundary segregation by the techniques of surface analysis such as AES [683]. In case of a 7050 aluminium-based alloy, magnesium segregation reduces the cohesion of the grain boundaries, while the opposite effect is observed for segregation of copper [691].

Recent study of the hydrogen-induced cracking in high-purity high-strength 4340 type steel showed that elimination of embrittling impurities does not eliminate the susceptibility of this material to hydrogen embrittlement under high yield strength if the strain rate is small enough [692].

### *7.1.3 Other Types of Non-reactive Environmental Decohesion*

Hydrogen is typical example of intergranular decohesion of polycrystalline materials caused by the impurity atoms localised at the grain boundaries. Similar effect was observed when liquid metal penetrates along the grain boundaries into the material volume (liquid metal embrittlement (LME)) or by penetration of the atoms adsorbed at the surface (either from the environment or from the material volume) to the free surface in the grain boundary micro-cracks (dynamic embrittlement). Let us briefly touch both these effects.

#### 7.1.3.1 Liquid Metal Embrittlement

Presence of a liquid metal on the surface of a polycrystalline material may be dangerous for its integrity because the liquid metal may penetrate along the grain boundaries and reduce substantially their cohesion. Classical example of this LME is the grain boundary penetration of mercury to β-brass at room temperature [7]. Other examples are listed in Table 7.1.

LME can have very dangerous consequences for service of some metallic parts of the machines. Quite frequently, this problem is observed in case of rotating parts of the gas turbines. A wide intergranular degradation was observed in case of a turbine rotor bolt made of IN 718, UNS N07718 after its service operation at about 540°C. High level of cadmium at the fracture surfaces indicated that this element present in cadmium-rich smear on the treated flank was melted (melting point of cadmium is 321°C) and penetrated quickly along the grain boundaries into the rotor bolt [693].

**Table 7.1** Metals exhibiting LME. According to [47]

| Matrix | Embrittling solute |
|---|---|
| Fe | Bi, Pb, Sn, Zn |
| Zn | Bi, Cd, Ga, Hg, In, Sn |
| Cu | Bi, Cd, Hg, In, Pb |
| Al | Cd, Ga, Hg, In, Na, Sn, Zn, |
| Ag | Ga, Hg |
| Cd | Ga, In, Sn |
| β-brass | Hg |

Similarly to other interfacial phenomena, LME is anisotropic, that is its velocity depends on the grain boundary structure. Kargol and Albright [694] showed that the $\Sigma = 3, 70°[110]$ and $\Sigma = 11, 130°[110]$ grain boundaries in aluminium are highly resistant to LME compared to other [110] characterised interfaces. Detailed study of structural dependence of fracture stress of zinc bicrystals deformed under effect of liquid gallium showed maximum resistance of the $\Sigma = 9, 56.6°[10\bar{1}0]$ grain boundary [695]. In polycrystalline bcc β-brass, the cracks of LME were nucleated and propagated preferentially at general grain boundaries [696].

Grain boundary segregation of some impurities can significantly enhance the susceptibility of the material to LME. For example, tin and antimony segregation in steel increase the LME by lead while it is reduced under phosphorus and arsenic segregation [47, 697]. Increased phosphorus segregation also reduces intergranular embrittlement of Monel 400 LME by mercury [698] and tin segregation at copper grain boundaries – although it does not cause materials embrittlement itself – increases susceptibility of copper to LME by mercury [699].

A special case of the LME is the grain boundary wetting resulting in grain boundary phase transitions (e.g. [72]). In this case the liquid metal may penetrate very quickly along the grain boundaries causing their "melting" and thus loss of cohesion. As examples of this effect, we can mention the grain boundary wetting in systems Al–Sn [700], Zn–Sn [701], Cu–In [702] or (Fe–Si)–Zn [703]. Here, we will not discuss this problem, but we refer to special publications in this field (e.g. [72]).

#### 7.1.3.2 Dynamic Embrittlement

In contrast to the low-temperature intergranular damage caused by equilibrium segregation and sudden brittle fracture, a relatively slow, stepwise brittle intergranular decohesion process can occur at enhanced temperatures in some materials under stress or when they are stressed in a specific environment. This process is called *stress-relief cracking* or *dynamic embrittlement* [704–706]. The mechanism of this process is controlled by adsorption of specific atoms at the free surface of the crack tip under external loading. Thus, the cohesion of the grain boundary changes and may result in cracking of the grain boundary bond closest to the surface in the crack tip. The atoms then adsorb again at the surface of the crack tip and the process of decohesion continues. The diffusive nature of this penetration controls the process, and therefore, the crack propagation is relatively slow [707]. The source of the embrittling element can be either external (adsorption from the environment) or internal (surface segregation from the material volume) [708].

The dynamic embrittlement can be described on basis of combined diffusion and straining processes. The unidirectional flux of atoms $I$ along the grain boundary, $J_I$, due to the stress-induced potential gradient under diffusion is [709]

$$J_I = -D_I^{\Phi}\frac{\partial X_I^{\Phi}}{\partial x} + \left(\frac{D_I^{\Phi} X_I^{\Phi} \Omega_I^{\Phi}}{kT}\frac{\partial \sigma}{\partial x}\right) + \left[\frac{D_I^{\Phi} X_I^{\Phi}\left(1 - X_I^{\Phi}\right)}{kT}\left(\varphi_M(\delta) - \varphi_I'(\delta)\right)\frac{\partial \delta}{\partial x}\right], \tag{7.7}$$

where the gradient of the chemical potential, $\nabla\mu_I^\Phi = \Omega_I^\Phi \nabla\sigma$, $\Omega_I^\Phi$ is the atom volume of $I$ and $\nabla\sigma$ is the relevant stress gradient, $D_I^\Phi$ and $X_I^\Phi$ are the grain boundary diffusivity and the concentration of $I$, respectively, $\delta$ is the grain boundary thickness and $\phi_J$ ($J = I, M$) are the interfacial "cohesive functions" of pure $I$ and $M$ interfaces. The continuity equation with constant $D_I^\Phi$ provides [709]

$$\frac{\partial X_I^\Phi}{\partial t} = D_I^\Phi \frac{\partial^2 X_I^\Phi}{\partial x^2} - \frac{D_I^\Phi \Omega_I^\Phi}{kT}\left(\frac{\partial X_I^\Phi}{\partial x}\frac{\partial \sigma}{\partial x}\right) - \frac{D_I^\Phi}{kT}\left[\frac{\partial\left(X_I^\Phi\left(1 - X_I^\Phi\right)\right)}{\partial x}\left(\varphi_M(\delta) - \varphi_I'(\delta)\right)\frac{\partial^2 \delta}{\partial x^2}\right], \tag{7.8}$$

where suffix $\Phi$ relates the variable/parameter to the grain boundary [709]. The kinetic parameter relating the diffusion and deformation rate, $\Re$, is under bulk diffusion control

$$\Re = \frac{t_\mathrm{d}}{t_\mathrm{r}} = \frac{\dot{a}\delta_\mathrm{c}^2}{D_I(\delta_\mathrm{c} - \delta_0)} \tag{7.9}$$

where $t_\mathrm{d}$ and $t_\mathrm{r}$ are the characteristic diffusion time and the time to rupture, $\delta_\mathrm{c}$ and $\delta$ are the critical and equilibrium separation distances, respectively, and $\dot{a}$ is the separation rate [707].

The dynamic embrittlement was observed in numerous systems, for example in bicrystals of Cu–Sn alloys in vacuum [710], polycrystalline IN 718 nickel-based superalloy [711], MnMoNiCr steel [712], $Ni_3Al$ [713] and Cu–Be alloy [714] in air.

Similarly to other examples of material embrittlement, dynamic embrittlement is also sensitive to grain boundary structure. As a result, the resistance of the material against dynamic embrittlement increases with increasing frequency of special grain boundaries (cf. Sect. 7.4). Oxygen-induced dynamic embrittlement of IN 718 nickel-based superalloy showed that general and/or the less coincident grain boundaries crack easily than the $\Sigma = 3$ and $\Sigma = 5$ grain boundaries although neither this boundary nor the low-angle interfaces are completely immune from cracking [708, 715].

## 7.2 Grain Boundary Corrosion

Since the grain boundaries are regions of enhanced energy, they are prone to localised corrosion. This is the basis for visualisation of the grain structure by chemical etching of polycrystalline samples in specific agents or by thermal etching during vacuum annealing. The corrosion attack can be accelerated or sometimes completely induced by stressing the material in aggressive environment resulting in intergranular brittle degradation of the material. In such case, the stress-corrosion cracking occurs. In the present part, we will discuss the pure corrosion and stress-corrosion cracking of the materials.

### 7.2.1 Corrosion

During chemical or thermal etching, the grain boundaries are attacked and the grooves are formed. If the period of the attack is long enough, polycrystalline material can be completely separated along the grain boundaries in the absence of applied stress. The mechanism of this environmentally assisted cracking is preferential attack of grain boundaries outcoming to a non-protected surface. The crack then propagates into the material along the grain boundaries due to electrochemical processes (e.g. anodic dissolution) [47]. Grain boundary segregation can play an important role in promoting this process in various ways. Either, a less noble layer is formed at the segregated grain boundary which is more prone to the above-mentioned dissolution (e.g. copper in silver, see below), or a segregated element can react with another one and change the anticorrosive ability of the grain boundary such as carbon with chromium in austenitic stainless steels [47].

Due to the anisotropy of grain boundary energy and segregation, there also exists the orientation dependence of intergranular corrosion. For example, minima of the penetration depth or the width of the etched boundary groove corresponding to the minima of the measured boundary energy were found at certain [100] tilt grain boundaries in aluminium [57, 716], Cr–Ni austenitic stainless steel [717], niobium [718] and copper and α-Cu–Al alloy [719]. Low values of the etch groove depth were also found at the low-$\Sigma$ grain boundaries in a polycrystalline Fe–Ni–Cr alloy [529, 720]. It was shown that a special etching agent containing picric acid as an active component attacks only the austenitic grain boundaries in a Ni–Cr steel that were segregated by phosphorus [721]. The treatment of polycrystalline stabilised 310 stainless steel in boiling nitric acid revealed the highest corrosion resistance of the $\Sigma = 3$, $\Sigma = 11$ and $\Sigma = 13$ grain boundaries [722]. Surprisingly, the maxima of the penetration depth were reported for the low energy {013}, {012} and {023} symmetrical tilt grain boundaries in a 17Cr–13Ni austenitic stainless steel containing silicon [533]. The orientation dependence of the penetration depth copies analogous dependence of the level of silicon segregation at individual grain boundaries for both 0.3 and 0.8mass.% of silicon in the bulk. The differences in corrosion behaviour of both symmetrical and asymmetrical tilt grain boundaries in one sample were studied on curved grain boundaries corresponding to 36.9°[100] and 45°[100] orientation relationships [87, 723]. As it is apparent in Fig. 7.5a, the width, $w$, of the etched groove changes with the grain boundary inclination. There exist two minima of $w$ at the inclinations 18.5° and 26.5° from original {013} grain boundary corresponding to the (001)/(034) and (011)/(017) asymmetrical grain boundaries, respectively (Fig. 7.5b) [723]. This finding is in very good agreement with the orientation dependence of the standard enthalpy of solute segregation (Fig. 5.16), and with the finding that the (011)/(017) asymmetrical grain boundary is special and the (001)/(034) one exhibits vicinal behaviour [94, 95]. Similarly, a single minimum of $w$ was found at the curved 45°[100] grain boundary corresponding to the special (001)/(011) grain boundary [87, 94, 95].

Another interesting example of anisotropic interfacial corrosion can be shown in case of the archaeological buried objects. The necklace of a Ag–1mass%Cu–0.3mass.%Au alloy, excavated from a grave and dated to the tenth century, exhibited

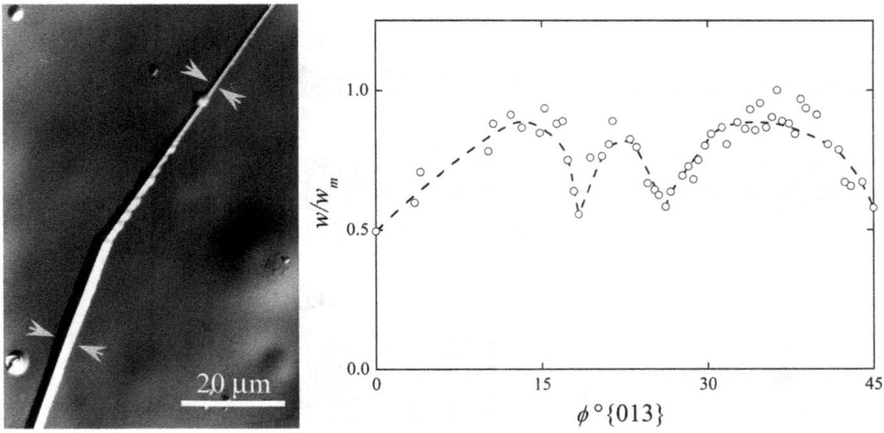


**Fig. 7.5** Corrosion of the curved 36.9°[100] grain boundary (**a**) micro-structure of the etched grove in Nomarski interference contrast (conical objects on the grain surfaces are dislocation etch pits); (**b**) orientation dependence of its width, $w$, normalised by the maximum value of the width, $w_m$ (according to [723])

an extended intergranular brittle failure [724]. The analysis of the parts of this necklace proved that this failure is caused by intergranular corrosion of less noble copper-segregated grain boundary regions under the conditions of decomposing human body. A thorough crystallographic investigation proved that all failed grain boundaries were non-coincidence and no separation was observed along the low-$\Sigma$ grain boundary. In addition, about 20% of non-coincidence grain boundaries were also found to be resistant to the corrosion attack. Some of these boundaries were surrounded exclusively by special twin boundaries preventing penetration of the electrolyte to them. However, some non-coincidence grain boundaries were found to be resistant to the corrosion although they were in contact to the strongly corroded interfaces. This is the case of the grain boundaries with the surface traces characterised by the $\langle 126\rangle/\langle 011\rangle$ and $\langle 112\rangle/\langle 237\rangle$ directions measured in respect to the adjoined grains parallel to the two respective grain boundaries (Fig. 7.6). These interfaces are joined with another non-coincidence grain boundaries characterised as $\langle 115\rangle/\langle 035\rangle$, and $\langle 116\rangle/\langle 023\rangle$ directions, that are heavily corroded.

We may deduce the character of these grain boundaries by accounting for the orientations of the planes belonging to the crystallographic zones of these directions. The crystallographic zones corresponding to at least one of the measured directions of both resistant grain boundaries, $\langle 011\rangle$ and $\langle 112\rangle$, can lay in the densest plane of the $\{111\}$ type. On the other hand, the $\langle 115\rangle$, $\langle 035\rangle$, $\langle 116\rangle$ and $\langle 023\rangle$ directions of the corroded grain boundaries cannot lay in the $\{111\}$ type plane. It is highly probable from energetic reasons that the untouched $\langle 126\rangle/\langle 011\rangle$ and $\langle 112\rangle/\langle 237\rangle$ grain boundaries in Fig. 7.6 are formed by the planes of the $\{111\}$ type. This deduction follows our model of classification of grain boundaries in bcc iron according to which

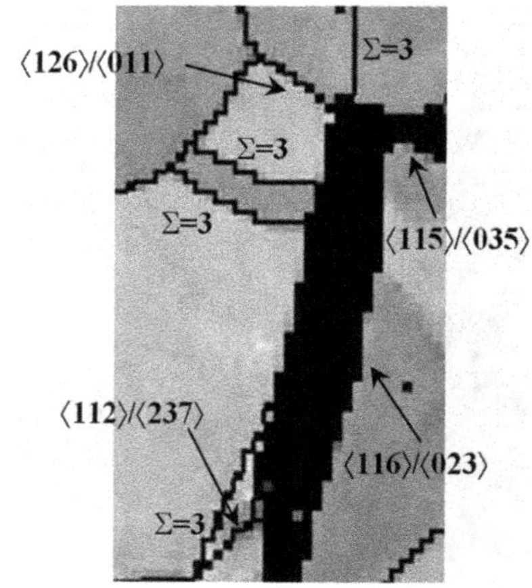


**Fig. 7.6** Example of resistant and corroded non-coincidence grain boundaries. The Miller symbols give the directions parallel to the grain boundary plane in respect to both adjoining grains [724]

there exists special asymmetrical tilt grain boundaries formed by the bcc densest {110} plane on one side [91, 94]. It is highly probable that the grain boundaries formed by the {111} fcc densest planes are special [94, 724].

## 7.2.2 *Stress Corrosion Cracking*

Simultaneous effect of corrosive environment and stressing the sample with a susceptible micro-structure can often result in brittle fracture also in case of the materials ductile under acting these effects separately. This phenomenon is called *stress corrosion cracking*. Stress corrosion cracking can pass transgranularly as well as intergranularly. Intergranular stress corrosion cracking is more frequent and also more dangerous for the behaviour of the parts in service [725], mainly in case of austenitic stainless steels in commercial boiling water reactor nuclear power plants [726].

Grain boundaries play an important role in localised corrosion as mentioned in the previous part. It is obvious that – similarly to simple corrosion – the grain boundary structure and dominantly its chemistry will also affect the course of intergranular stress corrosion cracking [725]. The initiation of the cracks during the stress corrosion is closely related to (a) the activated dislocation slip system, (b) the grain misorientation, (c) the misorientation of the Burgers vectors of the primary slip systems and the grain boundary plane and (d) the deformation behaviour in vicinity of the grain boundary [727]. Room-temperature intergranular stress corrosion cracking studied thoroughly in bicrystals of an $\alpha$-Cu–9at.%Al alloy containing specific [100] and [110] tilt grain boundaries in solution of $NH_4OH$ and NaOH under constant stresses in range of 0.8–2 related to the yield stress of the material was found to exhibit a pronounced anisotropy. The susceptibility minima of the interfaces to stress corrosion cracking (i.e. inverse time-to-fracture) were found for the {122}, {111} and {113}, [110] symmetrical tilt grain boundaries, and for {015}, {013}, {012}, {025} and {037}, [100] symmetrical tilt grain boundaries [725, 728]. From the point of view of the stress corrosion cracking, these grain boundaries are special. Note that the susceptibility cusps found for the [100] fit well with the minima of absolute values of the standard enthalpy of solute segregation for the symmetrical [100] tilt grain boundaries in bcc Fe–Si alloys [542] and this anisotropy well correlates with the structural dependence of the grain boundary energy [725].

Intergranular stress corrosion cracking of polycrystalline 304 austenitic stainless steel containing a distribution of general, low-energy and twin boundaries showed that $\Sigma = 3$ are more crack-resistant interfaces compared to other special grain boundaries ($\Sigma=9$ and $\Sigma=27$) [726]. The main reason for this degradation is chromium depletion of the grain boundaries of sensitised steels [729]. 3D characterisation of polycrystalline austenitic stainless steel enabling to determine the orientation of the grain boundary planes showed that the majority of the grain boundaries found on the fracture surface were characterised by high $\{hkl\}$ index planes: If there was present a low-index grain boundary, it failed in a ductile manner. This is direct proof of the resistance of individual grain boundaries to intergranular stress corrosion cracking and their role in crack-bridge development [730].

Classical experiments of Gleiter et al. [731, 732] with the corrosion remove of small single-crystalline spheres produced of copper or silver containing either $10^{-1}\%$ of bismuth or units of percents of gold, from the single-crystalline plate under ultrasonic irradiation in a bath of demineralised water can also be considered as example of the stress corrosion cracking. These experiments revealed that only the grain boundaries between the spheres and the plate exhibiting special orientation relationship remained untouched by corrosion while the spheres arranged in a general relationship were removed under the above treatment. The results well prove the close relationship between the strength of the grain boundaries and the level of bismuth segregation [731,732].

The effect of the grain boundary structure in intergranular stress corrosion cracking was also documented for example of a pipeline steel. Low-angle and special grain boundaries are more crack-resistant than general grain boundaries and the crack initiation and propagation of intergranular stress corrosion cracking is suppressed in material with large portion of special and low-angle grain boundaries. It also means that the strength of the texture of the material enhances its cohesion [733].

The intergranular stress corrosion cracking was found to propagate predominantly along general grain boundaries of 304L type austenitic stainless steels. If the portion of the general grain boundaries was reduced by thermomechanical treatment, the initiation and propagation of the intergranular stress corrosion cracking was substantially suppressed [734]. Although stress corrosion cracking of this alloy can be promoted by grain boundary segregation of phosphorus, it seems that this segregation alone is not sufficient to cause intergranular stress corrosion cracking [735]. The effects of decarburisation on the grain boundary segregation of phosphorus and primary water stress corrosion cracking have been investigated in a low-alloy steel. After decarburisation at 900°C, the low-alloy steel showed intergranular brittleness, arising from phosphorus segregation at grain boundaries. A drastic decrease in intergranular fracture strength was observed with increasing grain boundary concentration of phosphorus, e.g. 610 MPa after decarburisation alone and 320 MPa after subsequent holding at 490°C for 10 days. Oxide films to the depths of about 80 μm were formed along the grain boundaries of the decarburised surface region and, at slower strain rate, these acted as pre-cracks for intergranular stress corrosion cracking in a primary water environment [736].

High-resolution analytical electron microscopy study of stress corrosion cracking of 316 austenitic stainless steel and nickel-based alloy 600 revealed that active segregation and precipitation play a major role in this process. This is documented by presence of deeply attacked grain boundaries outside the main crack. The attacked grain boundaries contained layers of oxides maximum 10-nm thick indicating the role of the grain boundary character and water chemistry. Impurities (lead) were found in a nanometre wide layer along the grain boundaries are involved in the stress corrosion cracking process [214]. The stress corrosion cracking of nickel-based superalloys is also promoted by phosphorus segregation at and chromium depletion from the grain boundaries [737]. Austenitic stainless steels and nickel-based alloys exhibit generally greater susceptibility for intergranular stress corrosion cracking than ferritic/martensitic alloys [738]. Decarburisation of a low-alloy steel results in enhanced grain boundary segregation of phosphorus and consequently, in increased intergranular stress corrosion cracking in water. Under such conditions, the strength of the material was reduced by factor approximately 2, that is from 610 to 320 MPa after decarburisation. As a result of the corrosion attack, the oxide films were found along the grain boundaries [736].

Intergranular stress corrosion cracking can be also promoted by the segregation effects at the grain boundaries induced by irradiation, such as enrichment of nickel or depletion of chromium. However, quantitative relationship between radiation-induced segregation and intergranular stress corrosion cracking of the materials has not been derived till now [739].

## 7.3 Grain Boundary Diffusion

Grain boundary diffusion is the movement of atoms along the grain boundaries under a driving force, which is mainly the difference in chemical potentials. In case of different solute and matrix atoms, it induces concentration changes at the grain boundary and therefore, is in close relationship to the grain boundary segregation. Let us now give the brief description of this phenomenon with focus to its relationship to grain boundary segregation.

### *7.3.1 Fundamentals of Grain Boundary Diffusion*

The description of grain boundary diffusion is usually based on classical Fisher model [740, 741]. The grain boundary $\Phi$ is represented by a uniform and isotropic slab of the thickness, $\delta$, characterised by high diffusivity, $D_I^{\Phi}$, that is embedded in an isotropic crystal perpendicular to its surface with low diffusivity, $D_I \ll D_I{}^{\Phi}$

(Fig. 7.7a). Before the diffusion measurements, a layer of the solute or of the isotope of the same material is put at the surface. During annealing of such system at a constant temperature $T$ for a time $t$, the atoms diffuse from the surface into the material. This diffusion occurs in two ways (a) via the volume diffusion into the grains and (b) via the grain boundary diffusion along the interfaces. Since $D_I \ll D_I^{\Phi}$, the grain boundary diffusion is much faster. The atoms moving along the grain boundary may either continue in the grain boundary diffusion or to diffuse perpendicularly to the interface into the grain regions adjacent to the boundary. As a result, a zone of the volume diffusion occurs in vicinity of the boundary. Mathematically, this complex diffusion is described by a set of two coupled equations,

$$\frac{\partial X_I}{\partial t} = D_I \left( \frac{\partial^2 X_I}{\partial x^2} + \frac{\partial^2 X_I}{\partial y^2} \right), \tag{7.10}$$

if $|x| > \delta/2$, and

$$\frac{\partial X_I^{\Phi}}{\partial t} = D_I^{\Phi} \frac{\partial^2 X_I^{\Phi}}{\partial y^2} + \frac{2D_I}{\delta} \left( \frac{\partial^2 X_I}{\partial x^2} \right) \tag{7.11}$$

for $x = \delta/2$. In (7.10) and (7.11), $X_I(x, y, t)$ is the bulk concentration of the solute and $X_I^{\Phi}(y, t)$ is its grain boundary concentration. The solution of (7.10) and (7.11) should fulfil certain surface conditions specified below and natural initial and boundary conditions at $x$ and $y \to \infty$. A simple condition is keeping the constant source at the surface [740],

$$X_I(x, 0, t) = X_{I,0} = \text{const.} \tag{7.12}$$

This condition can be established by depositing a thick surface layer of the solute, the thickness $h$ of which exceeds the diffusion path, $h \gg (Dt)^{1/2}$. Presently, more frequently used experiments are performed with $h \ll (Dt)^{1/2}$. Therefore, another condition of so-called instantaneous source or thin layer should be applied,

$$X_I(x, y, 0) = M\delta_{\mathrm{D}}(y) \quad \text{and} \quad (\partial X_I/\partial y)_{y=0} = 0. \tag{7.13}$$

In (7.13) $M$ is the amount of the solute per unit area deposited at the surface, and $\delta_{\mathrm{D}}$ is the Dirac delta-function. This condition suggests that the initial layer of the solute is completely consumed by the specimen during the diffusion experiment.

The relationship between functions $X_I(x, y, t)$ and $X_I^{\Phi}(y, t)$ depends on both the character of the diffusing species (self-diffusion or impurity diffusion) and the type of the matrix material (pure metal or alloy). In case of the impurity diffusion in a pure metal, which is interesting from the point of view of the grain boundary segregation, the condition involves the equilibrium segregation factor $s$[1]

$$X_I^{\Phi}(y, t) = sX_I(y, t) \tag{7.14}$$

[1] Let us note that he segregation factor $s$ is identical to the grain boundary enrichment ratio $\beta_I^{\Phi}$ defined by (4.19).

which is independent of both, $y$ and $t$. This equation reflects two assumptions (a) the impurity atoms in the grain boundary are in local thermodynamic equilibrium with the atoms in the lattice adjacent to the boundary, that is the grain boundary segregation is locally at equilibrium at any depth; and (b) the grain boundary segregation follows the relationship $X_I^{\Phi} = sX_I$ supposing $s$ is the function of temperature only. Indeed, this relationship only holds when both $X_I^{\Phi}$ and $X_I$ are small enough, so that the thermodynamic activities of the solute in the grain boundary and in the bulk are proportional to the respective concentrations (Henry law). Another interesting situation is the grain boundary self-diffusion in concentrated alloys [742, 743]. If the grain boundary diffusion of the tracer B* in a binary A–B alloy occurs, (7.14) can be applied supposing (a) an equilibrium grain boundary segregation of B establishes before and persists during the diffusion annealing; (b) the amount of B* diffused into the sample is very small and does not affect the distribution of component B in the sample; and (c) the isotope equilibrium with respect to B is established between the grain boundary and the adjacent lattice at any depth. The factor $s$ in (7.14) equals to the ratio $X_B^{\Phi}/X_B$ of the corresponding net concentrations of B. This ratio characterises the equilibrium grain boundary segregation in the alloy. Let us mention that the concentrations of B do not need to be small and the grain boundary segregation can be either in the linear (Henry) type or in the saturation regime.

Let us now focus to the measurements of the grain boundary diffusion. The majority of these experiments are carried out using radiotracers and the serial sectioning technique [743]. After the deposition and the diffusion annealing, the concentration of the isotope is measured in removed thin layers of the material parallel to the surface (e.g. mechanically) using a crystalline $\gamma$-detector or a liquid scintillation counter. In this way, the average layered concentration of the diffusant, $\bar{c}$, is measured in dependence on the penetration depth $y$ (Fig. 7.7b). If the concentration profile calculated using the exact analytical solution [744] is plotted as $\log \bar{c}$ vs. $y^n$, it results in an almost straight line for $n = 6/5$ [745, 746]. Further, the linear

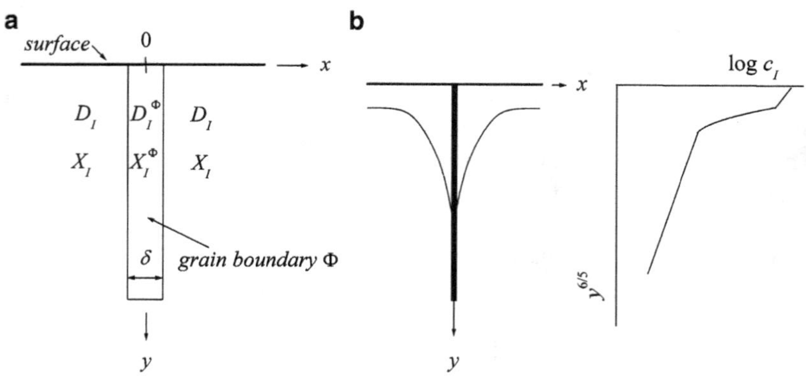


**Fig. 7.7** Schematic representation of the grain boundary diffusion (**a**) geometry; (**b**) penetration profile. If $a \ll 1$ and $\beta \gg 1$ (cf. (7.17) and (7.18)), the tail of the profile is a straight line in the coordinates $\log \bar{c}$ vs. $y^{6/5}$ (according to [740])

part of the profile varies with the reduced depth $\omega$,

$$\omega = \frac{y}{\sqrt{s\delta D_I^\Phi}}\left(\frac{4D_I}{t}\right)^{1/4} \tag{7.15}$$

with approximately constant slope $-\partial \ln \bar{c}/\partial y^{6/5} \approx 0.78$. Knowing the slope of the linear part of the profile, one can thus calculate the product $s\delta D_I^\Phi$ as

$$s\delta D_I^\Phi = 1.322\sqrt{\frac{D_I}{t}}\left(\frac{\partial \ln \bar{c}}{\partial y^{6/5}}\right)^{-5/3} \tag{7.16}$$

for constant source. The volume diffusion coefficient $D_I$ is supposed to be known from independent measurements. The application of (7.16) is only possible if the following conditions are fulfilled (a) the parameter $\beta$ (here the Le Claire parameter) [747] defined as

$$\beta = \frac{s\delta D_I^\Phi}{2D_I\sqrt{D_I t}} \tag{7.17}$$

must be large, $\beta > 10$; and (b) the parameter $\alpha$ defined as

$$\alpha = \frac{s\delta}{2\sqrt{D_I t}} \tag{7.18}$$

must be small, $\alpha < 0.1$.

The product $s\delta D_I^\Phi$ can be also determined from the linear part of the plot $\log \bar{c}$ vs. $y^n$, in case of an instantaneous source where $-\partial \ln \bar{c}/\partial y^{6/5} \approx 0.775$, and for $\alpha < 0.1$ and $\beta > 10^4$

$$s\delta D_I^\Phi = 1.308\sqrt{\frac{D_I}{t}}\left(\frac{\partial \ln \bar{c}}{\partial y^{6/5}}\right)^{-5/3}. \tag{7.19}$$

The numerical constant in (7.19) should be slightly modified for $\beta < 10^4$ [748].

Only one part of the profile dominated by the grain boundary diffusion is considered in these relations although the penetration profile has two parts as shown schematically in Fig. 7.7b. In the surface region, the volume diffusion runs from the surface. This part can be used for evaluation of the coefficient of volume diffusion. On the other hand, the region far from the surface involves simultaneous grain boundary and lateral volume diffusion from the boundary to the adjacent grains. This profile should become a straight line of the $\log \bar{c}$ vs. $y^{6/5}$ dependence and its slope can be then used to determine the grain boundary diffusivity.

Equations (7.16) and (7.19) are used to correlate the results of the grain boundary diffusion measurements and provide us with the product $s\delta D_I^\Phi$. Individual values of $s$, $\delta$ and $D_I^\Phi$ remain unknown. The assumption that $\delta = 0.5$ nm is a good approximation [740]. In the diffusion studies under the condition of the dilute limit, it is

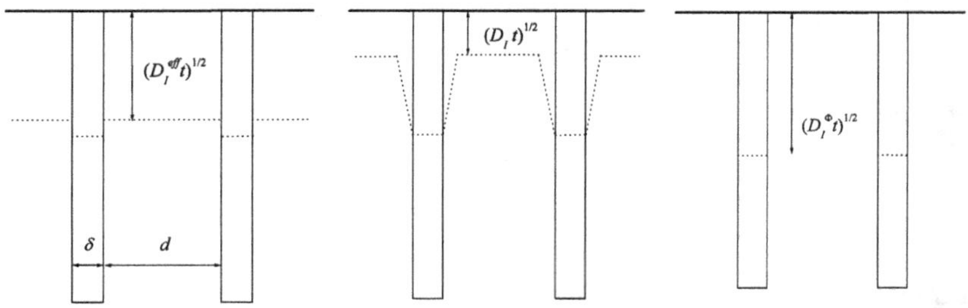


**Fig. 7.8** Schematic representation of regimes A, B and C of the diffusion kinetics in system containing grain and grain boundaries (according to [740])

possible to distinguish so-called A, B and C kinetic regimes of the diffusion process as measured on identical polycrystalline material [746] (Fig. 7.8).

The *A regime* is observed in the limiting case combining the conditions of high temperatures, very long annealing periods and small grain sizes $d$. In such conditions, the volume diffusion length, $(D_I t)^{1/2} \gg d$, and the volume diffusion dominates, although the grain boundary diffusion also occurs, which affects the total measured concentrations in different depths beneath the surface thus providing an effective diffusion coefficient $D_I^{\mathrm{eff}}$ according to the Fick law (Fig. 7.8).

At lower temperatures, shorter annealing periods and/or larger grain size than that applied in A regime, the diffusion exhibits so-called *B regime*. In this regime, the diffusion length should fulfil the condition $s\delta \ll (D_I t)^{1/2} \ll d$. It means that the grain boundary diffusion takes place simultaneously with the volume diffusion but in contrast to A regime individual grain boundaries are far from each other so that the expressions derived for an isolated grain boundary are valid. The condition for B regime implies $\alpha \ll 1$ and $\beta \gg 1$ supposing much deeper penetration of the solute along the grain boundaries compared to the volume [746]. Then, the penetration profile has a two-step shape shown schematically in Fig. 7.8b. The part of the profile related to the grain boundary diffusion depends on the dimensionless variable $\omega$ only defined by (7.15) suggesting the quasi-steady character of the grain boundary diffusion [748]. The triple product $s\delta D_I^{\Phi}$ is the only quantity that can be determined in B regime according to (7.16) and (7.19). B regime is the most commonly used for the measurements of grain boundary diffusion.

Much lower temperatures and/or shorter annealing times than applied in regime B result in almost "freezing" of the volume diffusion and in domination of the grain boundary diffusion without any essential leakage to the volume (Fig. 7.8c). This regime is called *C regime* and it is controlled by the condition $(D_I t)^{1/2} \ll s\delta$. Consequently, $a \gg 1$ (practically starting from $\alpha > 10$). The concentration profile in this regime is either a Gaussian function (instantaneous source) or an error function (constant source) with the diffusion coefficient $D_I^{\Phi}$. If the profile is measured experimentally, $D_I^{\Phi}$ can be determined separately from $s$ and $\delta$ [740].

### 7.3.2 Relationship Between Grain Boundary Diffusion and Segregation

Because there appears the segregation factor $s$, which is nearly identical with the grain boundary enrichment ratio $\beta^{\Phi}$ (4.19) in the expressions describing grain boundary diffusion, the thermodynamic parameters of the grain boundary segregation can also be obtained from the diffusion experiments.

As was mentioned above, the studies of the grain boundary diffusion are most frequently performed in B regime and for determination of the diffusion parameters, (7.16) and (7.19) are applied. For impurity diffusion, this method only provides the triple product $s\delta D_I^{\Phi}$. While the grain boundary width, $\delta$, can be approximated by a constant value $\delta = 0.5$ nm, the grain boundary segregation factor $s$ and the grain boundary diffusion coefficient $D_I^{\Phi}$ are still to be determined. Parameters $s$ and $D_I^{\Phi}$ can be separated on basis of the measurements of the grain boundary diffusion in a wide temperature range under conditions of both B and C regimes, because C regime can directly determine $D_I^{\Phi}$. Knowing the values of $D_I^{\Phi}$, $s$ can be determined from the values of the product $s\delta D_I^{\Phi}$ measured in B regime measurements as $s\delta = \left(s\delta D_I{}^{\Phi}\right)/D_I{}^{\Phi}$ [740].

The product $s\delta D_I^{\Phi}$ fulfils the Arrhenius law,

$$s\delta D_I^{\Phi} = \left(s\delta D_I^{\Phi}\right)_0 \exp\left(-\frac{Q_I^{\mathrm{app}}}{RT}\right), \tag{7.20}$$

where $Q_I^{\mathrm{app}}$ is the apparent activation energy. Similarly, $s$ is dependent on temperature. In the infinitesimal dilution limit when $X_I \ll X_I^{\Phi} \ll 1$,

$$s = s_0 \exp\left(-\frac{\Delta H_I^0}{RT}\right), \tag{7.21}$$

where $\Delta H_I^0$ is the standard enthalpy of segregation. As $D_I^{\Phi}$ also follows the Arrhenius law, $D_I^{\Phi} = D_{I,0}^{\Phi} \exp\left(-H_I^{\Phi}/RT\right)$,

$$Q_I^{\mathrm{app}} = H_I^{\Phi} + \Delta H_I^0. \tag{7.22}$$

If $Q_I^{\mathrm{app}}$ and $H_I^{\Phi}$ are determined from the diffusion measurements in B and C regimes, respectively, $\Delta H_I^0$ can be obtained [740]. From the temperature dependence of $D_I^{\Phi}$, the entropic contributions can also be obtained and similarly to $\Delta H_I^0$, $\Delta S_I^0$ can also be evaluated $\Delta S_I^0 = R \ln s_0$.

The possibility to determine the thermodynamic parameters of segregation from the measurements of grain boundary diffusion is very interesting. Despite the fact that two phenomena and thus their interaction can be studied simultaneously, it is possible to determine the thermodynamic parameters of the grain boundary segregation also for systems where other methods of interfacial analysis fail, such as AES or ESCA. This is especially important for non-brittle and less-segregating solutes.

Main problem is that the measurements in the C regime are extremely difficult and, therefore, they have been performed only rarely. Only recently, reliable and systematic measurements in the C regime have been facilitated by using of both the carrier-free radioisotope layers and the extremely sensitive detectors with a large counting efficiency and low background [749]. Up to now, combined measurements in B and C regime were realised in the systems such as tellurium [750], selenium [751] and nickel [752] in silver and gold [753], selenium [754] and silver [755] in copper.

In some systems at low temperatures, the grain boundary segregation is quite strong and the grain boundaries may reach their saturation with the solute. In this case, a non-linear relation between $X_I^\Phi$ and $X_I$ is observed, that is $s \neq$ const. It means that $X_I^\Phi$ changes with the changes of the depth $y$. The dependence of $X_I^\Phi$ on $y$ may alter the shape of the profile. In this case, Langmuir–McLean segregation isotherm (4.61) should be used instead of the linear relationship between $X_I^\Phi$ and $X_I$, that is the conditions of the Henry law is no more fulfilled. This replacement causes slight changes of the concentration profiles. Just below the surface, a saturation region of the grain boundary exists in which $\bar{c}$ rapidly decreases with $\omega$ and the profile has a strong upward curvature because $X_I^\Phi \approx 1$ while $X_I$ rapidly reduces. Under the saturation layer, the region of linear segregation region ($\omega > 1$) exists where the profile is linear with $\omega$ because both $X_I^\Phi$ and $X_I$ are small and the Henry isotherm is a good approximation. This part of kinetics can be used for determination of $s\delta D_I^\Phi$ [740]. For example, this situation arises in case of silver segregation in copper [754,755]. From this type of measurement, the segregation energy (enthalpy) of silver in copper was determined to be $-29 \pm 13\,\text{kJ/mol}$. This value is in good agreement with the AES data, $\Delta H_{\text{Ag}} \geq -40\,\text{kJ/mol}$ [756].

To obtain reliable results on grain boundary diffusion, appropriate measurements have to be performed at relatively high temperatures. In case of very pure polycrystalline metals with soluble solutes, recrystallisation or grain growth can occur during annealing the samples. Then the effect of moving grain boundaries must be taken into account. Non-linear segregation of poorly soluble elements like silver or iron in copper leads to strong curving the diffusion profiles which can be effectively used as a tool to measure segregation enthalpies in grain boundaries. Even in these cases, however, quantitative analysis of heterodiffusion profiles shows that grain boundary diffusion remains one of the best ways to obtain information on grain boundary chemistry [743].

The values of the standard enthalpy and entropy of grain boundary segregation determined on basis of the diffusion experiments are also included in Appendix A.

## 7.4 Grain Boundary Engineering

The knowledge of structural dependence of various grain boundary phenomena including solute segregation can also be used in practice for production of polycrystalline materials with optimum properties. As shown for many examples throughout

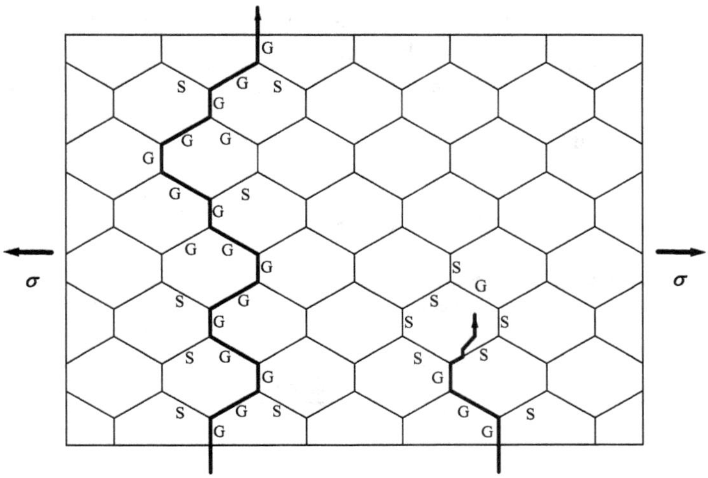


**Fig. 7.9** Schematic representation of the fracture in a polycrystalline material. The continuous path of general grain boundaries (*left side*) facilitates an easy fracture. On the other hand, the fracture along the path containing special grain boundaries interrupting the network of general interfaces (*right side*) is more difficult and needs larger stress and/or the material fractures in transcrystalline manner (According to [32])

this book, the behaviour of special and general grain boundaries differs substantially in many cases. One very important property is brittle fracture.

As mentioned above, the highly segregated general grain boundaries are more prone to intergranular brittle fracture than the special interfaces [757]. It means that the brittle separation of a polycrystalline material under stress will proceed preferentially along the general interfaces. If the general grain boundaries form a continuous path throughout the material (left side of Fig. 7.9), relatively low applied stress can cause this separation. If however, the path of the general grain boundaries is interrupted by special grain boundaries (right side of Fig. 7.9), the brittle fracture is not so easy as in previous case. To crack such material it is necessary to increase the stress and then, the fracture will also proceed in transrystalline manner. It follows from this consideration that the properties of a polycrystal will depend on the character and distribution of individual grain boundaries in the polycrystal. Watanabe [32] formulated the concept of the Grain Boundary Design of polycrystalline materials, the basic idea of which is to design and produce the polycrystalline material through controlling the distribution of the types of the grain boundaries in order to obtain polycrystalline materials with desirable properties and performance. At present, this concept is widely known as *grain boundary engineering (GBE)*.

Successful GBE is based on detail knowledge of the properties of individual grain boundaries. However, a polycrystalline material contains huge number of grain boundaries interconnected with each other. A small piece of polycrystal of the size of 1 $cm^3$ with the average grain size of 0.1 mm in diameter contains $10^6$ grains, each of them surrounded by 8–12 grain boundaries! It is clear that the design of the

**Table 7.2** Grain boundary control applicable to grain boundary engineering [759]

| Phenomenon | Controlling parameter | Application |
|---|---|---|
| Grain boundary density | Grain size, grain boundary volume | Enhancement of strength and ductility, new properties |
| Grain boundary geometry | Grain boundary inclination, dihedral angle | Enhancement of creep ductility and superconductivity |
| Grain boundary morphology | Precipitate shape, size and density | Enhancement of corrosion resistance and ductility |
| Grain boundary chemistry | Segregation level, width of precipitate-free zone | Reduction of embrittlement, enhancement of corrosion resistance |
| Grain boundary structure | Grain boundary type and character distribution | Enhancement of strength and ductility, new properties |

polycrystalline material and the manipulation with the grain boundaries is extremely complicated even in such a small piece of material. We must take into account that the grain boundaries are buried in the material and adjustment of a grain boundary to a desirable orientation may mean simultaneous change of the character of other boundaries of this single grain. In addition to structural parameters of grain boundaries, such as their character, structure and connectivity, there are many possible parameters that have to be considered for successful GBE. Geometrical parameters (grain size and shape, boundary area, boundary junctions), morphological parameters (boundary inclination and faceting, grain boundary phases, grain boundary width), composition parameters (grain boundary segregation and precipitation) and energetic parameters (grain boundary energy, electronic charge and magnetic state) belong among them [758]. In Table 7.2, the types of the grain boundary phenomena, controlling parameters and applications for the improvement of material performance are listed.

Besides all above-mentioned phenomena and parameters affecting the structure of the planned polycrystal, it is also necessary to consider formation of new recrystallised grain boundaries and their migration. In this way, the initially formed grain boundaries may change their structure and thus energy and migration characteristics. These processes can substantially control the micro-structure of the polycrystal during recrystallisation and following grain growth [758, 760].

One of the most important features necessary for successful GBE is the grain boundary classification, that is the specification of the grain boundary type, because the portion of the special grain boundaries is decisive for the properties of optimised polycrystal. As the low value of $\Sigma$ is the simplest geometric characteristics for the special grain boundaries, the CSL concept is widely used for characterisation of the grain boundaries and increased portion of low-$\Sigma$ grain boundaries is a measure of improved polycrystal [759–761]. The low-$\Sigma$ grain boundaries can also well multiply at the triple junctions. If the frequency of $\Sigma = 3$ grain boundaries is high, their meeting in the grain boundary triple junctions, namely results in the so-called multiple twinning, which seems to be a key factor of GBE in materials with low value of stacking fault energy. Such multiple twinning can be represented either by joining or by dissociation [762],

$$\Sigma A + \Sigma B \rightarrow \Sigma(A \times B) \quad \text{or} \quad \Sigma A + \Sigma B \rightarrow \Sigma(A/B). \qquad (7.23)$$

The relationship on the right-hand side of (7.23) is valid if A/B is an integer and A > B. Therefore, if two $\Sigma = 3$ grain boundaries meet at the triple junction, a $\Sigma = 9$ grain boundary occurs, and if $\Sigma = 3$ and $\Sigma = 9$ grain boundaries meet there, the resulting one is either another $\Sigma = 3$ or $\Sigma = 27$ [762].

Anisotropy of grain boundary segregation affects strongly many properties and processes occurring in polycrystals, for example grain boundary energy and mobility. In this way, the grain boundary character distribution will also be affected by presence of solutes and impurities. For example, increased bulk concentration of sulphur in pure nickel from $3 \times 10^{-1}$ ppm to 10 ppm reduces the portion of the $\Sigma = 3$ grain boundaries from more than 90% to about 55% (Fig. 7.10) [757]. Similarly, increasing bulk concentration of sulphur in α-iron reduces the fraction of special grain boundaries. Sulphur is active in this sense only in the absence of carbon: Under presence of carbon, the frequency of appearance of special grain boundaries remains similar to that observed in pure iron [763]. The existence of the compensation effect (cf. Chap. 5) and its consequence in reversing the type of anisotropy of grain boundary properties (e.g., the mobility [48] and solute segregation [566]) can also strongly affect the mechanism of formation of the structure and the final distribution of the grain boundaries.

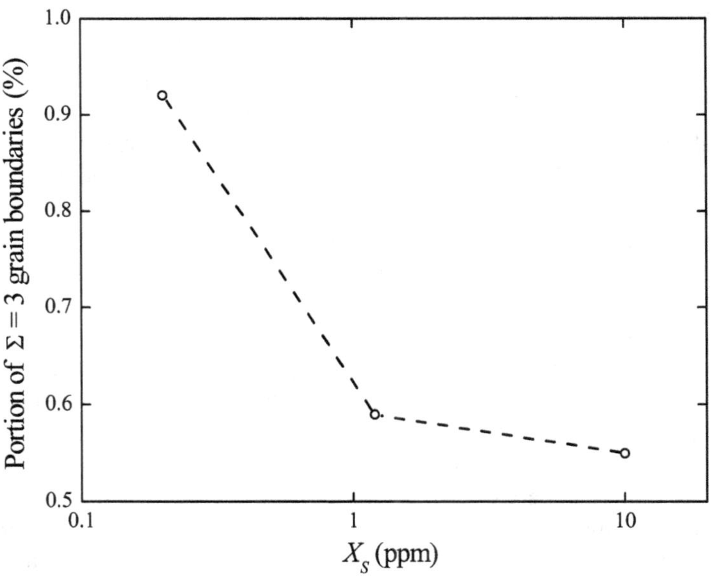


**Fig. 7.10** Dependence of the portion of $\Sigma = 3$ grain boundaries in pure nickel on the bulk concentration of sulphur (according to [757])

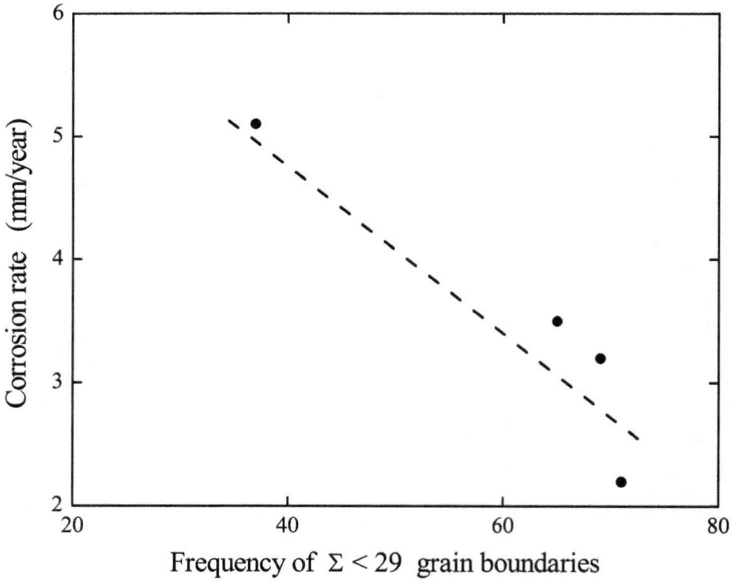


**Fig. 7.11** Dependence of the rate of grain boundary corrosion in a nickel-based alloy on the frequency of $\Sigma < 29$ grain boundaries. According to [768]

In many cases, a correlation between the density of low-$\Sigma$ grain boundaries and improved properties of materials are reported. For example, the ductility of an Fe–6.5mass.%Si alloy [764] and boron-free $Ni_3Al$ [765] was enhanced when the portion of low-$\Sigma$ grain boundaries was substantially increased. Optimisation of the grain boundary character distribution can improve the superplasticity in an Al–Li alloy [766] or magnetostriction-induced strain and shape memory effect in an Fe–Pd alloy [574]. Increased frequency of low-$\Sigma$ grain boundaries in 304 austenitic stainless steel suppressed boron segregation and carbide precipitation at grain boundaries [767] and suppressed grain boundary corrosion in a nickel-based alloy [762, 768] as is shown in Fig. 7.11. Corrosion and growth resistance of positive grids used in lead-acid batteries was increased by 20 and 75%, respectively, by increasing the portion of the CSL grain boundaries from 12 to more than 65% [769].

As was already mentioned in Chap. 2, the CSL approach reflects only the three DOFs related to the misorientation of the two adjoined grains but does not specify directly the crystallography of the boundary plane itself [769, 770]. It was already shown for many examples that the grain boundary plane orientation plays an important role in grain boundary classification. As mentioned above, the study of solute segregation at individual grain boundaries in a carefully chosen set of bicrystals of an Fe–Si alloy suggests that there should exist at least one special grain boundary for any misorientation of the two grains that is characterised by low-index plane [94, 100, 762, 771]. In case of pure nickel, it was shown that only 2/3 of $\Sigma = 3$ and $\Sigma = 9$ grain boundaries possessed "special" orientation in a polycrystal [772]. This

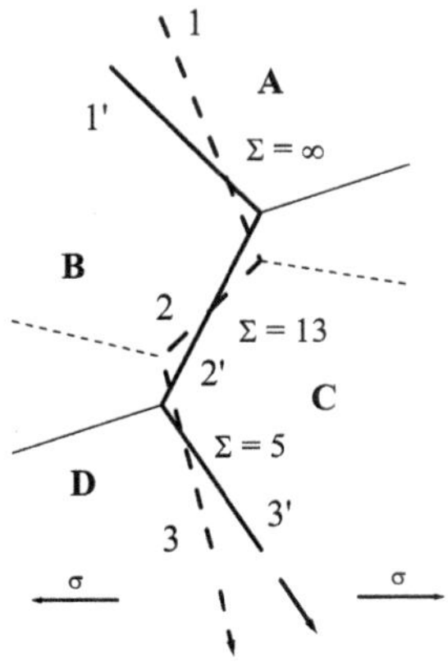


**Fig. 7.12** Model ensemble of four grains, A, B, C and D, with mutual orientation relationships A–B: 45°[100], B–C: 22.6°[100], and C–D: 36.9°[100]. Two alternative fracture paths (occurring under stress $\sigma$) along grain boundaries are depicted (**a**) 1–2–3, formed by general interfaces $\{0k_sl_s\}$, (023)/(015) and (0 1 14)/(0 20 23) (*full-line path*), and (**b**) $1'$–$2'$–$3'$ formed by special interfaces (001)/(011), {015} and {013} (*dashed-line path*). Two configurations are inclined by 22.5°[100]. According to [94]

idea modifies the approaches to the technologies of production of polycrystals with optimum properties in GBE: It is not necessary to consider exclusively the reorientation of the grains constrained in the material during recrystallisation or to locally deform the lattice but additionally, also to take into account the grain boundary reorientation during grain growth when the interface inclines into an energetically advantageous special orientation [94, 100, 773, 774]. This is schematically shown in Fig. 7.12 [94, 100]. It is obvious that the grain boundary inclination is much easier to be realised compared to the rotation of individual grains buried in a polycrystalline material.

Let us mention for completeness that in the GBE one has to deal not only with grain boundaries alone. They are constrained in triple junctions of three interfaces or quadruple points of four grain boundaries. The chemistry of these features as well as their migration ability is presently under intensive investigation. Similarly to grain boundaries, the triple junctions exhibit the segregation effects [195]. The importance of the triple junctions led to formulation of an alternative approach to the GBE: the grain boundary junction engineering [774, 775].

# Appendix A
# Parameters of Grain Boundary Segregation

## A.1 Enthalpy, Entropy and Gibbs Energy of Segregation

Matrix: α-Fe

| Element | $\Delta H_I^0$ ($\Delta H_I$) (kJ/mol) | $\Delta S_I^0$ ($\Delta S_I$) (J/(mol K)) | Ref. | Comments |
|---|---|---|---|---|
| P | −34.3 | +21.5 | [328] | Polycrystal |
| | −15 | +33 | [488] | Polycrystal, paramagnetic state |
| | (−38) | (−) | [343] | Polycrystal $\left(\Delta G_I^{\mathrm{eff}}\right)$ |
| | −50 | 0 | [340] | Polycrystal |
| | −21.2 | +37.3 | [354] | Polycrystal |
| | −32 | +22 | [551] | Polycrystal |
| | (−13.3) | (?) | [776] | Polycrystal, 12Cr–Mo–V steel. $\left(\Delta G_I^{\mathrm{eff}}\right)$ |
| | −22 | +28 | [329] | Polycrystal, low-alloy steel |
| | (−57) | (−12.4) | [400] | Polycrystal, Cr–Mo–V steel |
| | −7.9 | +42.7 | [271] | Bicrystal Fe–Si {112} |
| | −13.3 | +45.2 | [350] | Bicrystal Fe–Si {013} |
| | −10.9 | +42.5 | [111] | Bicrystal Fe–Si {012} |
| | −31 | +17 | [111] | Bicrystal Fe–Si {016} |
| | −16 | +38 | [111] | Bicrystal Fe–Si {015} |
| | −35 | +19 | [111] | Bicrystal Fe–Si {014} |
| | −37 | +18 | [111] | Bicrystal Fe–Si {0*kl*} 45°[100] |
| | −31 | +25 | [111] | Bicrystal Fe–Si {0 7 15} |
| | −34 | +20 | [111] | Bicrystal Fe–Si {059} |
| | −37 | +16 | [111] | Bicrystal Fe–Si {058} |
| | −32 | +19 | [86] | Bicrystal Fe–Si (018)/(047) |
| | −25 | +29 | [86] | Bicrystal Fe–Si (001)/(034) |
| | −14.5 | +39.3 | [86] | Bicrystal Fe–Si (017)/(011) |
| | −32 | +21 | [86] | Bicrystal Fe–Si (0 3 11)/(097) |

(*continued*)

| Element | $\Delta H_I^0$ ($\Delta H_I$) (kJ/mol) | $\Delta S_I^0$ ($\Delta S_I$) (J/(mol K)) | Ref. | Comments |
|---|---|---|---|---|
| | −19 | +38 | [87] | Bicrystal Fe–Si (001)/(011) 45°[100] |
| | −14 | +25 | [507] | Bicrystal Fe–Si (001)/(011) 90°[110] |
| | −22 | +32 | [94] | Bicrystal Fe–Si (001)/(015) |
| | −19 | +35 | [94] | Bicrystal Fe–Si (001)/(013) |
| | −26 | +28 | [94] | Bicrystal Fe–Si (001)/(012) |
| | −18 | +37 | [94] | Bicrystal Fe–Si (011)/(015) |
| | −17 | +37 | [94] | Bicrystal Fe–Si (011)/(013) |
| | −17 | +36 | [94] | Bicrystal Fe–Si (011)/(012) |
| C | (−80) | (0) | [339] | Polycrystal |
| | (−37.7) | (+43.2) | [355] | Polycrystal |
| | (−57) | (+21.5) | [589] | Polycrystal |
| | (−79) | (−13) | [395] | Polycrystal |
| | −40 | +12 | [271] | Bicrystal Fe–Si {013} |
| | −35.0 | +12 | [271] | Bicrystal Fe–Si {012} |
| | −49 | +1 | [111] | Bicrystal Fe–Si {016} |
| | −43 | +7 | [111] | Bicrystal Fe–Si {015} |
| | −50 | +2 | [111] | Bicrystal Fe–Si {014} |
| | −51 | +6 | [111] | Bicrystal Fe–Si {0*kl*} 45°[100] |
| | −45 | +6 | [111] | Bicrystal Fe–Si {0 7 15} |
| | −48 | +4 | [111] | Bicrystal Fe–Si {059} |
| | −53 | −1 | [111] | Bicrystal Fe–Si {058} |
| | −50 | +3 | [86] | Bicrystal Fe–Si (018)/(047) |
| | −44 | +6 | [86] | Bicrystal Fe–Si (001)/(034) |
| | −36 | +14 | [86] | Bicrystal Fe–Si (017)/(011) |
| | −48 | +3 | [86] | Bicrystal Fe–Si (0 3 11)/(097) |
| | −39 | +11 | [87] | Bicrystal Fe–Si (001)/(011) 45°[100] |
| | −33 | +8 | [507] | Bicrystal Fe–Si (001)/(011) 90°[110] |
| | −41 | +10 | [94] | Bicrystal Fe–Si (001)/(015) |
| | −38 | +14 | [94] | Bicrystal Fe–Si (001)/(013) |
| | −47 | +4 | [94] | Bicrystal Fe–Si (001)/(012) |
| | −40 | +9 | [94] | Bicrystal Fe–Si (011)/(015) |
| | −36 | +14 | [94] | Bicrystal Fe–Si (011)/(013) |
| | −34 | +16 | [94] | Bicrystal Fe–Si (011)/(012) |
| S | −51 | 0 | [381] | Polycrystal |
| Sn | −13 | +45 | [371] | Polycrystal |
| | −22.5 | +26.1 | [592] | Polycrystal |
| | −13.1 | +39.1 | [371] | Polycrystal, low-alloy steel |

(*continued*)

| Element | $\Delta H_I^0$ $(\Delta H_I)$ (kJ/mol) | $\Delta S_I^0$ $(\Delta S_I)$ (J/(mol K)) | Ref. | Comments |
|---|---|---|---|---|
| Sb | (−13) | (−) | [396] | Polycrystal $\left(\Delta G_I^{\text{eff}}\right)$ |
| | (−19) | (+28) | [335] | Polycrystal |
| | −23 | +37 | [338] | Polycrystal |
| Si | −4 | −4 | [331] | Bicrystal Fe−13%Si {112} |
| | −6.1 | −6.9 | [331] | Bicrystal Fe−13%Si {013} |
| | −3 | −3.8 | [271] | Bicrystal Fe−Si {112} |
| | −8.5 | −3 | [111] | Bicrystal Fe−Si {013} |
| | −4.1 | +0.2 | [111] | Bicrystal Fe−Si {012} |
| | −16 | −15 | [111] | Bicrystal Fe−Si {016} |
| | −12 | −9 | [111] | Bicrystal Fe−Si {015} |
| | −14 | −9 | [111] | Bicrystal Fe−Si {014} |
| | −17 | −13 | [111] | Bicrystal Fe−Si {0*kl*} 45°[100] |
| | −12 | −3 | [111] | Bicrystal Fe−Si {0 7 15} |
| | −12 | −5 | [111] | Bicrystal Fe−Si {059} |
| | −16 | −11 | [111] | Bicrystal Fe−Si {058} |
| | −10 | −8 | [86] | Bicrystal Fe−Si (018)/(047) |
| | −9 | −3 | [86] | Bicrystal Fe−Si (001)/(034) |
| | −6.1 | −2.2 | [86] | Bicrystal Fe−Si (017)/(011) |
| | −11 | −5 | [86] | Bicrystal Fe−Si (0 3 11)/(097) |
| | −6 | +2 | [87] | Bicrystal Fe−Si (001)/(011) 45°[100] |
| | −7 | +2 | [507] | Bicrystal Fe−Si (001)/(011) 90°[110] |
| | −8 | −3 | [94] | Bicrystal Fe−Si (001)/(015) |
| | −7 | −2 | [94] | Bicrystal Fe−Si (001)/(013) |
| | −10 | −5 | [94] | Bicrystal Fe−Si (001)/(012) |
| | −7 | −2 | [94] | Bicrystal Fe−Si (011)/(015) |
| | −6 | −1 | [94] | Bicrystal Fe−Si (011)/(013) |
| | −6 | −2 | [94] | Bicrystal Fe−Si (011)/(012) |
| Al | −12 | −8 | [587] | Polycrystal $Fe_3Al$+Cr |
| **Cr** | −8 | −4 | [587] | Polycrystal $Fe_3Al$+Cr |

| Element | $(\Delta G_I)$ (kJ/mol) | $T$(K) | Ref. | Comments |
|---|---|---|---|---|
| P | (−50) | 853 | [590] | Polycrystal, Mo−Cr low-alloy steel |
| | (−44 to −33.5) | 773 | [497] | Polycrystal, Fe−Mo−P |
| | (−53.1) | 873 | [402] | Polycrystal, low-alloy steel |
| | (−55.2) | 973 | [402] | Polycrystal, low-alloy steel |
| | (−55.5) | ? | [552] | Polycrystal, low-alloy steel. $\left(\Delta G_I^{\text{eff}}\right)$ |

*(continued)*

| Element | $(\Delta G_I)$ (kJ/mol) | $T$(K) | Ref. | Comments |
|---|---|---|---|---|
| | (−43.1) | 873 | [777] | |
| | (−46.6 ± 2) | 823 | [778] | |
| | (−47) | 723–1,173 | [360] | Polycrystal. $\left(\Delta G_I^{\mathrm{eff}}\right)$ |
| Si | (−17) | 1,073 | [340] | Polycrystal Fe–Si |
| | (−15.1) | 823 | [437, 779] | Polycrystal Fe–Si |
| | (−17) | 823 | [779] | Polycrystal Fe–Si |
| S | (−51.5) | 823 | [381] | (Polycrystal) |
| | (−141.1) | 873 | [393] | Polycrystal, low-alloy steel |
| | (−163.9) | 973 | [393] | Polycrystal, low-alloy steel |
| | (−75) | 1,143 | [370] | Polycrystal, Fe–S–C |
| Sb | (−32.7) | 823 | [416] | Polycrystal Fe–Ni–Sb |
| | (−13) | 723–923 | [320] | Polycrystal, Sb-doped C–Mn steel. $\left(\Delta G_I^{\mathrm{eff}}\right)$ |
| Sn | (−44) | 823 | [123] | Polycrystal Fe–Sn |
| | (−45.3) | 823 | [306] | Polycrystal Fe–Sn |
| | (−103) | 1,693 | [371] | Polycrystal Fe–Sn |
| Ni | (−14.4) | 823 | [416] | Polycrystal Fe–Ni–Sb |
| | (−3 ± 3) | 823 | [552] | Polycrystal, low-alloy steel |
| Mn | −8 ± 4 | 823 | [780] | Polycrystal, low-alloy steel |
| | (−5.4) | 723–923 | [320] | Polycrystal, Sb-doped C–Mn steel. $\left(\Delta G_I^{\mathrm{eff}}\right)$ |
| Mo | (−11.8 to −8.8) | 773 | [497] | Polycrystal, Fe–Mo–P |
| | (0) | 823 | [780] | Polycrystal, low-alloy steel |
| | (−50) | ? | [328] | Polycrystal, Fe–Cr–P. $\left(\Delta G_I^{\mathrm{eff}}\right)$ |
| | (−28) | 1,073 | [234] | Polycrystal, low-alloy steel |
| Cr | (0) | 823 | [780] | Polycrystal, low-alloy steel |
| | (−20) | 853 | [590] | Polycrystal, Mo–Cr low-alloy steel |
| Nb | (−38) | 1,073 | [233] | Polycrystal, low-alloy steel |
| B | (−39) | 673 | [781] | Polycrystal Fe–40Al |
| | (−44) | 673 | [347] | Polycrystal Fe–40Al |
| | (−30 to −34) | 673 | [128] | Polycrystal Fe–40Al |
| | (−49) | 673 | [347] | Polycrystal Fe–40Al |
| | (−100) | 1,073 | [781] | |

(*continued*)

| Element | $(\Delta G_I)$ (kJ/mol) | $T$(K) | Ref. | Comments |
|---|---|---|---|---|
| C | (−75) | 873 | [429] | |
| | (−92.4) | ? | [780] | Polycrystal, low-alloy steel. $(\Delta G_I^{\text{eff}})$ |
| | (−76) | 873 | [328] | Polycrystal Fe−P−C |
| Te | (−80) | 1,023 | [308] | Polycrystal Fe−Te, data from [439] |
| Se | (−67) | 1,023 | [308] | Polycrystal Fe−Se, data from [439] |
| Si | (−9) | 823 | [591] | Twist boundaries, Fe−3at.%Si |
| H | (−59) | ? | [406] | $(\Delta G_I^{\text{eff}})$ |

Matrix: Mo

| Element | $(\Delta H_I)$ (kJ/mol) | $(\Delta S_I)$ (J/(mol K)) | Ref. | Comments |
|---|---|---|---|---|
| C | (−69.5) | (+20.6) | [782] | |
| O | (−118.6) | (+0.08) | [782] | |
| | (−37.6) | | [776] | |

Matrix: W

| Element | $\Delta G_I$ (kJ/mol) | $T$(K) | Ref. | Comments |
|---|---|---|---|---|
| **Fe** | (−53) | 1,400 | [783] | Polycrystal |

Matrix: Cr

| Element | $\Delta G_I$ (kJ/mol) | $T$(K) | Ref. | Comments |
|---|---|---|---|---|
| Fe | (−11) | 973 | [784] | Polycrystal |

Matrix: Nb

| Element | $\Delta G_I$ (kJ/mol) | $T$(K) | Ref. | Comments |
|---|---|---|---|---|
| C | (−49.9) | 800–950 | [785] | Diffusion measurements. $(\Delta G_I^{\text{eff}})$ |

Matrix: Ir

| Element | $\Delta G_I$ (kJ/mol) | $T$(K) | Ref. | Comments |
|---|---|---|---|---|
| Lu | (−118) | 1,573–2,073 | [778] | Polycrystal Ir−W (3 mass%). $(\Delta G_I^{\text{eff}})$ |
| Y | (−150) | 1,573–2,073 | [778] | Polycrystal Ir−W (3 mass%). $(\Delta G_I^{\text{eff}})$ |
| Th | (−121) | | [786] | $(\Delta G_I^{\text{eff}})$ |

Matrix: $\gamma$-Fe

| Element | $(\Delta H_I)$ (kJ/mol) | $(\Delta S_I)$ (J/(mol K)) | Ref. | Comments |
|---|---|---|---|---|
| P | (−32) | (+17) | [787] | Polycrystal Fe−P−C−X (X = Mn,Ni,Cr) |
| | (−19.9) | (+22.8) | [377] | Polycrystal Fe−P |
| | (−14.1) | (+15.0) | [379] | Polycrystal 17Cr12Ni SS |
| Sn | (+0.7) | (+56.9) | [371] | |

| Element | $(\Delta G_I)$ (kJ/mol) | $T$(K) | Ref. | Comments |
|---|---|---|---|---|
| P | (−49.8) | 1,273 | [788] | Polycrystal, Ni−Cr SS |
| | (−51.9) | 1,173 | [377] | Polycrystal, Fe−P−X (X = C,Cr,Mn,Ni) |
| | (−53) | 1,273 | [377] | Polycrystal, Fe−P−X (X = C,Cr,Mn,Ni,V) |
| | (−50.5) | 1,273 | [411] | Polycrystal, Mn−Cr steel |
| | (−47) | 1,173–1,373 | [344] | Polycrystal, Fe−P−C. $(\Delta G_I^{\text{eff}})$ |
| | | 1,023–1,073 | [428] | |
| C | (−30) | 1,173–1,373 | [344] | Polycrystal, Fe−P−C. $(\Delta G_I^{\text{eff}})$ |
| B | (−57) | ? | [789] | Polycrystal, low-alloy steel |
| | (−63) | ? | [410] | Polycrystal, SS, $(\Delta G_I^{\text{eff}})$ |
| | (−97) | 1,023–1,373 | [378] | Polycrystal, Fe−10Mn−P−B. $(\Delta G_I^{\text{eff}})$ |
| | (−100) | 1,273 | [123] | |
| S | (−88.4) | 1,393 | [790] | Polycrystal, low-alloy steel |

Matrix: Ni

| Element | $\Delta H_I^0$ $(\Delta H_I)$ (kJ/mol) | $\Delta S_I^0$ $(\Delta S_I)$ (J/(mol K)) | Ref. | Comments |
|---|---|---|---|---|
| In | −38 | −0.05 | [524] | Bicrystal Ni−In {115} |
| | −39 | −0.05 | [524] | Bicrystal Ni−In {1110} |
| S | (−98) | (−) | [791] | Polycrystal $(\Delta G_I^{\text{eff}})$ |
| | (−70) | (−) | [792] | Polycrystal $(\Delta G_I^{\text{eff}})$ |
| | (−72) | (−) | [793] | Polycrystal $(\Delta G_I^{\text{eff}})$ |
| | −88 | −10 | [794] | Bicrystal Ni−Sb {133} |
| | −70 | +12 | [131] | Polycrystal Ni−Sb |
| | (−120) | | [795] | Polycrystal $(\Delta G_I^{\text{eff}})$ |
| Sb | −4 | +32 | [794] | Bicrystal Ni−Sb {133} |
| | −63 | −20 | [131] | Polycrystal Ni−Sb |
| B | (−10) | (+9) | [127] | Polycrystal $Ni_3Al$ + 0.48%B |

(*continued*)

Matrix: Ni

| Element | $\Delta H_I^0$ $(\Delta H_I)$ (kJ/mol) | $\Delta S_I^0$ $(\Delta S_I)$ (J/(mol K)) | Ref. | Comments |
|---|---|---|---|---|
| | (−11) | (+22) | [127] | Polycrystal $Ni_3Al$ + 0.048%B |
| | −30 | +1.3 | [347] | Polycrystal $Ni_3Al$ |
| | −19 | +39 | [347] | Polycrystal $Ni_3Al$ |

| Element | $\Delta G_I$ (kJ/mol) | $T$ (K) | Ref. | Comments |
|---|---|---|---|---|
| S | (−72.8) | 1,100 | [793] | Polycrystal |
| | (−51.5) | 823 | [451] | Polycrystal |
| In | (−50) | 970 | [524] | Polycrystal |
| B | (−32) | 1,323 | [347] | Polycrystal $Ni_3Al$ |
| | (−71) | 1,323 | [347] | Polycrystal $Ni_3Al$ |

Matrix: Co

| Element | $\Delta H_I^0$ $(\Delta H_I)$ (kJ/mol) | $\Delta S_I^0$ $(\Delta S_I)$ (J/(mol K)) | Ref. | Comments |
|---|---|---|---|---|
| Sb | (−5) | (+38) | [488] | Polycrystal Co–Ni50(para) |
| S | (−32) | (+47) | [488] | Polycrystal Co–Ni50(para) |

Matrix: Cu

| Element | $\Delta H_I^0$ $(\Delta H_I)$ (kJ/mol) | $\Delta S_I^0$ $(\Delta S_I)$ (J/(mol K)) | Ref. | Comments |
|---|---|---|---|---|
| S | (−65) | (?) | [752] | Polycrystal $(\Delta G_I^{eff})$ |
| Au | −9.7 | 0 | [754] | Polycrystal |
| Ag | (−39.5) | (−34.2) | [755] | Polycrystal, diffusion measurements |
| | (−25.7) | (+8.6) | [753] | Polycrystal, diffusion measurements |
| Bi | (−53.4) | (+8.8) | [796] | Polycrystal, diffusion measurements |
| | (−53.4) | (−24.1) | [797] | Polycrystal, diffusion measurements |

| Element | $(\Delta G_I)$ (kJ/mol) | $T$ (K) | Ref. | Comments |
|---|---|---|---|---|
| Bi | (−66.2) | 723–973 | [750] | Polycrystal $(\Delta G_I^{eff})$ |
| Ag | (−39) | 670 | [798] | Polycrystal |

Matrix: Ag

| Element | $\Delta H_I^0$ ($\Delta H_I$) (kJ/mol) | $\Delta S_I^0$ ($\Delta S_I$) (J/(mol K)) | Ref. | Comments |
|---|---|---|---|---|
| Te | (−43.3) | (−25.6) | [799] | Polycrystal, diffusion measurements |
| Se | (−21.4) | (−13.4) | [751] | Polycrystal, diffusion measurements |
| Ni | (−39.7) | (+3.3) | [750] | Polycrystal, diffusion measurements |

Matrix: Pt

| Element | ($\Delta G_I$) (kJ/mol) | $T$(K) | Ref. | Comments |
|---|---|---|---|---|
| Au | (−40 to −48) | 1,373–1,673 | [380] | Polycrystal Pt–Rh–Pd–Au ($\Delta G_I^{\text{eff}}$) |

Matrix: Al

| Element | ($\Delta G_I$) (kJ/mol) | $T$(K) | Ref. | Comments |
|---|---|---|---|---|
| Ga | (−11 to −18) | ? | [800] | ($\Delta G_I^{\text{eff}}$) |

Matrix: Zr

| Element | $\Delta H_I^0$ ($\Delta H_I$) (kJ/mol) | $\Delta S_I^0$ ($\Delta S_I$) (J/(mol K)) | Ref. | Comments |
|---|---|---|---|---|
| Cr | (−12) | (?) | [801] | Polycrystal, diffusion measurements. ($\Delta G_I^{\text{eff}}$) |

## A.2 Interaction Coefficients

Matrix: α-Fe

| Elements | $\alpha_I$ (kJ/mol) | $Q_{I,J}$[a] (kJ/mol) | $T$(K) | Ref. | Comments |
|---|---|---|---|---|---|
| C | −5 | | | [111] | Bicrystals |
| | 33.4 | | | [802] | |
| P | 8.9 | | | [802] | |
| | −4.4 | | | [552] | Polycrystal, low-alloy steel |
| | −2.2 | | | [777] | |
| B | 110–160 | | 673 | [128] | Polycrystal Fe−40Al |
| Sb | 10.8 | | 823 | [393] | Polycrystal Fe−Ni−Sb |

(*continued*)

| Elements | $\alpha_I$ (kJ/mol) | $Q_{I,J}$[a] (kJ/mol) | $T$(K) | Ref. | Comments |
|---|---|---|---|---|---|
| P–S | | 78.5 | 773 | [393] | Polycrystal, low-alloy steel |
| | | 60.5 | 873 | [393] | Polycrystal, low-alloy steel |
| | | 53.6 | 973 | [393] | Polycrystal, low-alloy steel |
| P–Si | | 92.4 | | [350] | Bicrystals |
| P–C | | 7 | | [350] | Bicrystals |
| | | 18 | | [802] | |
| | | 9 | | [777] | |
| Si–C | | −3 | | [350] | Bicrystals |
| P–Cr | | −5 to −2 | | [402] | Polycrystal, low-alloy steel |
| | | −102.6 | | [803] | |
| | | −100.3 | | [313] | |
| | | −17 ± 2.4 | | [777] | |
| | | −34 ± 5 | | [551] | Polycrystal, low-alloy steel |
| P–Mo | | −24 to −22 | | [402] | Polycrystal, low-alloy steel |
| | | −75 ± 13 | | [589] | Polycrystal, low-alloy steel |
| | | −25 to −17 | | [498] | |
| | | −184.0 | | [313] | |
| | | −180.5 | | [803] | |
| | | −37.5 ± 6.5 | | [777] | |
| P–Mn | | −21 ± 6 | | [551] | Polycrystal, low-alloy steel |
| | | −66.9 | | [313] | |
| | | −10.5 ± 3 | | [777] | |
| P–Ni | | −21 ± 4 | | [552] | Polycrystal, low-alloy steel |
| | | −26 | | [364] | Polycrystal |
| | | −75.2 | | [320] | |
| | | −30.1 | | [320, 803] | |
| | | −10.5 ± 2 | | [777] | |
| P–V | | −12 to −11 | | [402] | Polycrystal, low-alloy steel |
| | | −143.8 | | [313] | |
| P–W | | −170.7 | | [313] | |
| P–Ti | | −221.0 | | [313] | |
| P–Zr | | −256.4 | | [313] | |
| Mn–Sb | | −81.9 | | [320] | |
| | | −80.3 | | [320] | |
| Cr–Sb | | −33.5 | | [803] | |
| | | −33.6 | | [313] | |

(*continued*)

| Elements | $\alpha_I$ (kJ/mol) | $Q_{I,J}$[a] (kJ/mol) | $T$(K) | Ref. | Comments |
|---|---|---|---|---|---|
| Ni–Sb | | −113.7 | | [320] | |
| | | −19.6 | | [416] | |
| | | −56.9 | | [803] | |
| | | −56.8 | | [313] | |
| | | −24 | | [313] | |
| Ni–Se | | +66.9 | | [320] | |
| Ni–Te | | +59.4 | | [320] | |
| Mn–S | | −67 | | [803] | |
| | | −220 | | [313] | |
| Mn–N | | −150 | | [313] | |

[a]For the meaning of $Q_{IJ}$, see Table 4.3

# Appendix B
# Predicted Values of Enthalpy and Entropy of Segregation

| Solute | Solubility [551] | | Site | $\Delta H_I^0$ (kJ/mol)/$\Delta S_I^0$ (J/(mol K)) | | |
|---|---|---|---|---|---|---|
| | $X^*$ | $T$(K) | | 36.9°[100] {013} (special) $\Delta H^* =$ +6 kJ/mol | 36.9°[100] (001)/(034) (vicinal) $\Delta H^* =$ +2 kJ/mol | 45°[100] {0*kl*} (general) $\Delta H^* =$ −6 kJ/mol |
| Ag | 0.000002 | 1,185 | S | −94/−96 | −98/−100 | −106/−108 |
| Al | 0.2 | 718 | S̲ | −1/ + 4 | −5/−1 | −13/−9 |
| As | 0.095 | 1,173 | I | −12/ + 43 | −16/ + 39 | −24/ + 31 |
| Au | 0.01 | 1,141 | S | −28/−25 | −32/−32 | −40/−38 |
| B | 0.00005 | 1,185 | I̲ | −69/−20 | −73/−24 | −81/−33 |
| Be | 0.12 | 973 | I | −7/ + 48 | −11/ + 44 | −9/ + 36 |
| C | 0.001013 | 1,000 | I̲ | −38/ + 15 | −42/ + 11 | −50/ + 2 |
| Ca | 0.000003 | 1,115 | S? | −85/−86? | −89/−90? | −97/−99? |
| Ce | 0.0003 | 1,195 | S | −56/−55 | −60/−60 | −68/−68 |
| Co | 0.75 | 1,008 | S | 0/ + 5 | 0/ + 5 | −8/−4 |
| Cr | 0.37 | 1,073 | S̲ | −1/ + 4 | −5/0 | −13/−9 |
| Cu | 0.018 | 1,130 | S | −23/−20 | −27/−24 | −35/−33 |
| Er | 0.00015 | 1,183 | S | −61/−61 | −65/−65 | −73/−73 |
| Ga | 0.206 | 861 | S | −3/ + 2 | −7/−2 | −15/−11 |
| Gd | 0.007 | 1,205 | S | −32/−29 | −36/−34 | −44/−42 |
| Ge | 0.15 | 1,100 | I | −7/ + 48 | −11/ + 44 | −19/ + 35 |
| H | 0.0001 | 995 | I̲ | −53/−1 | −57/−5 | −65/−14 |
| Hf | 0.00196 | 1,173 | S | −41/−39 | −45/−43 | −53/−52 |
| In | 0.0037 | 873 | I | −25/ + 29 | −29/ + 25 | −37/ + 16 |
| Ir | 0.08 | 713 | S | −6/−1 | −10/−5 | −18/−14 |
| La | 0.001 | 1,053 | S | −41/−39 | −45/−43 | −53/−52 |

(*continued*)

| Solute | Solubility [551] | | Site | $\Delta H_I^0$ (kJ/mol)/$\Delta S_I^0$ (J/(mol K)) | | |
|---|---|---|---|---|---|---|
| | $X^*$ | $T$(K) | | 36.9°[100] {013} (special) $\Delta H^* =$ +6 kJ/mol | 36.9°[100] (001)/(034) (vicinal) $\Delta H^* =$ +2 kJ/mol | 45°[100] {0*kl*} (general) $\Delta H^* =$ −6 kJ/mol |
| Li | 0.00063 | 1,185 | I? | −50/+2? | −54/−2? | −62/−11? |
| Mg | 0.0004 | 1,185 | S | −53/−53 | −57/−57 | −65/−65 |
| Mn | 0.03 | 873 | S | −14/−10 | −18/−14 | −26/−23 |
| Mo | 0.055 | 1,173 | S | −16/−12 | −20/−16 | −28/−25 |
| N | 0.004 | 864 | I | −25/+30 | −29/+25 | −37/+17 |
| Nb | 0.007 | 1,234 | S | −33/−31 | −37/−35 | −45/−44 |
| Nd | 0.015 | 1,208 | S | −26/−23 | −30/−28 | −38/−36 |
| Ni | 0.058 | 748 | S | −8/−4 | −12/−8 | −20/−16 |
| O | 0.000008 | 1,185 | I | −83/−33 | −87/−38 | −95/−46 |
| Os | 0.027 | 893 | S | −15/−11 | −19/−15 | −27/−24 |
| P | 0.033 | 1,173 | I | −20/+35 | −24/+31 | −32/+22 |
| Pd | 0.0353 | 1,088 | S | −17/−14 | −21/−18 | −29/−26 |
| Pt | 0.04 | 1,000 | S | −15/−11 | −19/−15 | −27/−24 |
| Pu | 0.0002 | 1,135 | S | −56/−55 | −60/−59 | −68/−68 |
| Re | 0.195 | 1,168 | S | −6/−1 | −10/−6 | −18/−14 |
| Rh | 0.53 | 1,000 | S | 0/+5 | −2/+3 | −10/−6 |
| Ru | 0.048 | 773 | S | −9/−5 | −13/−9 | −21/−18 |
| S | 0.00033 | 1,200 | I | −56/−4 | −60/−8 | −68/−17 |
| Sb | 0.0419 | 1,173 | I | −18/+37 | −22/+33 | −30/+24 |
| Sc | 0.02 | 1,208 | S | −24/−21 | −28/−25 | −36/−34 |
| Se | 0.02 | 1,149 | I | −23/+32 | −27/+27 | −35/+19 |
| Si | 0.305 | 1,313 | S | −4/+1 | −8/−4 | −16/−12 |
| Sn | 0.095 | 1,183 | I | −12/+43 | −16/+39 | −24/+30 |
| Ta | 0.0071 | 1,238 | S | −33/−31 | −37/−35 | −45/−44 |
| Tc | 0.01 | 1,153 | S | −28/−25 | −32/−29 | −40/−38 |
| Ti | 0.0308 | 1,173 | S | −20/−17 | −24/−21 | −32/−29 |
| V | 0.25 | 1,178 | S | −4/0 | −8/−4 | −16/−13 |
| W | 0.02 | 1,173 | S | −23/−20 | −27/−24 | −35/−33 |
| Zn | 0.21 | 920 | S? | −3/+2? | −7/−3? | −15/−11? |
| Zr | 0.0016 | 1,173 | S | −42/−41 | −46/−45 | −54/−53 |

Model of Lejček and Hofmann [98, 375]. S substitutional segregation, I interstitial segregation. The mark "?" means that the segregation position and thus the value of segregation entropy is not clear. Underlined symbols represent the experimentally or theoretically proven segregation position at the grain boundaries

# References


1. R.F. Tylecote, *A History of Metallurgy* (Institute of Metals, London, 1989)
2. V.S. Arunachalam, MRS. Bull. **25**(1), 55 (2000)
3. G. Agricola, *De Re Metallica* (J. Froben & N. Episcopius, Basel, 1556, translated by H.C. Hoover, L.H. Hoover, London, 1912, reprinted by Dover Publications, New York, 1950)
4. P.G. Shewmon, Metall. Mater. Trans. A **29**, 1535 (1998)
5. W. Hampe, Z. Berg- Hütten- und Salinenwesen **23**, 93 (1874)
6. W.C. Roberts-Austen, Phil. Trans. Soc. A **179**, 339 (1888)
7. H.K. Huntington, J. Inst. Met. **11**, 108 (1914)
8. J.H. Westbrook, R.L. Fleischer (eds.), *Intermetallic Compounds* (Wiley, Chichester, vols. 1 and 2, 1995, vol. 3, 2002)
9. K. Otsuka, C.M. Wayman (eds.), *Shape Memory Alloys*, (Cambridge University Press, Cambridge, 1998)
10. A. Cottrell, MRS. Bull. **25**(2), 43 (2000)
11. A.L. Dragoo, Acta Mater. **48**, 4629 (2000)
12. A.P. Sutton, R.W. Balluffi, *Interfaces in Crystalline Materials* (Clarendon Press, Oxford, 1995)
13. E.D. Hondros, M.P. Seah, S. Hofmann, P. Lejček, in *Physical Metallurgy*, 4th edn., ed. by R.W. Cahn, P. Haasen (North-Holland, Amsterdam, 1996), pp. 1201–1289
14. J.Y. Laval, T.S. Orlova, Supercond. Sci. Technol. **16**, 1139 (2003)
15. R. Schafrik, R. Sprague, Adv. Mater. Proc. **162**(4), 27 (2004)
16. H. Biloni, W.J. Boetinger, Solidification, in *Physical Metallurgy*, 4th edn., ed. by R.W. Cahn, P. Haasen (North-Holland, Amsterdam, 1996), pp. 669–842
17. P. Lejček, J. Janovec, R. Konečná, Acta Metall. Slov. **13**, 117 (2007)
18. R.G. Faulkner, Int. Mater. Rev. **41**, 198 (1996)
19. D. McLean, *Grain Boundaries in Metals* (Clarendon Press, Oxford, 1957)
20. P. Lejček, S. Hofmann, Crit. Rev. Sol. State. Mater. Sci. **20**, 1 (1995)
21. H. Gleiter, B. Chalmers, High-Angle Grain Boundaries, in *Progress in Materials Science*, vol. 16, ed. by B. Chalmers, J.W. Christian, T.B. Massalski (Pergamon Press, Oxford, 1972) pp. 1–272
22. V. Randle, *The Measurement of Grain Boundary Geometry* (Institute of Physics Publications, Bristol, 1993)
23. D. Wolf, J. Appl. Phys. **68**, 3221 (1990)
24. D. Wolf, S. Yip (eds.), *Materials Interfaces: Atomic-level Structure and Properties* (Chapman & Hall, London, 1992)
25. W.T. Read, W. Shockley, Phys. Rev. **78**, 275 (1950)
26. V. Randle, Grain-boundary geometry: measurement, in *Concise Encyclopedia of Materials Characterization*, 2nd edn., ed. by R.W. Cahn (Elsevier, Amsterdam, 2005) pp. 293–297
27. D. Wolf, Atomic-Level geometry of crystalline interfaces, in *Materials Interfaces: Atomic-Level Structure and Properties*, ed. by D. Wolf, S. Yip (Chapman & Hall, London, 1992) pp. 1–57

28. R.C. Pond, Proc. Roy. Soc. (Lond.) A **357**, 463 (1977)
29. D. Wolf, Acta Metall. Mater. **38**, 781 (1990)
30. L. Priester, *Les Joints de Grains: De la Théorie à l'Ingénierie* (EDP Sciences, Les Ulis, 2006)
31. D. Wolf, J.F. Lutsko, Z. Kristallogr. **189**, 239 (1989)
32. T. Watanabe, Res. Mech. **11**, 47 (1984)
33. H. Mykura,A checklist of cubic coincidence site lattice relations, in *Grain-Boundary Structure and Kinetics*, ed. by R.W. Balluffi (ASM, Metals Park, 1980) pp. 445–456
34. C.T. Forwood, L.M.C. Clarebrough, *Electron Microscopy of Interfaces in Metals and Alloys* (Adam Hilger, Bristol, 1991)
35. V. Randle, Acta Mater. **46**, 1459 (1997)
36. D. Wolf, K.L. Merkle, Correlation between the structure and energy of grain boundaries in metals, in *Materials Interfaces: Atomic-Level Structure and Properties*, ed. by D. Wolf, S. Yip (Chapman & Hall, London, 1992) pp. 87–150
37. W. Rosenhain, J.C.W. Humphrey, J. Iron Steel Inst. **87**, 219 (1917)
38. P. Keblinski, D. Wolf, S.R. Philpot, H. Gleiter, Phys. Lett. A **226**, 205 (1997)
39. R.W. Cahn, Nature. **390**, 344 (1997)
40. D. Wolf, Curr. Opin. Sol. State. Mater. Sci. **5**, 435 (2001)
41. S. von Alfthan, P.D. Haynes, K. Kaski, A.P. Sutton, Phys. Rev. Lett. **96**, 055505 (2006)
42. N.F. Mott, Proc. Phys. Soc. **60**, 391 (1948)
43. A.P. Sutton, R.W. Balluffi, Acta Metall. **35**, 2177 (1987)
44. R.W. Balluffi, M. Rühle, A.P. Sutton, Mater. Sci. Eng. **89**, 1 (1987)
45. A. Bourret, Atomic structure of grain boundaries, in *Polycrystalline Semiconductors*, ed. by G. Harbeke (Springer, Berlin, 1985) pp. 2–17
46. D.A. Smith, Grain boundary structure and migration, in *Materials Interfaces: Atomic-Level Structure and Properties*, ed. by D. Wolf, S. Yip (Chapman & Hall, London, 1992) pp. 212–227
47. P.E.J. Flewitt, R.K. Wild, *Grain Boundaries: Their Microstructure and Chemistry* (Wiley, Chichester, 2001)
48. G. Gottstein, L.S. Shvindlerman, *Grain Boundary Migration in Metals: Thermodynamics, Kinetics, Applications* (CRC Press, Boca Raton, 1999)
49. M. Finnis, M. Rühle, Structures of interfaces in crystalline solids, in *Materials Science and Technology: A Comprehensive Treatment*, ed. by R.W. Cahn, P. Haasen, E.J. Kramer, vol. 1, *Structure of Solids*, ed. by V. Gerold (VCH, Weinheim, 1993) pp. 533–605
50. J.M. Burgers, Proc. Phys. Soc. **52**, 23 (1940)
51. G.I. Taylor, Proc. Roy. Soc. A **145**, 362 (1934)
52. D.G. Brandon, Acta Metall. **14**, 1479 (1966)
53. M.E. Glicksman, C.L. Vold, Surf Sci. **31**, 50 (1972)
54. M. Winnig, G. Gottstein, L.S. Shvindlerman, Acta Mater. **50**, 353 (2002)
55. G.H. Bishop, B. Chalmers, Scripta Metall. **2**, 133 (1968)
56. G.C. Hasson, J.B. Guillot, B. Baroux, C. Goux, Phys. Stat. Sol. (a). **2**, 551 (1972)
57. G.C. Hasson, J.-Y. Boss, I. Herbeuval, M. Biscondi, C. Goux, Surf. Sci. **31**, 115 (1972)
58. A.P. Sutton, Philos. Mag. A **46**, 171 (1982)
59. A.P. Sutton, V. Vitek, Philos. Trans. R. Soc. Lond. A **309**, 1 (1983)
60. A.P. Sutton, V. Vitek, Philos. Trans. R. Soc. Lond. A **309**, 37 (1983)
61. A.P. Sutton, V. Vitek, Philos. Trans. R. Soc. Lond. A **309**, 55 (1983)
62. G.J. Wang, A.P. Sutton, V. Vitek, Acta Metall. **32**, 1093 (1984)
63. G.J. Wang, V. Vitek, Acta Metall. **34**, 951 (1986)
64. V. Vitek, J. Phys. France. **49**, C5–115 (1988)
65. P.D. Bristowe, R.W. Balluffi, J. Phys. France. **46**, C4–155 (1985)
66. M.F. Ashby, F. Spaepen, S. Williams, Acta Metall. **26**, 1647 (1978)
67. C.L. Briant, Acta Metall. **31**, 257 (1983)
68. D. Wolf, J.A. Jaszczak, Computer simulation of the elastic behavior of thin films and superlattices, in *Materials Interfaces: Atomic-Level Structure and Properties*, ed. by D. Wolf, S. Yip (Chapman & Hall, London, 1992) pp. 364–406
69. T. Vystavěl, PhD Thesis, Université J. Fourier-Grenoble I (1999)

70. S.M. Foiles, D.N. Seidman, Atomic resolution study of solute-segregation at grain boundaries: experiments and Monte Carlo simulations, in *Materials Interfaces: Atomic-Level Structure and Properties*, ed. by D. Wolf, S. Yip (Chapman & Hall, London, 1992) pp. 497–515
71. E.I. Rabkin, L.S. Shvindlerman, B.B. Straumal, Int. J. Mod. Phys. B **5**, 2989 (1991)
72. B.B. Straumal, *Fazovye Perekhody na Granitsakh Zeren* (*Grain Boundary Phase Transitions*) (Nauka, Moscow, 2003) In Russian
73. M.L. Kronberg, F.H. Wilson, Trans. AIME. **185**, 501 (1949)
74. N.H. Fletcher, Crystal interface models – a critical survey, in *Advances in Materials Research*, vol. 5, ed. by H. Herman (Wiley, New York, 1971), pp. 281–314
75. W. Bollmann, *Crystal Defects and Crystalline Interfaces* (Springer, Heidelberg, 1970)
76. G. Friedel, *Leçons de Cristallographie* (Berger Levrault, Paris, 1926)
77. V. Paidar, Czech. J. Phys. A **38**, 131 (1988) (In Czech)
78. G. Palumbo, K.T. Aust,Special properties of Σ grain boundaries, in *Materials Interfaces: Atomic-Level Structure and Properties*, ed. by D. Wolf, S. Yip (Chapman & Hall, London, 1992), pp. 191–211
79. M.W. Finnis, Phys. Stat. Sol. (a) **166**, 397 (1998)
80. D.G. Brandon, B. Ralph, S. Ranganathan, M.S. Wald, Acta Metall. **12**, 813 (1964)
81. Y. Ishida, M. McLean, Philos. Mag. **27**, 1125 (1973)
82. M. Deschamps, F. Baribier, A. Marrouche, Acta Metall. **35**, 101 (1987)
83. G. Palumbo, K.T. Aust, E.M. Lehockey, U. Erb, P. Lin, Scripta Mater. **38**, 1985 (1998)
84. V. Paidar, Acta Metall. **35**, 2035 (1987).
85. V. Randle, *The Role of Coincidence Site Lattice in Grain Boundary Engineering* (The Institute of Materials, London, 1996)
86. P. Lejček, J. Adámek, S. Hofmann, Surf. Sci. **246**, 449 (1992)
87. P. Lejček, V. Paidar, J. Adámek, S. Hofmann, Acta Mater. **45**, 3915 (1997)
88. G. Palumbo, E.M. Lehockey, P. Lin, J. Metals. **50**, 40 (1998)
89. W. Lojkowski, H.-J. Fecht, Prog. Mater. Sci. **45**, 339 (2000)
90. L.S. Shvindlerman, B.B. Straumal, Acta Metall. **33**, 1735 (1985)
91. A.P. Sutton, Prog. Mater. Sci. **36**, 167 (1992)
92. R.W. Balluffi, Grain boundary structure and segregation, in *Interfacial Segregation*, ed. by W.C. Johnson, J.M. Blakely (ASM, Metals Park, 1979) pp. 193–237
93. D.A. Smith, R.C. Pond, Int. Met. Rev. **205**, 61 (1967)
94. P. Lejček, S. Hofmann, V. Paidar, Acta Mater. **51**, 3951 (2003)
95. P. Lejček, V. Paidar, Mater. Sci. Technol. **21**, 393 (2005)
96. S. Yip, D. Wolf, Mater. Sci. Forum **46**, 77 (1989)
97. V. Paidar, Philos. Mag. A **66**, 41 (1992)
98. P. Lejček, S. Hofmann,Grain boundary segregation, anisotropy and prediction, in *Encyclopedia of Materials: Science and Technology – Updates*, ed. by K.H.J. Bunshaw, R.W. Cahn, M.C. Flemings, E.J. Kramer, S. Mahajan (Pergamon, Amsterdam, 2002) #200115, pp 1–7
99. P. Lejček, V. Paidar, S. Hofmann, Mater. Sci. Forum **294–296**, 103 (1999)
100. P. Lejček, S. Hofmann, V. Paidar, Structure/property relationship for grain boundary engineering of polycrystals, in *Proceedings of the 1st International Conference on Advanced Materials Processing*, Rotorua, New Zealand, 19–23 November 2000, ed. by D.L. Zhang, K.L. Pickering, X.Y. Xiong (Institute of Material Engineering Australasia Ltd., Auckland, 2000) pp. 615–620
101. J.W. Gibbs, Transactions of the Connecticut Academy **III**, **108**, 343 (1875–1878), *The Scientific Papers of J. Willard Gibbs. vol. I: Thermodynamics* (Longmans, Green & Co., London, 1906)
102. J.W. Cahn, Thermodynamics of solid and fluid surfaces, in *Interfacial Segregation*, ed. by W.C. Johnson, J.M. Blakely (ASM, Metals Park, 1979) pp. 3–23
103. J.W. Cahn, J. Phase. Equilib. **15**, 373 (1994)
104. See any handbook on thermodynamics, e.g. K. Denbigh: *The Principles of Chemical Equilibrium* (Cambridge University Press, Cambridge, 1961); S. Stølen. T. Grande, *Chemical Thermodynamics of Materials* (Wileys, Chichester, 2005)

105. L.S. Chang, E. Rabkin, B.B. Straumal, S. Hofmann, B. Baretzky, W. Gust, Defects. Diff. Forum **156**, 135 (1998)
106. J. Janovec, *Nature of Alloy Steel Intergranular Embrittlement* (Veda, Bratislava, 1999)
107. J. Bruley, R. Brydson, H. Müllejans, J. Mayer, G. Gutekunst, W. Mader, D. Knauß, M. Rühle, J. Mater. Res. **9**, 2574 (1994)
108. P.S. Ho, B.D. Silverman, S.-L. Chiu, Microscopic nature of metal-polymer interfaces, in *Materials Interfaces: Atomic-Level Structure and Properties*, ed. by D. Wolf, S. Yip (Chapman & Hall, London, 1992) pp. 616–638
109. E.D. Hondros, Grain boundary segregation assessment of investigative techniques, in *Grain Boundary Structure and Properties*, ed. by G.A. Chadwick, D.A. Smith (Academic, London, 1976) pp. 265–298
110. E.L. Hall, C.L. Briant, Metall. Trans. A **16**, 1225 (1985)
111. P. Lejček, Anal. Chim. Acta **297**, 165 (1994)
112. P. Lejček, Surf. Interface Anal. **30**, 321 (2000)
113. P. Lejček, S. Hofmann, Crit. Rev. Sol. State. Mater. Sci. **33**, 133 (2008)
114. D. Briggs, M.P. Seah (eds.), *Practical Surface Analysis, vol. 1, Auger and X–ray Photoelectron Spectroscopy*, 2nd edn. (Wiley, Chichester, 1990)
115. D. Briggs, J.C. Rivière, Spectral interpretation, in *Practical Surface Analysis*, 2nd edn., ed. by D. Briggs, M.P. Seah (Wiley, Chichester, 1990) pp. 85–142
116. S. Hofmann, Mikrochim. Acta **1**, 321 (1987)
117. C.L. Briant, R.P. Messmer (eds.), *Auger Electron Spectroscopy* (Academic, London, 1988)
118. J.C. Walmsley, G. Jones, B.J. Lee, R.K. Wild, Appl. Surf. Sci. **108**, 289 (1997)
119. F. Reniers, C. Tewell, J. Electron. Spectrosc. Relat. Phenom. **142**, 1 (2005)
120. S. Hofmann, Vacuum **40**, 9 (1990)
121. D. Kalderon, Proc. Inst. Mech. Eng. **186**, 341 (1972)
122. D. Briggs, J.T. Grant (eds.), *Surface Analysis by Auger and X–Ray Photoelectron Spectroscopy*, (IM Publications, Chichester, 2003)
123. H.J. Grabke, Grain boundary segregation of impurities in iron and steels and effects on steel properties, in *Impurities in Engineering Materials*, ed. by C.L. Briant (Marcel Dekker Inc., New York, 1999) pp. 143–192
124. C.L. Briant, The effect of grain boundary segregation on intergranular failures, in *Impurities in Engineering Materials*, ed. by C.L. Briant (Marcel Dekker Inc., New York, 1999) pp. 193–224
125. C.L. Briant, Metall. Trans. A **19**, 137 (1988)
126. J.X. Dong, M.C. Zhang, X.S. Xie, R.G. Thompson, Mater. Sci. Eng. A **328**, 8 (2002)
127. A. Choudhury, C.L. White, C.R. Brooks, Acta Metall. Mater. **40**, 57 (1992)
128. A. Fraczkiewicz, A.-S. Gay, M. Biscondi, Mater. Sci. Eng. A **258**, 108 (1998)
129. B.S. Bokstein, A.N. Smirnov, Mater. Lett. **57**, 4501 (2003)
130. A.V. Krajnikov, A.S. Drachinskii, V.N. Slyunyaev, Refr. Met. Hard. Mater. **11**, 175 (1992)
131. P. Lejček, A. Rar, S. Hofmann, Surf. Interface Anal. **34**, 375 (2002)
132. P. Lejček, M. Koutník, J. Brádler, V. Paidar, A. Potměšilová, Appl. Surf. Sci. **44**, 75 (1990)
133. P. Lejček, S. Hofmann, J. Mater. Sci. Lett. **7**, 646 (1988)
134. P. Lejček, J. Brádler, V. Paidar, M. Koutník, J. Mater. Sci. **22**, 3974 (1987)
135. P. Lejček, J. Adámek, S. Hofmann, Surf. Sci. **264**, 449 (1992)
136. H. Fukushima, H.K. Birnbaum, Scripta Metall. **16**, 753 (1982)
137. J.M. Walls (ed.), *Methods of Surface Analysis* (Cambridge University Press, Cambridge, 1989)
138. M.P. Seah, J. Phys. F. Metal Phys. **10**, 1043 (1980)
139. M. Menyhard, B. Rothman, C.J. McMahon Jr., P. Lejček, V. Paidar, Acta Metall. Mater. **39**, 1289 (1991)
140. A. Benninghoven, F.G. Rudenauer, H.W. Werner, *Secondary Ion Mass Spectrometry* (Wiley, Chichester, 1987)
141. D. Briggs, M.P. Seah (eds.), *Practical Surface Analysis, vol. 2, Ion and Neutral Spectroscopy*, 2nd edn. (Wiley, Chichester, 1992)
142. A. Benninghoven, Surf. Sci. **53**, 596 (1975)

143. J.C. Vickerman, Static secondary ion mass spectrometry, in *Methods of Surface Analysis*, ed. by J.M. Walls (Cambridge University Press, Cambridge, 1989) pp. 169–215
144. D. Briggs, Surf. Interface Anal. **4**, 151 (1982)
145. H. Fukushima, H.K. Birnbaum, Acta Metall. **32**, 851 (1984)
146. M.T. Bernius, G.H. Morrison, Rev. Sci. Instrum. **58**, 1789 (1987)
147. R.L. Gavrilov, S.J. Bennison, K.R. Mikesa, R. Levi-Setti, Acta Mater. **47**, 4031 (1999)
148. P. Williams, C.A. Evans Jr., M.L. Grossbeck, H.K. Birnbaum, Anal. Chem. **48**, 964 (1976)
149. L. Karlsson, H.–O. Andrén, H. Nordén, Scripta Metall. **16**, 297 (1982)
150. X. Huang, M.C. Chatuverdi, N.L. Richards, J. Jackman, Acta Mater. **45**, 3095 (1997)
151. B. Ladna, H.K. Birnbaum, J. Physique. IV. Colloq. **49**, C5–441 (1988)
152. J. Cho, C.M. Wang, H.M. Chan, J.M. Rickman, M.P. Harmer, Acta Mater. **47**, 4197 (1999)
153. K.V. Hansen, K. Norrman, M. Mogensen, Surf. Interface Anal. **38**, 911 (2006)
154. K.K. Soni, A.M. Thompson, M.P. Harmer, D.B. Williams, J.M. Chabala, R. Levi–Setti, Appl. Phys. Lett. **66**, 2795 (1995)
155. H. Niehus, W. Heiland, E. Taglauer, Surf. Sci. Rep. **17**, 252 (1993)
156. W.A. Grant, Rutherford back-scattering spectroscopy, in *Methods of Surface Analysis*, ed. by J.M. Walls (Cambridge University Press, Cambridge, 1989) pp. 299–337
157. P. Lejček, O. Schneeweiss, A. Fraczkiewicz, Surf. Sci. **566–568**, 826 (2004)
158. A. Benninghoven, Surf. Sci. **53**, 596 (1975)
159. R.K. Marcus, J.A.C. Broekaert (eds.), *Glow Discharge Plasmas in Analytical Spectroscopy*, (Wiley, Chichester, 2003)
160. E.M. Skelly Frame, Determination of trace impurities in materials, in *Impurities in Engineering Materials*, ed. by C.L. Briant (Marcel Dekker Inc., New York, 1999) pp. 91–142
161. H.W. Werner, R.P.H. Garten, Trends. Anal. Chem. **4**, 11 (1985)
162. A. Joshi, D.F. Stein, J. Inst. Metals. **99**, 178 (1971)
163. B.D. Powell, H. Mykura, Acta Metall. **21**, 1151 (1973)
164. J.P. Coad, J.C. Riviere, M. Guttmann, P.R. Krahe, Acta Metall. **25**, 151 (1977)
165. A. Šólyom, P. Marko, V. Frič, V. Kavečanský, V. Karel, Acta Phys. Slov. **35**, 266 (1985)
166. P. Ho, D.F. Mitchell, M.J. Graham, Appl. Surf. Sci. **15**, 108 (1983)
167. M.B. Hintz, A.A. Heldt, S.P. Clough, J.F. Moulder, Scripta Metall. **17**, 1415 (1983)
168. K.R. Hallam, R.K. Wild, Surf. Interface Anal. **23**, 133 (1995)
169. G.M. Ingo, E. Angelini, G. Bultrini, T. De Caro, L. Pandolfi, A. Mezzi, Surf. Interface Anal. **34**, 328 (2002)
170. G.M. Ingo, E. Angelini, T. De Caro, G. Bultrini, Appl. Phys. A **79**, 171 (2004)
171. D.G. Armour, Ion scattering techniques, in *Methods of Surface Analysis*, ed. by J.M. Walls (Cambridge University Press, Cambridge, 1989) pp. 263–298
172. H.H. Brongersma, T.M. Buck, Surf Sci. **53**, 649 (1975)
173. E.D. Hondros, M.P. Seah, Int. Metals. Rev. **22**, 867 (1977)
174. J. Bernardini, P. Gas, E.D. Hondros, M.P. Seah, Proc. Roy. Soc. Lond. A **379**, 611 (1982)
175. M. Guttmann, P.R. Krahe, F. Abel, G. Amsel, M. Bruneaux, C. Cohen, Metall. Trans. **5**, 167 (1974)
176. E.V. Barrera, S.M. Heald, H.L. Marcus, Scripta Metall. **21**, 1633 (1987)
177. M. Gülgün, M. Rühle, Key. Eng. Mater. **171–174**, 793 (2000)
178. B. Soenen, A.K. De, S. Vandeputte, B.C. De Cooman, Acta Mater. **52**, 3483 (2004)
179. P. Lejček, O. Schneeweiss, Surf. Sci. **487**, 210 (2001)
180. Y. Sugimoto, P. Pou, M. Abe, P. Jelínek, R. Peréz, S. Morita, Ó. Custance, Nature. **446**, 64 (2007)
181. R. Castaing, J. Descamps, J. Phys. Radium. **16**, 304 (1955)
182. F. Christien, R. Le Gall, Surf. Sci. **602**, 2463 (2008)
183. J.L. Pouchou, F. Pichoir, Anal. Chim. Acta **283**, 81 (1993)
184. P. Doig, P.E.J. Flewitt, Acta Metall. **29**, 1831 (1981)
185. J. Bruley, I. Tanaka, H.–J. Klebe, M. Rühle, Anal. Chim. Acta **297**, 97 (1994)
186. P.E. Batson, N.D. Browning, D.A. Muller, MSA. Bull. **24**, 371 (1994)
187. V.J. Keast, D.B. Williams, Cur. Opin. Sol. State. Mater. Sci. **5**, 23 (2001)
188. J.C.H. Spence, J. Taftø, J. Microsc. **130**, 147 (1983)

189. S.J. Pennycook, N.D. Browning, M.M. McGibbon, A.J. McGibbon, D.E. Jesson, M.F. Chisholm, Philos. Trans. R. Soc. Lond. A **354**, 2619 (1996)
190. V.J. Keast, P.A. Midgley, S.J. Lloyd, P.J. Thomas, M. Weyland, C.B. Boothroyd, C.J. Humphreys, Composition of grain boundaries and interfaces: A comparison of modern analytical techniques using a 300-kV FEGSTEM, in *Electron Microscopy and Analysis*, ed. by C.J. Kiely (Institute of Physics, Bristol, 1999) pp. 35–38
191. A.D. Romig Jr., J.C. Lippold, M.J. Cieslak, Met. Trans. A **19**, 35 (1988)
192. E.L. Hall, D. Imeson, J.B. Vander Sande, Philos. Mag. A **43**, 1569 (1981)
193. V.J. Keast, D.B. Williams, J. Microsc. **199**, 45 (2000)
194. A.J. Papworth, M. Watanabe, D.B. Williams, Ultramicroscopy **88**, 265 (2001)
195. F. Sorbello, G.M. Hughes, P. Lejček, P.J. Heard, P.E.J. Flewitt, Ultramicroscopy **109**, 147 (2009)
196. P. Doig, P.E.J. Flewitt, Metall. Trans. A **13**, 1397 (1982)
197. P. Doig, D. Lonsdale, P.E.J. Flewitt, Metal. Sci. **16**, 335 (1982)
198. D.A. Muller, S. Subramanian, P.E. Batson, J. Silcox, S.L. Sass, Acta, Mater. **44**, 1637 (1996)
199. V.J. Keast, J. Bruley, P. Rez, J.M. MacLaren, D.B. Williams, Acta Mater. **46**, 481 (1998)
200. D.B. Williams, A.D. Romig Jr, Ultramicroscopy **30**, 38 (1989)
201. R.G. Faulkner, T.S. Morgan, E.A. Little, X-Ray Spectrom. **23**, 195 (1994)
202. D. Bouchet, C. Colliex, P. Flora, O. Krivanek, C. Mory, M. Tencé, Microsc. Microanal. Microstruct. **1**, 443 (1990)
203. T. Katoh, T. Ichimori, H. Ichinose, K. Ito, Y. Ishida, HRTEM study of segregation at grain boundary in Al, in *Interface Science and Materials Interconnections*, ed. by Y. Ishida, M. Morita, T. Suga H. Ichinose, O. Ohashi, J. Echigoya (JIM, Tokyo, 1996) pp. 253–256.
204. E.L. Hall, J. Phys. **43**, C6–239 (1982)
205. C.S. Pande, N. Suenaga, B. Vyas, S. Issacs, D.F. Harling, Scripta Metall. **11**, 687 (1977)
206. S.J. Pennycook, M.F. Chisholm, Y. Yan, G. Duscher, S.T. Pantelides, Phys. B **273**, 453 (1999)
207. C.M. Wang, G.S. Cargill III, H.M. Chan, M.P. Harmer, Acta Mater. **48**, 2579 (2000)
208. P. Gruffel, C. Carry, J. Europ. Ceram. Soc. **11**, 189 (1993)
209. S. Lartigue-Korinek, C. Carry, L. Priester, J. Eur. Ceram. Soc. **22**, 1525 (2002)
210. Y.-M. Zhu, Y.L. Corcoran, M. Suenaga, Interface Sci. **1**, 361 (1993)
211. D.E. Luzzi, Ultramicroscopy **37**, 180 (1991)
212. D.E. Luzzi, M. Yan, M. Šob, V. Vitek, Phys. Rev. B **67**, 1894 (1991)
213. M. Yan, M. Šob, D.E. Luzzi, V. Vitek, G.J. Ackland, M. Methfessel, C.O. Rodriguez, Phys. Rev. B **47**, 5571 (1993)
214. S.M. Bruemmer, L.E. Thomas, Surf. Interface Anal. **31**, 571 (2001)
215. P.J. Turner, J.M. Papazian, Metal. Sci. J. **7**, 81 (1973)
216. M.K. Miller, Int. Mater. Rev. **32**, 221 (1987)
217. T.T. Tsong, *Atom-Probe Field-Ion Microscopy* (Cambridge University Press, Cambridge, 1990)
218. D.N. Seidman, B.W. Krakauer, D.K. Chan, MSA. Bull. **24**, 375 (1994)
219. J.D. Rittner, PhD Thesis, Nortwestern University, Evanston (1996)
220. F. Pérez-Willard, D. Wolde-Giorgis, T. Al-Kassab, G.A. López, E.J. Mittermeijer, R. Kirchheim, D. Gerthsen, Micron **39**, 45 (2008)
221. D. Blavette, A. Bostel, J.M. Sarrau, B. Deconihout, A. Menand, Nature **363**, 432 (1993)
222. A. Cerezo, G.D.W. Smith, Field-ion microscopy and atom probe microanalysis, in *Materials Science and Technology: Characterisation of Materials*, ed. by E. Lifshin (VCH, Weinheim, 1994) pp. 513–539
223. Y.S. Zang, J.G. Bae, C.G. Park, Mater. Sci. Eng. A **508**, 148 (2009)
224. D. Blavette, B. Deconihout, A. Bostel, J.M. Sarrau, M. Bouet, A. Menand, Rev. Sci. Instrum. **64**, 2911 (1995)
225. D.J. Larson, A.K. Petford-Long, A. Cerezo, G.D.W. Smith, Acta Mater. **47**, 4019 (1999)
226. N. Masahashi, MRS. Symp. Proc. **364**, 749 (1995)
227. R. Jayaram, M.K. Miller, Surf. Sci. **266**, 310 (1992)
228. H. Nordén, H.O. Andrén, Surf. Interface Anal. **12**, 179 (1988)
229. M.K. Miller, P.A. Beaven, G.D.W. Smith, Surf. Interface Anal. **1**, 149 (1979)

230. R. Jayaram, M.K. Miller, Appl. Surf. Sci. **67**, 311 (1993)
231. B.W. Krakauer, D.N. Seidman, Rev. Sci. Instrum. **63**, 4071 (1992)
232. N. Maruyama, G.D.V. Smith, A. Cerezo, Mater. Sci. Eng. A **353**, 126 (2003)
233. M.K. Miller, I.M. Anderson, K.F. Russell, Appl. Surf. Sci. **94/95**, 288 (1996)
234. M. Thuvander, K. Stiller, Appl. Surf. Sci. **87/88**, 251 (1995)
235. M.K. Miller, S.S. Brenner, M.G. Burke, Metall. Trans. A **18**, 519 (1987)
236. A.R. Waugh, M.J. Southon, Surf. Sci. **68**, 79 (1977)
237. H.C. Eaton, H. Nordén, Scripta Metall. **17**, 1043 (1983)
238. R. Herschitz, D.N. Seidman, Scripta Metall. **16**, 849 (1982)
239. R. Herschitz, D.N. Seidman, Acta Metall. **33**, 1547 (1985)
240. R. Herschitz, D.N. Seidman, Acta Metall. **33**, 1565 (1985)
241. B.W. Krakauer, D.N. Seidman, Metal. Sci. Forum **126–128**, 161 (1993)
242. M.K. Miller, G.D.W. Smith, Appl. Surf. Sci. **87/88**, 243 (1995)
243. N. Maruyama, G.D.W. Smith, A. Cerezo, Mater. Sci. Eng. A **353**, 126 (2003)
244. S.S. Babu, S.A. David, J.M. Vitek, M.K. Miller, Appl. Surf. Sci. **87/88**, 207 (1995)
245. E. Cadel, A. Fraczkiewicz, D. Blavette, Annu. Rev. Mater. Res. **33**, 215 (2003)
246. D. Blavette, E. Cadel, A. Fraczkiewicz, A. Menand, Science **17**, 2317 (1999)
247. M.K. Miller, E.A. Kenik, M.S. Mousa, K.F. Russell, A.J. Bryhan, Scripta Mater. **46**, 299 (2002)
248. D. Wolde-Giorgis, T. Al-Kassab, R. Kirchheim, Surf. Interface Anal. **39**, 246 (2007)
249. M.K. Miller, J. Mater. Sci. **41**, 7808 (2006)
250. P. Pareige, A. Etienne, B. Radiguet, J. Nucl. Mater. **389**, 259 (2009)
251. N.L. Allan, Thermodynamics and materials modelling, in *Chemical Thermodynamics of Materials*, ed. by S. Stølen, T. Grande (Wiley, Chichester, 2005), pp. 337–376.
252. P. Hohenberg, W. Kohn, Phys. Rev. B **136**, 864 (1964)
253. G.S. Painter, F.W. Averill, Phys. Rev. Lett. **58**, 234 (1987)
254. S.P. Chen, A.F. Voter, R.C. Albers, D.J. Srolovitz, Scripta Metall. **20**, 1389 (1986)
255. Y. Zhang, G.H. Lu, T. Wang, S. Deng, M. Kohyama, R. Yamamoto, Mater. Trans. **47**, 2678 (2006)
256. J.M. Raulot, A. Fraczkiewicz, T. Cordonnier, H. Aourag, T. Grosdidier, J. Mater. Sci. **43**, 3867 (2008)
257. S. Tang, A.J. Freeman, G.B. Olson, Phys. Rev. B **50**, 1 (1994)
258. R. Wu, A.J. Freeman, G.B. Olson, Phys. Rev. B **50**, 75 (1994)
259. L. Zhong, R. Wu, A.J. Freeman, G.B. Olson, Phys. Rev. B **62**, 13938 (2000)
260. R. Wu, A.J. Freeman, G.B. Olson, Science **265**, 376 (1994)
261. R. Schweinfest, A.T. Paxton, M.W. Finnis, Nature **432**, 1008 (2004)
262. M. Čák, M. Šob, J. Hafner, Phys. Rev. B **78**, 054418 (2008)
263. R. Car, M. Parrinello, Phys. Rev. Lett. **55**, 2471 (1985)
264. M.S. Daw, M.I. Baskes, Phys. Rev. B **29**, 6443 (1984)
265. M.W. Finnis, J.E. Sinclair, Phil. Mag. A **50**, 45 (1984)
266. F. Ercolessi, E. Tosatti, M. Parrinello, Phys. Rev. Lett. **57**, 719 (1986)
267. D.G. Pettifor, Sol. State. Phys. **40**, 43 (1987)
268. A.B. Walker, R. Taylor, J. Phys. Cond. Matter. **2**, 9481 and 9501 (1990)
269. V. Heine, I.J. Robertson, M.C. Payne, Philos. Trans. R. Soc. Lond. A **334**, 1 (1990)
270. K. Masuda-Jindo, Phys. Stat. Sol. B **134**, 545 (1986)
271. P. Lejček, S. Hofmann, Surf. Interface Anal. **16**, 546 (1990)
272. J.E. Lennard-Jones, Proc. Roy. Soc. Lond. A **106**, 463 (1924)
273. P.M. Morse, Phys. Rev. **34**, 57 (1929)
274. B. Lezzar, O. Khalfallah, A. Larere, V. Paidar, O. Harduin Duparc, Acta Mater. **52**, 2809 (2004)
275. V. Vitek, G.J. Ackland, M. Menyhard, M. Yan, Atomic structure of grain boundaries in ordered and disordered binary alloys, in *Interfaces: Structure and Properties*, ed. by S. Ranganathan, C.S. Pande, B.B. Rath, D.A. Smith (Oxford & IBH Publication, New Delhi, 1993) pp. 3–19
276. G.L. Krasko, Scripta Metall. Mater. **28**, 1543 (1993)

277. O. Harduin Duparc, M. Torent, Interface Sci. **2**, 7 (1994)
278. M.I. Baskes, V. Vitek, Metall. Trans. A **16**, 1625 (1985)
279. P.C. Millett, R.P. Selvam, A. Saxena, Acta Mater. **55**, 2329 (2007)
280. V. Pontikis, J. Phys. **49**, C5–327 (1988)
281. M. Hashimoto, Y. Ishida, R. Yamamoto, M. Doyama, Acta Metall. **32**, 1 (1983)
282. N. Metropolis, M.N. Rosenbluth, A.W. Rosenbluth, A.H. Teller, E. Teller, J. Chem. Phys. **21**, 482 (1972)
283. D. Udler, D.N. Seidman, Phys. Stat. Sol. B **172**, 267 (1992)
284. A. Seki, D.N. Seidman, Y. Oh, S.M. Foiles, Acta Metall. Mater. **39**, 3167 and 3179 (1991)
285. D. Udler, D.N. Seidman, Acta Metall. Mater. **42**, 1959 (1994)
286. D. Udler, D.N. Seidman, Interface Sci. **3**, 41 (1995)
287. D. Udler, D.N. Seidman, J. Mater. Res. **10**, 1993 (1995)
288. S.M. Foiles, M.I. Baskes, M.S. Daw, MRS. Symp. Proc. **122**, 343 (1988)
289. S.M. Foiles, Phys. Rev. B **40**, 11502 (1989)
290. J. Creuze, F. Berthier, R. Tetot, B. Legrand, Phys. Rev. B **62**, 2813 (2000)
291. X. Xie, Y. Mishin, Acta Mater. **50**, 4303 (2002)
292. D. Udler, D.N. Seidman, Phys. Rev. Lett. **77**, 3379 (1996)
293. B. Jiang, X. Liu, L.P. Zheng, D.X. Li, Mater. Lett. **44**, 319 (2000)
294. J.D. Rittner, D. Udler, D.N. Seidman, Y. Oh, Phys. Rev. Lett. **74**, 1115 (1995)
295. R. Najafabadi, H.Y. Wang, D.J. Srolovitz, R. LeSar, Acta Metall. Mater. **39**, 3071 (1991)
296. H.Y. Wang, R. Najafabadi, D.J. Srolovitz, R. LeSar, Acta Metall. Mater. **41**, 2533 (1993)
297. H.Y. Wang, R. Najafabadi, D.J. Srolovitz, R. LeSar, Philos. Mag. A **65**, 625 (1992)
298. H.Y. Wang, R. Najafabadi, D.J. Srolovitz, R. LeSar, Interface Sci. **1**, 31 (1993)
299. S. Hofmann, Scanning Electron Microsc. **III**, 1071 (1985)
300. S. Hofmann, J. Chim. Phys. **84**, 141 (1987)
301. S. Hofmann, Segregation at grain boundaries, in *Surface Segregation Phenomena*, ed. by P.A. Dowben, A. Miller (CRC Press, Boca Raton, 1990) pp. 107–134
302. J. duPlessis, Sol. State Phenom. **11**, 1 (1990)
303. P. Wynblatt, D. Chatain, Metall. Mater. Trans. A **37**, 2595 (2006)
304. G.N. Lewis, M. Randall, *Thermodynamics* (McGraw–Hill, New York, 1923)
305. E.D. Hondros, Proc. Roy. Soc. Lond. A **286**, 479 (1965)
306. E.D. Hondros, M.P. Seah, Scripta Metall. **6**, 1007 (1972)
307. M.P. Seah, E.D. Hondros, Proc. Roy. Soc. Lond. A **335**, 191 (1973)
308. E.D. Hondros, M.P. Seah, Metall. Trans. A **8**, 1363 (1977)
309. J.-G. Hu, D.N. Seidman, Scripta Metall. Mater. **27**, 693 (1992)
310. J. duPlessis, G.N. van Wyk, J. Phys. Chem. Solids **49**, 1451 (1988)
311. J. duPlessis, G.N. van Wyk, J. Phys. Chem. Solids. **49**, 1441 (1988)
312. P. Lejček, Sol. State. Phenom. **138** 339 (2008)
313. M. Guttmann, D. McLean, Grain boundary segregation in multicomponent systems, in *Interfacial Segregation*, ed. by W.C. Johnson, J.M. Blakely (ASM, Metals Park, 1979) pp. 261–348
314. C.J. McMahon Jr., L. Marchut, J. Vac. Sci. Technol. **15**, 450 (1978)
315. K.L. Luthra, C.L. Briant, Metall. Trans. A **19**, 2091 (1988)
316. J. Cabané, F. Cabané, Sol. State. Phenom. **15–16**, 1 (1991)
317. C.L. Briant, Metall. Trans. A **21**, 2339 (1990)
318. A. Zangwill, *Physics at Surfaces* (Cambridge University Press, Cambridge, 1989)
319. M. Polak, L. Rubinovich, Surf. Sci. Rep. **38**, 127 (2000)
320. M. Guttmann, Surf. Sci. **53**, 213 (1975)
321. D.R. Gaskell, Metallurgical thermodynamics, in *Physical Metallurgy*, ed. by R.W. Cahn, P. Haasen, 4th edn. (North-Holland, Amsterdam, 1996) pp. 413–469
322. L. Rubinovich, M. Polak, Eur. J. Phys. B **22**, 267 (2001)
323. P. Lejček, S. Hofmann, Eur. Phys. J. B **31**, 439 (2003)
324. P. Wynblatt, R.C. Ku, Surf. Sci. **65**, 511 (1977)
325. P. Wynblatt, R.C. Ku, Surface segregation in alloys, in *Interfacial Segregation*, ed. by W.C. Johnson, J.M. Blakely (ASM, Metals Park, 1979) pp. 115–136

326. E. Rabkin, Mater. Lett. **25**, 199 (1995)
327. P. Lejček, Prog. Surf. Sci. **35**, 209 (1991)
328. H. Erhart, H.J. Grabke, Metal. Sci. **15**, 401 (1981)
329. J. Janovec, D. Grman, J. Perháčová, P. Lejček, J. Patscheider, P. Ševc, Surf. Interface Anal. **30**, 354 (2000).
330. L.S. Shvindlerman, G. Gottstein, V.A. Ivanov, D.A. Molodov, D. Kolesnikov, W. Lojkovski, J. Mater. Sci. **41**, 7725 (2006)
331. S. Hofmann, P. Lejček, J. Phys. France **51**, C1-C179 (1990)
332. D. Udler, D.N. Seidman, Acta Mater. **46**, 1221 (1998)
333. S. Hofmann, Depth profiling in AES and XPS, in *Practical Surface Analysis*, vol. 1, ed. by D. Briggs, M.P. Seah, 2nd edn. (Wiley, Chichester, 1990) pp. 143–199
334. T. Muschik, S. Hofmann, W. Gust, B. Predel, Appl. Surf. Sci. **37**, 439 (1989)
335. R. Mast, H. Viefhaus, H.J. Grabke, Steel. Res. **70**, 239 (1999)
336. C.L. Briant, Philos. Mag. Lett. **73**, 345 (1996)
337. P. Lejček, S. Hofmann, Interface Sci. **1**, 161 (1993)
338. P. Lejček, J. Alloys. Compd. **378**, 85 (2004)
339. S. Suzuki, M. Obata, K. Abiko, H. Kimura, Scripta Metall. **17**, 1325 (1983)
340. C.M. Liu, K. Abiko, H. Kimura, Effect of silicon on the grain boundary segregation of phosphorus and the phosphorus induced intergranular fracture in high purity Fe–Si–P alloys, in *Strength of Metals and Alloys*, vol. 2, ed. by P.O. Kettunen, T.K. Lepistö, M.E. Lehtonen, (Pergamon, Oxford, 1988) pp. 1101–1106
341. J. Rittner, D.N. Seidman, Acta Mater. **45**, 3191 (1997)
342. P. Lejček, Mater. Sci. Eng. A **185**, 109 (1994)
343. K. Tatsumi, N. Okumura, S. Funaki, Trans. JIM. **27**, 427 (1986) Suppl.
344. M. Paju, H. Viefhaus, H.J. Grabke, Steel. Res. **59**, 336 (1988)
345. P. Lejček, Surf. Interface Anal. **26**, 800 (1997)
346. B.S. Bokstein, A.N. Smirnov, Mater. Lett. **57**, 4501 (2003)
347. P. Lejček, A. Fraczkiewicz, Intermetallics **11**, 1053 (2003)
348. M. Guttmann, Interfacial segregation and temper embrittlement, in *Encyclopedia of Materials: Science and Technology*, ed. by K.H.J. Bunshaw, R.W. Cahn, M.C. Flemings, E.J. Kramer, S. Mahajan (Elsevier, Amsterdam, 2001) pp. 4141–4148
349. C.L. Briant, H.J. Grabke, Mater. Sci. Forum **46**, 253 (1989)
350. P. Lejček, S. Hofmann, Acta Metall. Mater. **39**, 2469 (1991)
351. G. Tauber, H.J. Grabke, Ber. Bunsen. Gesell. Phys. **82**, 298 (1978)
352. R.H. Jones, S.M. Bruemmer, M.T. Thomas, D.R. Baer, Metall. Trans. A **12**, 1621 (1981)
353. Y.–Q. Weng, C.J. McMahon Jr., Mater. Sci. Technol. **3**, 207 (1987)
354. S. Suzuki, S. Tanii, K. Abiko, H. Kimura, Metall. Trans. A **18**, 1109 (1987)
355. H. Hänsel, H.J. Grabke, Scripta Metall. **20**, 1641 (1986)
356. H. Kimura, K. Abiko, S. Suzuki, M. Obata, J. Kumagai, H. Kimura, Trans. JIM. **27**, 53 (1986) Suppl.
357. C.L. Briant, Acta. Metall. **36**, 1805 (1988)
358. Ph. Dumoulin, M. Guttmann, M. Foucault, M. Palmier, M. Wayman, M. Biscondi, Metal. Sci. **14**, 458 (1985)
359. H. Sato, K. Sato, J. Jpn. Inst. Met. **41**, 458 (1977)
360. A.I. Kovalev, V.P. Mishina, G.V. Shcherbedinskiy, Phys. Met. Metall. **62**, 112 (1986)
361. R.H. Jones, D.R. Baer, L.A. Charlot, M.T. Thomas, Metall. Trans. A **19**, 2005 (1988)
362. K. Abiko, S. Suzuki, H. Kimura, Trans. Iron. Steel. Inst. Jpn. **69**, 625 (1983)
363. R.D.K. Misra, Acta Mater. **44**, 4367 (1996)
364. A.I. Kovalev, V.P. Mishina, Metallofizika. **9**, 45 (1987) In Russian
365. R.R. de Avillez, P.R. Rios, Scripta Metall. **17**, 677 (1983)
366. R.D.K. Misra, Surf. Interface Anal. **31**, 509 (2001)
367. M. Guttmann, Grain boundary segregation and embrittlement in multicomponent systems: recent literature revisited, in *Structural Materials: Engineering Application Through Scientific Insight*, ed. by E.D. Hondros, M. McLean (Cambridge University Press, Cambridge, 1996) pp. 59–81

368. J.C. Kim, N.H. Heo, J.G. Na, J.S. Woo, G.M. Kim, Scripta Mater. **38**, 1071 (1998)
369. K.S. Shin, B.H. Tsao, Scripta Metall. **22**, 585 (1988)
370. C.L. Briant, Acta Metall. **35**, 149 (1987)
371. M.P. Seah, C. Lea, Philos. Mag. **31**, 627 (1975)
372. H. Hofmann, S. Hofmann, Scripta Metall. **18**, 77 (1984)
373. C.L. Briant, Metall. Trans. A. **18**, 691 (1987)
374. S.M. Bruemmer, R.H. Jones, M.T. Thomas, D.R. Baer, Metall. Trans. A **14**, 223 (1989)
375. P. Lejček, S. Hofmann, J. Janovec, Mater. Sci. Eng. A **462**, 76 (2006)
376. H.J. Grabke, W. Paulitschke, G. Tauber, H. Viefhaus, Surf. Sci. **63**, 377 (1977)
377. M. Paju, H.J. Grabke, Mater. Sci. Technol. **5**, 148 (1989)
378. M. Paju, H.J. Grabke, Steel. Res. **60**, 41 (1989)
379. P. Ševc, J. Janovec, P. Lejček, P. Záhumenský, J. Blach. Scripta Mater. **46**, 7 (2002)
380. YuP. Denisov, A.S. Drachinskii, YuN. Ivashchenko, A.V. Krainikov, A.V. Pronin, N.I. Timofeev, S.A. Firstov, Metallofizika. **9**, 122 (1987) In Russian
381. C.L. Briant, Acta Metall. **33**, 1241 (1985)
382. C.J. McMahon Jr., Y.Q. Weng, M.J. Morgan, M. Menyhard, Mater. Sci. Eng. **42**, 215 (1980)
383. E.V. Barrera, M. Menyhard, D. Bika, B. Rothman, C.J. McMahon Jr., Scripta Metall. Mater. **27**, 643 (1995)
384. I. Baker, Examination of grain-boundary structure and chemistry in ordered alloys, in *Structure and Property Relationships for Interfaces*, ed. by J.R. Walter, A.H. King, K. Tangri (ASM, Metals Park, 1991) pp. 67–119
385. H. Dosch, *Critical Phenomena at Surfaces and Interfaces* (Springer, Berlin, 1992)
386. H. Dosch, H. Reichert, Acta Mater. **48**, 4387 (2000)
387. D.N. Sieloff, S.S. Brenner, M.G. Burke, MRS. Symp. Proc. **81**, 87 (1987)
388. I. Baker, E.M. Schulson, J.R. Michael, Phil. Mag. B **57**, 379 (1988)
389. A. Choudhury, C.L. White, C.R. Brooks, Acta Metall. Mater. **40**, 57 (1992)
390. S.S. Subramanian, D.A. Muller, J. Silcox, S.L. Sass, Mater. Sci. Eng. A **239–240**, 297 (1997)
391. M.K. Miller, I.M. Anderson, K.F. Russel, Appl. Surf. Sci. **94/95**, 288 (1996)
392. C.T. Liu, E.P. George, W.C. Olivier, Intermetallics **4**, 77 (1996)
393. M. Guttmann, Interfacial segregation in multicomponent systems, in *Atomistics of Fracture*, ed. by R.M. Latanision, J.R. Pickens (Plenum Press, New York, 1983) pp. 465–491
394. Z. Zhang, Q.-Y. Lin, Z.–S. Yu, Mater. Sci. Eng. A **291**, 22 (2000)
395. J.M. Papazian, D.N. Beshers, Metall. Trans. **2**, 491 (1971)
396. M. Guttmann, Surf. Sci. **53**, 168 (1975)
397. H. Kimura, K. Abiko, S. Suzuki, MRS. Symp. Proc. **122**, 379 (1988)
398. R.E.E. Pulkinen, M. Lähdeniemi, J. Mater. Sci. **18**, 3421 (1983)
399. R.H. Jones, S.M. Bruemmer, M.T. Thomas, D.R. Baer, Scripta Metall. **16**, 615 (1982)
400. R. Möller, H.J. Grabke, Scripta Metall. **18**, 527 (1984)
401. J. Bernardini, P. Gas, Scripta Metall. **22**, 1165 (1988)
402. J. Janovec, D. Grman, V. Magula, J. Patscheider, P. Lejček, P. Ševc, J. Blach, J. Surf. Anal. **5**, 294 (1999)
403. P. Dumoulin, M. Guttmann, Mater. Sci. Eng. **42**, 249 (1980)
404. W.T. Nachtrab, Y.T. Chou, J. Mater. Sci. **19**, 2134 (1984)
405. J.P. Hirth, Metall. Trans. A **11**, 861 (1980)
406. L. Heatherly, E.P. George, Acta Mater. **49**, 289 (2001)
407. A.S. Drachinskii, Yu.N. Ivashchenko, A.V. Krajnikov, Poverkhnost'. Fiz. Khim. Mech. 5, pp.145–146 (1986) In Russian
408. M. Paju, H.P. Hougardy, H.J. Grabke, Scand. J. Metall. **18**, 235 (1989)
409. E.P. George, R.L. Kennedy, Trace elements effects on high temperature fracture, in *Impurities in Engineering Materials*, ed. by C.L. Briant (Marcel Dekker Inc., New York, 1999) pp. 225–258
410. L. Karlsson, H. Nordén, Acta Metall. **36**, 13 (1988)
411. M. Paju, R. Möller, Scripta Metall. **18**, 813 (1984)
412. J. Yu, C.J. McMahon Jr., Metall. Trans. A **11**, 291 (1980)
413. A. Biedermann, M. Schmid, B.M. Reichl, P. Varga, Fresenius. J. Anal. Chem. **353**, 259 (1995)

414. M. Militzer, J. Wieting, Scripta Metall. **21**, 1375 (1987)
415. M. Militzer, J. Wieting, Acta Metall. **35**, 2765 (1987)
416. P. Gas, M. Guttmann, J. Bernardini, Acta Metall. **30**, 1309 (1982)
417. H.J. Grabke, C.L. Briant, Scripta Metall. **22**, 1169 (1988)
418. J. Bernardini, P. Gas, Scripta Metall. **22**, 1188 (1988)
419. C.L. Briant, Scripta Metall. **15**, 1013 (1981)
420. H.J. Grabke, K. Hennesen, R. Möller, W. Wei, Scripta Metall. **21**, 1329 (1987)
421. D.Y. Lee, E.V. Barrera, J.P. Stark, H.L. Marcus, Metall. Trans. A **15**, 1415 (1984)
422. C.L. Briant, A.M. Ritter, Acta Metall. **32**, 2031 (1984)
423. C.J. McMahon Jr., Y.Q. Weng, M.J. Morgan, M. Menyhard, Trans. JIM. **27**, 53 (1986) Suppl.
424. C.L. Briant, Grain boundary segregation in ordered and disordered alloys, in *Structure and Property Relationships for Interfaces*, ed. by J.R. Walter, A.H. King, K. Tangri (ASM, Metals Park, 1991) pp. 43–63
425. S.-H. Chen, T. Takasugi, D.P. Pope, Metall. Trans. A **14**, 571 (1983)
426. YuN. Ivashchenko, A.V. Krajnikov, S.A. Firstov, M. Shaper, Metallofizika. **13**, 77 (1991) In Russian
427. T. Sakurai, Y. Kuk, A.K. Birchenall, H.W. Pickering, H.J. Grabke, Scripta Metall. **15**, 535 (1981)
428. J. Janovec, A. Výrostková, J. Perháčová, V. Homolová, H.J. Grabke, P. Ševc, M. Lucas, Steel. Res. **70**, 269 (1999)
429. H.J. Grabke, Surf. Interface Anal. **14**, 686 (1989)
430. F. Pavlyák, G. Hárs, Scripta Metall. **19**, 919 (1985)
431. Y. Iwasaki, K. Fujimoto, J. Physique.IV. Colloq. **42**, C5–C475 (1981)
432. H. Hänninen, E. Minni, Metall. Trans. A **13**, 2281 (1982)
433. H.J. Grabke, R. Möller, H. Erhart, S.S. Brenner, Surf. Interface Anal. **10**, 202 (1987)
434. J. Janovec, M. Bogyó, J. Blach, P. Záhumenský, A. Výrostková, Can. Metall. Quart. **40**, 97 (2001)
435. D.J. Nettleship, R.K. Wild, Surf. Interface Anal. **16**, 552 (1990)
436. E.L. Hall, C.L. Briant, MRS. Symp. Proc. **122**, 237 (1988)
437. D.N. Seidman, B.W. Krakauer, D. Udler, J. Phys. Chem. Sol. **55**, 1035 (1994)
438. R.H. Fowler, E.A. Guggenheim, *Statistical Thermodynamics* (Cambridge University Press, Cambridge, 1939)
439. C. Pichard, M. Guttmann, J. Rieu, C. Goux, J. Phys. **36**, C4–151 (1975)
440. M. Menyhard, B. Rothman, C.J. McMahon Jr., Scripta Metall. Mater. **29**, 1005 (1993)
441. M.P. Seah, E.D. Hondros, Scripta Metall. **7**, 735 (1973)
442. J.R. Rellick, C.J. McMahon Jr., H.L. Marcus, R.W. Palmberg, Metall. Trans. **2**, 1492 (1971)
443. A.S. Gay, A. Fraczkiewicz, M. Biscondi, J. Phys. IV. Colloq. **9**, Pr4–75 (1999)
444. L.Z. Mezey, J. Giber, Physica. Scripta **28**, 97 (1983)
445. L.Z. Mezey, Surf. Sci. **162**, 510 (1985)
446. L.Z. Mezey, J. Giber, Surf. Sci. **162**, 514 (1985)
447. L.Z. Mezey, J. Giber, W. Hofer, P. Varga, Surf. Sci. **234**, 197 (1990)
448. L.Z. Mezey, W. Hofer, Surf. Sci. **269–270**, 1135 (1992)
449. W. Hofer, L.Z. Mezey, Vacuum. **43**, 601 (1992)
450. L.Z. Mezey, W. Hofer, Surf. Interface Anal. **19**, 618 (1992)
451. M. Menyhard, M. Yan, V. Vitek, Acta Metall. Mater. **42**, 2783 (1994)
452. S. Brunauer, P.H. Emmett, E. Teller, J. Am. Chem. Soc. **60**, 309 (1938)
453. H.J. Butt, K. Graf, M. Kappl, *Physics and Chemistry of Interfaces* (Wiley-VCH, Weinheim, 2003)
454. R.H. Jones, Grain boundary segregation and environmentally induced fracture of materials, in *Mechanical Properties and Phase Transformations in Engineering Materials*, ed. by S.D. Antolovich, R.O. Ritchie, W.W. Gerberich (TMS. AIME. Conf. Proc. 1986) pp. 227–249
455. C.L. White, W.A. Coghlan, Metall. Trans. A **8**, 1403 (1977)
456. C.L. White, D.F. Stein, Metall. Trans. A **9**, 13 (1978)
457. T.N. Nowicki, M. Biscondi, J. Phys. France **49**, C5–409 (1988)
458. T.N. Nowicki, J.-C. Joud, M. Biscondi, J. Phys. France **51**, C1–293 (1990)

459. R. Kirchheim, Prog. Mater. Sci. **32**, 261 (1988)
460. R. Kirchheim, Physics and chemistry of segregation at internal interfaces, in *Materials Interfaces: Atomic-Level Structure and Properties*, ed. by D. Wolf, S. Yip (Chapman and Hall, London, 1992) pp. 481–496
461. S. Suzuki, ISIJ. Int. **30**, 1000 (1990)
462. F.L. Williams, D. Nason, Surf. Sci. **45**, 377 (1974)
463. R. Defay, I. Prigogine, A. Bellmans, D.H. Everett, *Surface Tension and Adsorption* (Wiley, New York, 1996)
464. J. Fridel, Adv. Phys. **3**, 446 (1954)
465. A.R. Miedema, Z. Metallkde. **69**, 455 (1978)
466. J.W. Christian, *The Theory of Transformations in Metals and Alloys* (Pergamon Press, Oxford, 1975)
467. J.S. Sundaraman, P. Wynblatt, Surf. Sci. **52**, 569 (1975)
468. M.A. Hoffmann, P. Wynblatt, Metall. Trans. A **20**, 215 (1989)
469. R.C. McCune, P. Wynblatt, J. Am. Ceram. Soc. **66**, 111 (1983)
470. R.H. Ewing: Acta Metall. **19**, 1359 (1971)
471. J.R. Chelikowsky, Surf. Sci. **139**, L197 (1984)
472. V. Kumar, Phys. Rev. B **23**, 3756 (1981)
473. S. Mukherjee, J.L. Morán-López, Surf. Sci. **188**, L742 (1987)
474. S. Mukherjee, J.L. Morán-López, Prog. Surf. Sci. **25**, 139 (1987)
475. G. Bozzolo, J. Ferrante, J.R. Smith, Phys. Rev. B **45**, 493 (1992)
476. A.M. Rodriguez, G. Bozzolo, J. Ferrante, Surf. Sci. **307–309**, 625 (1994)
477. G. Bozzolo, B. Good, J. Ferrante, Surf. Sci. **289**, 169 (1993)
478. G. Bozzolo, J. Ferrante, Mater. Sci. Forum **126–128**, 225 (1993)
479. G. Tréglia, B. Legrand, F. Ducastelle, A. Saúl, C. Gallis, I. Meunier, C. Mottet, A. Senhaji, Comput. Mater. Sci. **15**, 196 (1999)
480. G. Tréglia, B. Legrand, F. Ducastelle, Europhys. Lett. **7**, 575 (1988)
481. F. Berthier, B. Legrand, G. Tréglia, Acta Mater. **47**, 2705 (1999)
482. F. Berthier, J. Creuze, R. Tétot, B. Legrand, Phys. Rev. B **65**, 195413 (2002)
483. J. Creuze, F. Berthier, R. Tétot, B. Legrand, Phys. Rev. B **62**, 2813 (2000)
484. V. Rosato, M. Guillopé, B. Legrand, Philos. Mag. A **59**, 321 (1989)
485. R. Haydock, V. Heine, M.J. Kelly, J. Phys. C. **5**, 2845 (1972)
486. J.R. Lee, Y.M. Chiang, Mater. Sci. Forum **207**, 129 (1996)
487. W. Lojkowski, E. Rabkin, B. Straumal, L.S. Shvindlerman, W. Gust, Def. Dif. Forum **156**, 163–174 (1998)
488. Y. Ishida, S. Yokoyama, T. Nishizawa, Acta Metall. **33**, 225 (1985)
489. S. Tsurekawa, K. Kawahara, K. Okamoto, T. Watanabe, R. Faulkner, Mater. Sci. Eng. A **387–389**, 442 (2003)
490. S. Tsurekawa, T. Watanabe, X. Zhao, L. Zuo, Scripta Mater. **54**, 969 (2006)
491. S. Tsurekawa, K. Okamoto, K. Kawahara, T. Watanabe, J. Mater. Sci. **40**, 895 (2005)
492. K.E. Sklarz, M.L. Wayman, Acta Metall. **29**, 341 (1981)
493. M.R. Fitzsimmons, A. Röll, E. Burkel, K.E. Sikafus, M.A. Nastasi, G.S. Smith, R. Pynn, J. Appl. Phys. **76**, 6295 (1994)
494. M.R. Fitzsimmons, A. Röll, E. Burkel, K.E. Sikafus, M.A. Nastasi, G.S. Smith, R. Pynn, Nanostr. Mater. **6**, 539 (1995)
495. M. Šob, I. Turek, V. Vitek, Mater. Sci. Forum **294–296**, 17 (1999)
496. B.D. Powell, D.P. Woodruff, Philos. Mag. A **34**, 169 (1976)
497. M. Menyhard, C.J. McMahon Jr., Acta Metall. **37**, 2287 (1989)
498. K. Tatsumi, N. Okumura, M. Yamamoto, J. Phys. France. **49**, C5–699 (1988)
499. S. Suzuki, M. Obata, K. Abiko, H. Kimura, Scripta Metall. **17**, 1325 (1983)
500. J. Kameda, C.J. McMahon Jr., Metall. Trans. A **12**, 31 (1981)
501. A. Joshi, Scripta Metall. **9**, 251 (1975)
502. C.T. Liu, C.L. White, J.A. Horton, Acta Metall. **33**, 213 (1985)
503. C.L. Briant, R.P. Messmer, Acta Metall. **32**, 2043 (1984)

504. W. Losch, Intergranular fracture, in *Chemistry and Physics of Fracture*, ed. by R.M. Latanision, R.H. Jones (Martinus Nijhoff, Dordrecht, 1987) pp. 461–490
505. M. Polak, Interfacial segregation in steel and steel-related alloys, in *Surface Segregation Phenomena*, ed. by P.A. Dowben, A. Miller (CRC Press, Boca Raton, 1990) pp. 291–325
506. P. Lejček, V. Paidar, J. Adámek, S. Hofmann, Segregation behaviour of two (001)/(011) asymmetrical tilt grain boundaries in iron, in *Interface Science and Materials Interconnections*, ed. by Y. Ishida, M. Morita, T. Suga H. Ichinose, O. Ohashi, J. Echigoya (JIM, Tokyo, 1996) pp. 245–248
507. P. Lejček, S. Hofmann, Surf. Sci. **307–309**, 793 (1994)
508. C. Li, D.B. Williams, Interface Sci. **11**, 461 (2003)
509. H. Gleiter, Acta Metall. **18**, 117 (1970)
510. P. Lejček, V. Paidar, M. Koutník, Scripta Metall. **22**, 1379 (1988)
511. C.R.M. Grovenor, C.M.F. Rae, Scripta Metall. **15**, 1305 (1981)
512. W. Swiatnicki, S. Lartigue-Korinek, A. Dubon, J.Y. Laval, Mater. Sci. Forum **126–128**, 193 (1993)
513. M. Pierantoni, B. Aufray, F. Cabané, J. Phys. France **46**, C4–517 (1985)
514. E. Smiti, P. Joufrey, A. Kobylanski, Scripta Metall. **18**, 673 (1984)
515. A. Roshko, W.D. Kingery, J. Am. Ceram. Soc. **68**, C–331 (1988)
516. M.H. Berger, J.Y. Laval, J. Phys. France **51**, C1–965 (1990)
517. J.Y. Laval, C. Delamarre, M.H. Berger, C. Cabanel, J. Phys. France **51**, C1–331 (1990)
518. L.T. Romano, P.R. Wilshaw, N.J. Long, C.R.M. Grovenor, Supercond. Sci. Technol. **1**, 285 (1989)
519. D.H. Shin, J. Silcox, S.E. Russel, D.K. Lathrop, B. Moeckly, R.A. Buhrman, Appl. Phys. Lett. **57**, 508 (1990)
520. Y. Zhu, Z.L. Wang, M. Suenaga, Philos. Mag. A **67**, 11 (1993)
521. Y. Yoshitomi, S. Suzuki, T. Ueda, S. Tsurekawa, H. Nakashima, H. Yoshinaga, Scripta Metall. Mater. **3**, 1067 (1995)
522. R. Herschitz, D.N. Seidman, Scripta Metall. **16**, 849 (1982)
523. R. Herschitz, D.N. Seidman, A. Brokman, J. Phys. France **46**, C4–451 (1985)
524. T. Muschik, W. Gust, S. Hofmann, B. Predel, Acta Metall. **37**, 2971 (1989)
525. W.R. Thomas, B. Chalmers, Acta Metall. **3**, 17 (1955)
526. T. Watanabe, S. Kitamura, S. Karashima, Acta Metall. **28**, 455 (1980)
527. T. Watanabe, T. Murakami, S. Karashima, Scripta Metall. **12**, 361 (1978)
528. M. Biscondi, J. Phys. France. **43**, C6–293 (1982)
529. T. Ogura, T. Watanabe, S. Karashima, T. Masumoto, Acta Metall. **35**, 1807 (1987)
530. El M'Rabat, L. Priester, Mater. Sci. Eng. A **101**, 117 (1988)
531. A. Fraczkiewicz, M. Biscondi, J. Phys. France **46**, C4–497 (1985)
532. R. Monzen, Mater. Sci. Eng. A **387**, 424 (2004)
533. J. Stolarz, J. LeCoze, J. Phys. France **51**, C1–641 (1990)
534. D. Bouchet, L. Priseter, Scripta Metall. **20**, 961 (1986)
535. D. Bouchet, L. Priseter, Scripta Metall. **21**, 475 (1987)
536. W. Swiatnicki, S. Lartigue, M. Biscondi, D. Bouchet, J. Phys. France **51**, C1–341 (1990)
537. M.C. Juhas, L. Priester, M. Biscondi, Mater. Sci. Eng. A **185**, 71 (1994)
538. M.C. Juhas, L. Priester, MRS. Symp. Proc. **229**, 319 (1991)
539. S. Suzuki, K. Abiko, H. Kimura, Scripta Metall. **15**, 1139 (1981)
540. J. Bernardini, F. Cabané, J. Cabané. Surf. Sci. **162**, 519 (1985)
541. Y. Pang, P. Wynblatt, J. Am. Ceram. Soc. **88**, 2286 (2005)
542. S. Hofmann, P. Lejček, J. Adámek, Surf. Interface Anal. **19**, 601 (1992)
543. D.N. Seidman, Mater. Sci. Eng. A **137**, 57 (1991)
544. J.-G. Hu, D.N. Seidman, Phys. Rev. Lett. **65**, 1615 (1990)
545. A.P. Sutton, V. Vitek, Acta Metall. **30**, 2011 (1982)
546. V. Vitek, G.J. Wang, J. Phys. France **43**, C6–147 (1982)
547. V. Vitek, G.J. Wang, Surf. Sci. **144**, 110 (1984)
548. G.J. Wang, V. Vitek, Acta Metall. Sin. A **1**, 7 (1988)

549. M. Hashimoto, Y. Ishida, S. Wakayama, R. Yamamoto, M. Doyama, T. Fujiwara, Acta Metall. **32**, 13 (1983)
550. S. Wakayama, M. Hashimoto, Y. Ishida, R. Yamamoto, M. Doyama: Acta Metall. **32**, 21 (1983)
551. S. Suzuki, K. Abiko, H. Kimura, Trans. ISIJ. **23**, 746 (1983)
552. M. Guttmann, Ph. Dumoulin, M. Wayman, Metall. Trans. A. **13**, 1693 (1982)
553. S.M. Foiles, Phys. Rev. B **32**, 7685 (1985)
554. M.I. Baskes, S.M. Foiles, M.S. Daw, Mater. Sci. Forum **46**, 187 (1989)
555. H.Y. Wang, R. Najafabadi, D.J. Srolovitz, R. LeSar, Metall. Trans. A **23**, 3105 (1992)
556. J.O. Vasseur, P.A. Deymier, B. Djafari-Rouhani, L. Dobrzynski, Interface Sci. **1**, 49 (1993)
557. J.O. Vasseur, P.A. Deymier, Interface Sci. **1**, 125 (1993)
558. R. Hultgren, P. Desai, D. Hawkins, M. Gleiser, K. Kelley (eds.), *Selected Values of the Thermodynamic Properties of Binary Alloys* (ASM, Metals Park, 1973)
559. S. Hofmann, P. Lejček, Scripta Metall. Mater. **25**, 2259 (1991)
560. O. Kubaschewski, *Iron – Binary Phase Diagrams* (Springer, Berlin, 1982); T.B. Massalski, *Binary Alloy Phase Diagrams* (ASM, Metals Park, 1987); B. Predel, *Phase Equilibria of Binary Alloys* (Springer, Berlin, 2003)
561. P. Lejček, Z. Metallkde **96**, 1129 (2005)
562. Ye.L. Maksimova, B.B. Straumal, V.Ye. Fradkov, L.S. Shvindlerman, Phys. Met. Metall **56**, 133 (1983)
563. L. Liu, Q.–X. Guo, Chem. Rev. **101**, 673 (2001)
564. J. Šesták, *Heat, Thermal Analysis and Society* (Nucleus, Hradec Králové, 2004)
565. P. Lejček, S. Hofmann, Interface Sci. **9**, 221 (2001)
566. P. Lejček, S. Hofmann, Rev. Adv. Mater. Sci. **21**, 27 (2009)
567. D.A. Steigerwald, S.J. Miller, P. Wynblatt, Surf. Sci. **155**, 79 (1985)
568. K. Ishida, J. Alloy Compd. **235**, 244 (1996)
569. R.W. Siegel, Nanophase materials: structure – property correlations, in *Materials Interfaces: Atomic-Level Structure and Properties*, ed. by D. Wolf, S. Yip (Chapman & Hall, London, 1992) pp. 431–460
570. C.D. Terwilliger, Y.–M. Chiang, Acta Metall. Mater. **43**, 319 (1995)
571. F.F.D. Chabert, F. Tancret, F. Christien, R. Le Gall, J.–F. Castagné, J. Mater. Sci. **42**, 9765 (2007)
572. F.F.D. Chambet, F. Tancret, F. Christien, R. Le Gall, J.-F. Castagné, J. Mater. Sci. **42**, 9765 (2007)
573. S.V. Divinski, F. Hisker, Y.–S. Kang, J.–S. Lee, Chr. Herzig, Acta Mater. **52**, 631 (2004)
574. S.F. Liu, B. Li, X.H. Wang, W. Su, H. Han, J. Mater. Proc. Technol. **209**, 3999 (2009)
575. J. Weissmüller, J. Mater. Res. **9**, 4 (1994)
576. R. Kirchheim, Acta Mater. **50**, 413 (2002)
577. F. Liu, R. Kirchheim, Scripta Mater. **51**, 521 (2004)
578. B Färber, E. Cadel, A. Menand, G. Schmitz, R. Kirchheim, Acta Mater. **48**, 789 (2000)
579. F. Liu, R. Kirchheim, J. Cryst. Growth. **264**, 385 (2004)
580. S.–G. Kim, Y.–B. Park, Acta Mat. **56**, 3739 (2008)
581. C.E. Krill, H. Erhardt, R. Birringer, Z. Metallkde. **96**, 1134 (2005)
582. K.A. Darling, R.N. Chan, P.Z. Wong, J.E. Semones, R.O. Scattergood, C.C. Koch, Scripta Mater. **59**, 530 (2008)
583. E. Botcharova, J. Freudenberg, L. Schulz, Acta Mater. **54**, 3333 (2006)
584. H. Gleiter, Prog. Mater. Sci. **33**, 223 (1989)
585. L.S. Shvindlerman, G. Gottstein, Scripta Mater. **54**, 1041 (2006)
586. J.R. Trelewicz, C.A. Schuh, Phys. Rev. B **79**, 094112 (2009)
587. P. Lejček, S. Hofmann, Surf. Interface Anal. **36**, 938 (2004)
588. S. Suzuki, M. Obata, K. Abiko, H. Kimura, Scripta Metall. **17**, 427 (1983)
589. H.J. Grabke, Steel Res. **57**, 401 (1986)
590. P. Ševc, J. Janovec, J. Katana, Scripta Metall. Mater. **31**, 1673 (1994)
591. B.W. Krakauer, D.N. Seidman, Acta Mater. **46**, 6145 (1998)

592. H.J. Grabke, Segregation at interfaces, in *Chemistry and Physics of Fracture*, ed. by R.M. Latanision, R.H. Jones (Nijhoff, Dordrecht, 1987) pp. 388–415
593. M.P. Seah, C. Lea, Philos. Mag. **31**, 1241 (1975)
594. U. Otterbein, S. Hofmann, Surf. Interface Anal. **24**, 263 (1996)
595. P. Lejček, R. Konečná, J. Janovec, Surf. Interface Anal. **40**, 503 (2008)
596. P. Wynblatt, Z. Shi, J. Mater. Sci. **40**, 2765 (2005)
597. P. Wynblatt, D. Chatain, Y. Pang, J. Mater. Sci. **41**, 7760 (2006)
598. G. Rowlands, D.P. Woodruff, Philos. Mag. A **40**, 459 (1979)
599. M.P. Seah, Acta Metall. **25**, 345 (1977)
600. M. Militzer, J. Wieting, Acta Metall. **34**, 1229 (1986)
601. M. Militzer, J. Wieting, Surf. Sci. **192**, L905 (1987)
602. M. Militzer, J. Wieting, Surf. Sci. **200**, 342 (1988)
603. M. Militzer, Y.N. Ivashchenko, A.V. Krajnikov, P. Lejček, J. Wieting, S.A. Firstov, Surf. Sci. **261**, 267 (1992)
604. S. Hofmann, J. Erlewein, Surf. Sci. **77**, 591 (1978)
605. M.F. Boudjemaa, A. Mosser, J. Less Common Metals **145**, 55 (1988)
606. K.J. Rowlings, S.D. Folias, B.J. Hopkins, Surf. Sci. **109**, 513 (1981)
607. K.T. Aust, S.J. Armijo, E.F. Koch, J.A. Westbrook, Trans. ASM. **60**, 360 (1967)
608. T.R. Anthony, Acta Metall. **17**, 603 (1969)
609. S.J. Bercovichi, C.E.L. Hunt, P. Niessen, J. Mater. Sci. **5**, 326 (1970)
610. P. Doig, P.E.J. Flewitt, Acta Metall. **29**, 1831 (1981)
611. R.G. Faulkner, Acta Metall. **35**, 2905 (1987)
612. T. Xu, B. Cheng, Prog. Mater. Sci. **49**, 109 (2004)
613. T. Xu, J. Mater. Sci. **7**, 241 (1988)
614. X.L. He, Y.Y. Chu, J.J. Jonas, Acta Metall. **37**, 2905 (1989)
615. T. Xu, K. Wang, S. Song, Sci. China E-Tech. Sci. **52**, 893 (2009)
616. C.L. Briant, Metall. Trans. A **16**, 2061 (1985)
617. P. Maier, R.G. Faulkner, Mater. Charact. **51**, 49 (2003)
618. L. Karlsson, Acta Metall. **36**, 25 (1988)
619. R.G. Faulkner, S. Song, P.E.J. Flewitt, J. Nucl. Mater. **212–215**, 608 (1994)
620. T. Shimoda, T. Nakamura, Acta Metall. **29**, 1631 and 1637 (1981)
621. T.R. Allen, J.T. Busby, G.S. Was, E.A. Kenik, J. Nucl. Mater. **255**, 44 (1998)
622. T.S. Duh, J.J. Kai, F.R. Chen, J. Nucl. Mater. **283–287**, 198 (2000)
623. E.P. Simonen, L.A. Chariot, S.M. Bruemmer, J. Nucl. Mater. **225**, 117 (1995)
624. L. Fournier, B.H. Spencer, G.S. Was, E.P. Simonen, S.M. Bruemmer, J. Nucl. Mater. **321**, 192 (2003)
625. S. Watanabe, N. Sakaguchi, N. Hashimoto, M. Nakamura, H. Takahashi, C. Namba, N.Q. Lain, J. Nucl. Mater. **232**, 113 (1996)
626. T. Ezawa, E. Wakai, R. Oshima, J. Nucl. Mater. **283–287**, 244 (2000)
627. N. Sakaguchi, S. Watanabe, H. Takahashi, J. Mater. Sci. **40**, 889 (2005)
628. N. Sakaguchi, S. Watanabe, H. Takahashi, Nucl. Instrum. Methods. Phys. Res. B **153**, 142 (1999)
629. A.Y. Lozovoi, A.T. Paxton, Phys. Rev. B **77**, 165413 (2008)
630. M.J. Hackett, J.T. Busby, G.S. Was, Metall. Mater. Trans. A **39**, 218 (2008)
631. M.J. Hackett, R. Najafabadi, G.S. Was, J. Nucl. Mater. **389**, 279 (2009)
632. G.S. Was, J.T. Busby, Philos. Mag. **85**, 443 (2005)
633. S.-H. Song, H. Zhuang, J. Wu, L.-Q. Weng, Z.-X. Yuan, T.-H. Xi, Mater. Sci. Eng. A **486**, 433 (2008)
634. R.G. Faulkner, J. Mater. Sci. **16**, 373 (1981)
635. M. Militzer, W.P. Sun, J.J. Jonas, Acta Metall. Mater. **42**, 133 (1994)
636. J.W. Rutter, K.T. Aust, Trans. AIME. **218**, 682 (1960)
637. P.-R. Cha, S.G. Kim, D.-H. Yeon, J.-K. Yoon, Acta Mater. **50**, 3817 (2002)
638. J. Kameda, Acta Metall. **34**, 1721 (1986)
639. J. Li, J. Wang, G. Yang, Acta Mater. **57**, 2108 (2009)
640. N. Ma, S.A. Dregia, Y. Wang, Acta Mater. **51**, 3687 (2003)

641. S.H. Zhang, X.L. He, T. Ko, J Mater. Sci. **29**, 2655 (1994)
642. S.H. Zhang, X.L. He, T. Ko, J Mater. Sci. **29**, 2663 (1994)
643. M. Lane, Annu. Rev. Mater. Res. **33**, 29 (2003)
644. C.J. McMahon Jr., V. Vitek, Acta Metall. **27**, 507 (1979)
645. M.L. Jokl, V. Vitek, C.J. McMahon Jr., Acta Metall. **28**, 1479 (1980)
646. E.D. Hondros, D. McLean, Grain boundary fragility, in *Grain-Boundary Structure and Properties*, ed. by G.A. Chadwick, D.A. Smith (Academic, London, 1976) pp. 353–381
647. M.P. Seah, Proc. Roy. Soc.Lond. A **349**, 535 (1976)
648. J.P. Hirth, J.R. Rice, Metall. Trans. A **11**, 1501 (1980)
649. J.R. Rice, J.–S. Wang, Mater. Sci. Eng. A **107**, 23 (1989)
650. U. Otterbein, S. Hofmann, M. Rühle, The temperature dependece of grain boundary segregation in B doped Ni3Al bicrystals, in *High-Temperature Ordered Intermetallic Alloys V*, vol. 288, ed. by I. Baker, R. Darolia, J.D. Whittenberger, M. H. Yoo (MRS Proc. Symp. MRS Pittsburgh 1993) pp. 183–188
651. P. Lejček, S. Hofmann, A. Krajnikov, Mater. Sci. Eng. A **234–236**, 283 (1997)
652. M.P. Seah, Acta Metall. **28**, 955 (1980)
653. E. Smiti, P. Joufrey, A. Kobylanski, C. R. Acad. Sci. Paris C **298**, 555 (1984)
654. E. Smiti, P. Joufrey, A. Kobylanski, Scripta Metall. **18**, 673 (1984)
655. S.P. Lynch, B.C. Muddle, T. Passang, Philos. Mag. A **82**, 3361 (2002)
656. C.L. White, J. Vac. Sci. Technol. A **4**, 1633 (1986)
657. S. Lozano-Perez, M. Schröder, T. Yamada, T. Terachi, C.A. English, C.R.M. Grovenor, Appl. Surf. Sci. **255**, 1541 (2008)
658. M.P. Seah, E.D. Hondros, Atomistic mechanisms of intergranular embrittlement, in *Atomistics of Fracture*, ed. by R.M. Latanision (Plenum Press, New York, 1983) pp. 855–887.
659. H. Kurishita, H. Yoshinaga, Mater. Forum **13**, 161 (1989)
660. C.L. Briant, R.P. Messmer, Acta Metall. **30**, 1181 (1982)
661. R.P. Messmer, C.L. Briant, Acta Metall. **30**, 457 (1982)
662. G.L. Krasko, G.B. Olson, Solid. State. Commun. **76**, 247 (1990)
663. M.P. Seah, J. Vac. Sci. Technol. **17**, 16 (1980)
664. R. Wu, A.J. Freeman, G.B. Olson, J. Mater. Res. **7**, 2403 (1992)
665. P. Rez, J.R. Alvare, Acta Mater. **47**, 2069 (1999)
666. Y. Ishida, M. Mori, J. Phys. France **46**, C4–465 (1985)
667. Y.–Q. Fen, C.–Y. Wang, Comput. Mater. Sci. **20**, 48 (2001)
668. Z.-Z. Chen, C.Y. Wang, J. Phys. Condens. Matter. **17**, 6645 (2005)
669. M.E. Eberhart, K.H. Johnson, R.M. Latanision, Acta Metall. **32**, 955 (1984)
670. M. Yamaguchi, M. Shiga, H. Kaburaki, Science **307**, 393 (2005)
671. W.T. Geng, J.–S. Wang, G.B. Olson, Science **309**, 1677c (2005)
672. M. Yamaguchi, M. Shiga, H. Kaburaki, Science **309**, 1677d (2005)
673. M. Yamaguchi, M. Shiga, H. Kaburaki, J. Phys. Condens. Matter **16**, 3933 (2004)
674. X.J. Wu, Q.Q. Zheng, B. Hu, Z. Zeng, F.X. Zhou, Z.Y. Chen, Effect of bismuth and silver on chemical bonding of grain boundaries for fcc metal copper, in *Strength of Metals and Alloys*, ed. by P.O. Kettunen, T.K. Lepistö, M.E. Lehtonen (Pergamon, Oxford, 1988) pp. 1369–1373
675. R. Haydock, J. Phys. C **14**, 3807 (1981)
676. G. Duscher, M. Chisholm, U. Alber, M. Rühle, Nat. Mater. **3**, 621 (2004)
677. A.Y. Lozovoi, A.T. Paxton, M.W. Finnis, Phys. Rev. B **74**, 155416 (2006)
678. U. Alber, H. Müllejans, M. Rühle, Acta Mater. **47**, 4047 (1999)
679. D.J. Srolovitz, W.H. Yang, R. Najafabadi, H.Y. Wang, R. LeSar, Microstructural and segregation effects in the fracture of polycrystals, in *Materials Interfaces: Atomic-Level Structure and Properties*, ed. by D. Wolf, S. Yip (Chapman & Hall, London, 1992) pp. 691–702
680. A. Kobylanski, C. Goux, C. R. Acad. Sci. Paris C **272**, 1937 (1971)
681. N. Gokon, S. Ohyama, M. Kajihara, Mater. Sci. Eng. A **488**, 252 (2008)
682. M.E. Eberhart, D.D. Vvedensky, Mater. Sci. Forum **46**, 169 (1989)
683. M. Aucouturier, J. Phys. France **43**, C6–175 (1982)
684. R.M. Latanision, H. Opperhauser, Metall. Trans. **5**, 483 (1974)
685. S.P. Lynch, Mater. Sci. Forum **46**, 1 (1989)

686. C.J. McMahon Jr., Mater. Charact. **26**, 269 (1991)
687. J. Kameda, C.J. McMahon Jr., Metall. Trans. A **14**, 903 (1983)
688. W.J. Mills, M.R. Lebo, J.J. Kearns, Metall. Mater. Trans. A **30**, 1579 (1999)
689. R.G. Faulkner, R.B. Jones, Z. Lu, S. Song, P.E.J. Flewitt, Philos. Mag. **85**, 2065 (2005)
690. C.J. McMahon Jr., Interface Sci. **12**, 141 (2004)
691. R.G. Song, M.K. Tseng, B.J. Zhang, Z.H. Jin, K.S. Shin, Acta Mater. **44**, 3241 (1996)
692. X.–Y. Liu, J. Kameda, J.W. Anderegg, S. Takaki, K. Abiko, C.J. McMahon Jr., Mater. Sci. Eng. A **492**, 218 (2008)
693. D. Cameron, Mater. Charact. **33**, 37 (1994)
694. J.A. Kargol, D.L. Albright, Metall. Trans. A **8**, 27 (1977)
695. T. Watanabe, S. Shima, S. Karashima, Misorientation dependence of liquid metal-induced intergranular fracture of zinc bicrystals, in *Embrittlement by Liquid and Solid Metals*, ed. by M.H. Kamdar (THS–AIME, New York, 1982) pp. 161–172
696. T. Watanabe, M. Tanaka, S. Karashima, Intergranular fracture caused by liquid gallium in polycrystalline beta brass with bcc structure, in *Embrittlement by Liquid and Solid Metals*, ed. by M.H. Kamdar (THS–AIME, New York, 1982) pp. 183–196
697. S. Dinda, W.S. Warke, Mater. Sci. Eng. **24** 199 (1976)
698. A.W. Funkenbusch, L.A. Heldt, D.F. Stein, Metall. Trans. A **13**, 611 (1982)
699. X.–Y. Liu, D. Tham, D. Yates, C.J. McMahon Jr., Mater. Sci. Eng. A **458**, 123 (2007)
700. B. Straumal, W. Gust, D. Molodov, Interface Sci. **3**, 127 (1995)
701. B.B. Straumal, W. Gust, T. Watanabe, Mater. Sci. Forum **294–296**, 411 (1999)
702. B. Straumal, T. Muschik, W. Gust, B. Predel, Acta Metall. **40**, 939 (1992)
703. B.B. Straumal, O.I. Noskovich, V.N. Semenov, L.S. Shvindlerman, W. Gust, B. Predel: Acta Metall. Mater. **40**, 795 (1992)
704. C.A. Hipsley, Acta Metall. **35**, 2399 (1987)
705. D. Bika, C.J. McMahon Jr., Acta Metall. Mater. **43**, 1909 (1995)
706. U. Krupp, Int. Mater. Rev. **50**, 83 (2005)
707. Y. Mishin, P. Sofronis, J.L. Bassani, Acta Mater. **50**, 3609 (2002)
708. U. Krupp, C.J. McMahon Jr., J. Alloy. Comp. **378**, 79 (2004)
709. Y. Mishin, C.J. McMahon Jr., J.L. Bassani, P. Sofronis, MRS. Symp. Proc. **586**, 27 (2000)
710. R.C. Muthiah, J.A. Pfandtner, C.J. McMahon Jr., P. Lejček, V. Paidar, MRS. Symp. Proc. **458**, 289 (1997)
711. J.A. Pfandtner, C.J. McMahon Jr., Acta Mater. **49**, 3369 (2001)
712. D. Bika, C.J. McMahon Jr., Acta Metall. Mater. **43**, 1895 (1995)
713. C.T. Liu, C.L. White, Acta Metall. **35**, 643 (1987)
714. R.C. Muthiah, C.J. McMahon Jr., A. Guha, Mater. Sci. Forum **207–209**, 585 (1996)
715. W.M. Kane, C.J. McMahon Jr., Mater. Sci. Eng. A **507**, 61 (2009)
716. O.P. Arora, M. Metzger, Trans. Met. Soc. AIME **236**, 1205 (1966)
717. L. Beaunier, M. Froment, C. Vignaud, J. Electroanal. Chem. **119**, 125 (1972)
718. X.R. Quian, Y.T. Chou, Philos. Mag. **45**, 1075 (1982)
719. M. Yamashita, T. Mimaki, S. Hashimoto, S. Miura, Philos. Mag. A **63**, 695 (1991)
720. T. Ogura, A. Makino, T. Masumoto, Metall. Trans. A **15**, 1563 (1984)
721. T. Ogura, C.J. McMahon Jr., H.C. Feng, V. Vitek, Acta Metall. **26**, 1317 (1978)
722. S. Visitserngtrakul, S. Hashimoto, S. Miura, M. Okubo, Intergranular corrosion and grain boundary structure in stabilized 310 stainless steel, in *Proceedings of EVALMAT 89* (ISIJ, Tokyo, 1989) pp. 365–372
723. P. Lejček, V. Paidar, Scripta Metall. **30**, 283 (1994)
724. P. Lejček, V. Gärtnerová, A. Jäger, J. Vaníčková, J. Děd, J. Haloda, Mater. Sci. Forum **638–642**, 2852 (2010)
725. M. Yamashita, T. Mimaki, S. Hashimoto, S. Miura, Philos. Mag. A **63**, 707 (1991)
726. S.M. Bruemmer, MRS. Symp. Proc. **819**, N2.2.1 (2004)
727. T. Mimaki, Y. Nakazawa, S. Hashimoto, S. Miura, Metall. Trans. A **21**, 2355 (1990)
728. M. Yamashita, M. Yoshioka, T. Mimaki, S. Hashimoto, S. Miura, Acta Metall. Mater. **38**, 1619 (1990)
729. S.M. Bruemmer, Mater. Sci. Forum **294–296**, 7 (1999)

730. A. King, G. Johnson, D. Engelberg, W. Ludwig, J. Marrow, Science **321**, 382 (2008)
731. H. Sautter, H. Gleiter, G Bäro, Acta Metall. **25**, 467 (1977)
732. A. Roy, U. Erb, H. Gleiter, Acta Metall. **30**, 1847 (1982)
733. M.A. Arafin, J.A. Szpunar, Corros. Sci. **51**, 119 (2009)
734. R. Ishibashi, T. Horiuchi, J. Kuniya, M. Yamamoto, S. Tsurekawa, H. Kokawa, T. Watanabe, T. Shoji, Mater. Sci. Forum **475**, 3863 (2005)
735. P. Muraleedharan, F. Schneider, K. Mummert, J. Nucl. Mater. **270**, 342 (1999)
736. N.H. Heo, Y.C. Jung, J.K. Lee, K.T. Kim, Scripta Mater. **59**, 1200 (2008)
737. Y.–I. Takeda, H. Yamauchi, Q. Peng, T. Shoji, Key. Eng. Mater. **297–300**, 986 (2005)
738. G.S. Was, P. Ampornrat, G. Gupta, S. Teysseyre, E.A. West, T.R. Allen, K. Sridharan, L. Tan, Y. Chen, X. Ren, C. Pister, J. Nucl. Mater. **31**, 176 (2007)
739. G.S. Was, S.M. Bruemmer, J. Nucl. Mater. **216**, 326 (1994)
740. Y. Mishin, Chr. Herzig, Mater. Sci. Eng. A **260**, 55 (1999)
741. J.C. Fisher, J. Appl. Phys. **22**, 74 (1951)
742. J. Bernardini, Def. Diff. Forum **66–69**, 667 (1989)
743. J. Bernardini, P. Gas: Def. Diff. Forum **95–98**, 393 (1993)
744. J. Philibert, *Atom Movements-Diffusion and Mass Transport in Solids* (Les Editions de Physique, Les Ulis, 1991)
745. R.T.P. Whipple, Philos. Mag. **45**, 1225 (1954)
746. H.S. Levine, C.J. MacCallum, J. Appl. Phys. **31**, 595 (1960)
747. A.D. Le Claire, Brit. J. Appl. Phys. **14**, 351 (1963).
748. I. Kaur, Y. Mishin, W. Gust, *Fundamentals of Grain and Interphase Boundary Diffusion* (Wiley, Chichester, UK, 1995)
749. S. Divinski, Chr. Herzig, J. Mater. Sci. **43**, 3900 (2008)
750. Chr. Herzig, J. Geise, Y.M. Mishin, Acta Metall. Mater. **41**, 1683 (1993)
751. T. Surholt, C. Minkwitz, Chr. Herzig, Acta Mater. **46**, 1849 (1998)
752. T. Surholt, Y.M. Mishin, Chr. Herzig, Phys. Rev. B **50**, 3577 (1994)
753. T. Surholt, Chr. Herzig, Def. Diff. Forum **143–147**, 1391 (1997)
754. S. Divinski, M. Lohmann, Chr. Herzig, Acta Mater. **49**, 249 (2001)
755. S.V. Divinski, Chr. Herzig, Arch. Metall. Mater. **49**, 305 (2004)
756. F. Christien, R. Le Gall, G. Saindrenan, Scripta Mater. **48**, 301 (2003)
757. G. Palumbo, K.T. Aust, Can. Met. Quart. **34**, 165 (1995)
758. T. Watanabe, Toward grain boundary design and control for advanced materials, in *Grain Boundary Engineering*, ed. by U. Erb, G. Palumbo (Can. Inst. Mining, Metall. Petrol., Montreal, 1993) pp. 57–87
759. T. Watanabe, Mater. Sci. Forum **11**, 284 (1989)
760. T. Watanabe, Scripta Metall. **27**, 1497 (1992)
761. M. Kumar, A.J. Schwartz, W.E. King, Acta Mater. **50**, 2599 (2002)
762. V. Randle, Grain boundary engineering, in *Encyclopedia of Materials: Science and Technology*, ed. by K.H.J. Bunshaw, R.W. Cahn, M.C. Flemings, E.J. Kramer, S. Mahajan (Elsevier, Amsterdam, 2005) doi:10.1016/B0-08-043152-6/02035
763. M. Tacikowski, M.W. Grabski, J. Driver, A. Kobylanski, Mater. Sci. Eng. A **205**, 133 (1996)
764. T. Watanabe, H. Fujii, H. Oikawa, K.I. Arai, Acta Metall. **37**, 941 (1989)
765. T. Hirano. Acta Metall. Mater. **38**, 2667 (1990)
766. T. Watanabe, S. Tsurekawa, Acta Mater. **47**, 4171 (1999)
767. M. Kurban, U. Erb, K.T. Aust, Scripta Mater. **54**, 1053 (2006)
768. P. Lin, G. Palumbo, U. Erb, K.T. Aust, Scripta Metall. Mater. **33**, 1387 (1995)
769. E.M. Lehockey, D. Limoges, G. Palumbo, J. Sklarchuk, K. Tomantschger, A. Vincze, J. Power Sources **78**, 79 (1999)
770. C.–S. Kim, A.D. Rollet, G.S. Rohrer, Scripta Mater. **54**, 1005 (2006)
771. V. Randle, Scripta Mater. **54**, 1011 (2006)
772. V. Randle, Mater. Sci. Technol. **15**, 246 (1999)
773. V. Randle, J. Metals **50**, 56 (1998)
774. V. Randle, Acta Mater. **46**, 1459 (1998)
775. G. Gottstein, D.A. Molodov, L.S. Shvindlerman, J. Mater. Sci. **41**, 7730 (2006)

776. A. Kumar, B.L. Eyre, Proc. Roy. Soc. Lond. A **370**, 431 (1980)
777. Yu.I. Ustinovshikov, Metal. Sci. **18**, 545 (1984)
778. L. Ben Mostefa, G. Saindrenan, N. Barbouth, A.M. Brass, J. Chêne: Scripta Metall. **24**, 773 (1990)
779. B.W. Krakauer, D.N. Seidman, Phys. Rev. B **48**, 6724 (1993)
780. K. Hoffmann, K.H. Sauer, H.J. Grabke, Steel. Res. **59**, 251 (1988)
781. A. Fraczkiewicz, Mater. Trans. JIM. **41**, 166 (2000)
782. A.S. Drachinskii, A.V. Krajnikov, V.N. Slyuniaev, Fyz. Met. Metalloved. **66**, 538 (1988)
783. L. Uray, M. Menyhard, Phys. Stat. Sol. (a) **84**, **65** (1984)
784. Yu.N. Ivashchenko, N.P. Korzhova, G.G. Kurdyumova, Y.N. Milman, Fyz. Met. Metalloved. **68**, 122 (1990)
785. B. Bokstein, I. Razumovskii, Interface Sci. **11**, 41 (2003)
786. J.J. Liao, PhD Thesis, Auburn University (1992)
787. H. Erhart, M. Paju, Scripta Metall. **17**, 171 (1983)
788. J.Q. Clayton, J.F. Knott, Metal. Sci. **16**, 145 (1982)
789. J.E. Morral, W.F. Jandenska Jr., Metall. Trans. A **11**, 1628 (1980)
790. C. Carbonnaux, R. Dessieux, G. Cizeron, A. Larere, T.T. Nguyen, G. Saindrenan, D. Roptin, J. Phys. France **51**, C1–789 (1990)
791. A. Larere, M. Guttmann, Ph. Dumoulin, C. Roques-Carmes, Acta Metall. **30**, 685 (1982)
792. S. Hofmann, P. Lejček, Interface Sci. **3**, 241 (1996)
793. R.A. Mulford, Metall. Trans. A **14**, 865 (1983)
794. S. Hofmann, P. Lejček, Segregation of sulphur and antimony at the {133} symmetrical tilt grain boundary in nickel, in *Interface Science and Materials Interconnections*, ed. by Y. Ishida, M. Morita, T. Suga H. Ichinose, O. Ohashi, J. Echigoya (JIM, Tokyo, 1996) pp. 229–232
795. B. Ladna, H.K. Birnbaum, Scripta metall. **22**, 223 (1988)
796. F. Moya, G.E. Moya–Goutier, J. Phys. France. **36**, C4–157 (1975)
797. S. Divinski, M. Lohmann, Chr. Herzig, Acta Mater. **52**, 3973 (2004)
798. M. Menyhard, Mater. Sci. Forum **126–128**, 205 (1993)
799. J.R. Michael, D.B. Williams, Metall. Trans. A **15**, 99 (1984)
800. S. Schmidt, W. Sigle, W. Gust, M. Rühle, Z. Metallkde. **93**, 428 (2002)
801. C. Corvalán Moya, M.J. Iribarren, N. Di Lalla, F. Dyment, J. Nucl. Mater. **382**, 35 (2008)
802. M. Guttmann, J. Phys. IV **5**, C7–85 (1995)
803. M. Guttmann, Metal. Sci. **10**, 337 (1976)

# Index


Activity, 54, 57, 59, 63, 66, 81, 84, 91, 108, 132, 133
Adsorption, 25, 33, 53, 54, 75, 92, 93, 143, 176, 184
Alkali metals, 33, 42
Alumina ($Al_2O_3$), 32, 34, 36, 113, 114
Aluminium, 10, 36, 41, 42, 48, 104, 177, 179, 183, 184, 186
Antimony, 1, 30, 33, 34, 70, 71, 75, 76, 83, 84, 86, 89, 93, 109, 112, 113, 130, 131, 149, 173, 177, 182, 184
Arrhenius equation, 195
Arsenic, 36, 173, 182, 184
Atomic structure of interfaces, 44
Austenitic steels
  Cr–Ni steels, 113, 186

Beryllium, 179
Bicrystal, 5, 7, 11, 14, 23, 30, 31, 39, 70, 73, 92, 113, 119
Bismuth, 1, 10, 33, 36–38, 41, 43, 48, 76, 86, 88, 93, 108, 112, 113, 118, 120, 125, 126, 130, 173, 178–181, 189
Boron, 32, 36, 38, 41, 43, 45, 81, 83, 87–89, 91, 112, 114, 126, 127, 147, 161, 171, 177–179, 181, 200

Calcium, 36, 114, 142, 143
Carbides, 85
Carbon, 21, 27, 33, 38, 72, 77, 84, 98, 112, 114, 116, 121–124, 137–139, 149, 158, 175, 177–180, 186, 199
Cast iron, 149
Chemical potentials, 23, 25, 46, 55, 57–60, 63, 76, 89, 111, 176, 190
Chlorine, 29
Chromium, 112, 147, 164, 165, 186, 189, 190
Cohesion, 41, 43, 48, 165, 173–184, 189
Coincidence site lattice (CSL), 13–16, 121, 198, 200
Compensation effect, 71, 96, 98, 118, 135–140, 146, 148, 199
Configurational entropy, 48, 74, 75, 127, 137, 138
Convolution, 69
Copper, 1, 29, 33, 36–38, 41–44, 47–49, 75, 76, 79, 86, 88, 93, 99, 111–113, 118, 120, 125–127, 130, 144, 173, 177, 179, 180, 183, 184, 186, 187, 189, 196
Crack tip, 174, 178, 181, 182, 184

Deconvolution, 36, 69
Density functional theory, 40, 41
Depletion, 36, 38, 63, 67, 83, 107, 114, 119, 165, 189, 190
Dislocations
  edge, 9
  screw, 12, 167
Ductile–brittle transition temperature (DBTT), 175, 176

Embedded atom method (EAM), 41, 43
Embrittlement
  dynamic, 79, 184, 185
  hydrogen, 28, 173, 181–183
  liquid metal, 2, 173, 183
  temper, 2, 29, 173
Energy
  Gibbs, 13, 24, 25, 38, 51, 55–63, 66, 68, 72, 73, 75, 76, 87, 92–94, 101, 108–111, 129, 130, 132, 140, 143–145, 150, 157
  Helmholtz, 23, 45–49, 66

internal, 23, 42–44, 49, 55, 60, 74, 75
Enrichment, 25, 38, 54, 62–64, 66, 73, 81, 90, 91, 110, 120, 125, 127–131, 154, 156, 159, 160, 171, 190, 191, 195
Equation
Schrödinger, 39, 40

Ferritic steels
low alloy, 78
Fick law, 156, 194
Field emission gun (FEG), 29, 36
Focussed ion beam (FIB), 36, 37
Fracture
ideal work of, 174–176
surface, 26, 28–31, 33, 71, 73, 112–114, 116, 117, 148, 183
Free energy minimisation, 48

Gallium, 41, 179, 184
Germanium, 173
Gettering, 85
Gold, 48, 49, 77, 99, 126–128, 140, 151, 152, 189, 196
Grain boundaries
adsorption, 143, 176
asymmetrical, 18, 19, 21, 113, 119, 121, 186
character distribution, 199, 200
classification, 13, 21, 66, 73, 121, 187, 198, 200
classification level, 17
crystallography, 6, 26, 200
delimiting, 11
diffusion
A regime, 194
B regime, 194, 195
C regime, 194–196
dislocations, 1, 2, 9, 10, 12, 14–16, 38, 51, 128, 156, 163–167, 171, 175, 178, 181
electrochemistry, 98
energetics, 38, 39, 43, 46, 146
energy, 3, 10, 12, 13, 23, 25, 42, 48, 53, 54, 56, 58, 59, 86, 96, 111, 118, 129, 143, 144, 173, 186, 189, 199
engineering (GBE), 15, 196–198, 201
enrichment ratio, 54, 91, 127, 130, 131, 160, 171, 191, 195
general, 11, 12, 19, 71, 74, 114, 115, 118, 123, 135, 142, 148, 149, 167, 180, 184, 189, 197
high-angle, 8, 10, 38, 47, 114–116, 123, 129
internal energy, 23, 44, 49, 55, 60, 74, 75
large-angle, 8
low-angle, 8, 9, 12, 15, 115, 116, 118
migration, 164, 167, 168, 170, 171
mixed, 7
mobility, 170, 180, 199
plane, 5, 6, 11, 14, 15, 17, 19–21, 41, 101, 114, 118, 121, 123, 150, 164, 180, 188, 200
plane engineering, 200
plane hierarchy, 17, 19
random, 7, 12, 13, 146
saturation, 59, 75, 80, 88, 89, 103, 110, 116, 118, 128, 131, 141, 144, 192, 196
segregation diagram, 131–135, 146
singular, 11–14, 17, 19, 21
small-angle, 8
special, 12–15, 18, 22, 67, 116, 118, 129, 135, 140, 167, 168, 179, 197–200
stress corrosion, 165, 182, 185, 188–190
structure, 8, 10, 38, 39, 41, 43–46, 103, 114, 115, 125, 129, 131, 180, 184, 189
symmetrical, 7, 8, 10–12, 14, 15, 17, 19–21, 41, 42, 44, 45, 47, 49, 64, 67, 94, 112, 113, 115, 118–123, 125–129, 134, 137, 138, 141, 176, 180, 186, 188
thermodynamics of, 22–25, 40, 44, 45, 47, 51, 53, 55, 57–64, 66–69, 73, 74, 90, 94, 95, 103, 107, 111, 112, 114, 121, 132, 136, 137, 141, 143, 144, 148, 149, 175, 192, 195
tilt, 7–15, 17, 19–22, 41, 42, 44, 45, 47, 49, 64, 67, 112–129, 134, 137, 138, 141, 176, 180, 186, 188, 189
twin, 14, 31, 41, 48, 113, 114, 118, 125, 128, 164, 187, 189, 198
twist, 7–9, 17, 46–49, 114, 116, 117, 125, 127, 128, 150–152, 178
vicinal, 11–13, 15, 16, 21, 22, 115, 121–123, 135, 149, 186
volume, 17, 145
Grain misorientnation, 6, 14

Hafnium, 164, 165
Helium, 32, 44
Hydrogen, 28, 31, 32, 41, 44, 173, 181–183

Ideal, 96
Impurity drag, 159, 165, 167, 170
Indium, 77, 104, 114
Interactions
  attractive, 78, 82, 83, 86, 106, 107, 147, 149
  binary, 52–54, 57, 59, 62, 65, 66, 72–77, 81, 83, 86, 90, 93, 94, 96, 98–100, 103–107, 116, 125, 130, 131, 144, 147, 148, 153
  repulsive, 42, 77, 82, 83, 85–87, 89, 105–107
  ternary, 26, 82, 83, 86, 91, 107, 108, 149, 155
Interatomic potentials
  Lennard–Jones, 42
  Morse, 42
  pairwise potentials, 42
Intermetallics
  FeAl, 38, 41, 89
  $Ni_3Al$, 36, 38, 47, 80, 91, 112, 176, 177, 185, 200
  NiAl, 38, 47
Interplanar spacing, 5, 16–18, 37
Iridium, 177
Iron, 1, 17, 18, 21, 22, 27, 29, 32, 36, 38, 41–43, 45, 54, 67, 68, 70–72, 75–79, 83–88, 91, 93, 94, 98, 109–115, 117, 118, 120–124, 126, 127, 130, 131, 133–135, 137–140, 146–149, 165, 173, 177–179, 182, 187, 196, 199

Lead, 113, 115, 180, 184, 200
Lithium, 177
Local density approximation, 41, 71

Magnesium, 32, 47, 48, 142, 183
Manganese, 33, 34, 36, 78, 112, 161, 164, 165, 173, 177
Methods
  ab-initio, 39, 41, 42, 48, 178
  first principles quantum mechanical, 39–41, 180
  molecular dynamics (MD), 37, 39, 43–46, 102, 125, 126, 128
  molecular statics (MS), 39
Microscopy
  analytical electron (AEM), 35–37, 113
  atom location channelling enhanced microanalysis (ALCHEMI), 36
  atom-probe field ion (AP FIM), 38, 54
  energy dispersive X–ray analysis (EDX), 35, 36
  field ion (FIM)
    3–D atom probe (3DAP), 37
    3–D tomography, 37, 38
    tomographic atom probe (TAP), 37
  high-resolution transmission electron (HRTEM), 9, 17, 35, 37
  scanning transmission electron (STEM), 35, 36
Misorientation angle, 6, 10, 21, 112, 115, 117, 118, 120, 122, 124, 128, 141, 180
Model
  BFS model, 100
  coincidence site lattice (CSL)
    coincidence site, 13–15
    reciprocal density of coincidence sites, 18
  different segregation sites, 92–94
  Hofmann and Erlewein, 156–158
  individual site models, 93, 94
  Kumar, 98, 99
  layer-by-layer model, 156–158
  Lejček and Hofmann, 146–149
  limited reaction rates, 158
  Luthra and Briant, 91, 92
  Miedema, 98
  Mukherjee and Morán-López, 99
  Seah, 97, 98
  SMA–TBIM, 101, 102
  Wynblatt and Ku, 95–97
  Wynblatt and Shi, 150–152
Modern thermodynamic calculation of interfacial properties, 90, 91
Molybdenum, 9, 12, 29, 36, 38, 75, 83, 85, 94, 114, 116, 118, 147, 177, 178, 180
Monte Carlo simulations (MC), 45–48
  overlapping distribution (ODMC), 46, 48

Nanocrystals, 143, 144
Neon, 37
Nickel
  base superalloys, 1, 29, 38, 83, 185
Niobium, 32, 38, 75, 114, 121, 144, 164
Nitrides, 178
Nitrogen, 28, 38, 43, 77, 116, 178, 179
Noble gases, 33

O-lattice, 15, 16
Osmium, 38
Oxides, 26, 34, 79, 190

Oxygen, 26, 27, 29, 33, 34, 36, 38, 41, 43, 79, 94, 112, 114, 116, 118, 119, 173, 177, 179, 180, 185

Palladium, 38, 49, 77, 128, 129, 140, 164
Parameter
  Sigma ($\Sigma$), 13–15, 17
Phase boundary, 5
Phosphorus, 17, 18, 21, 33, 36, 38, 41–43, 45, 54, 64, 67, 70, 72, 76–78, 83, 85–87, 91, 93, 94, 109, 112, 113, 117, 118, 120–124, 126, 127, 130, 137, 138, 140, 141, 148, 155, 161, 162, 173, 176, 178, 179, 182, 184, 186, 189, 190
Platinum, 48, 77, 91, 127, 128, 164, 165
Polonium, 115
Polycrystals, 3, 22, 70, 80, 92, 94, 126, 138, 141–143, 145, 146, 149, 197–201
Potassium, 32
Precipitation, 25, 26, 78, 79, 85, 149, 156, 178, 190, 198, 200

Rhodium, 77

Scavenging, 78, 85
Segregation
  anisotropy of, 135
  effective functions of, 68
  effect of
    bulk composition, 54, 61, 77, 85, 99, 103, 170
    grain boundary energy, 111
    grain boundary plane orientation, 180
    grain size, 141
    magnetic, 108
    nature of segregating element, 129
    pressure, 107
  enthalpy of, 41, 68, 72, 94, 96, 123, 126, 127, 131, 137, 148, 149, 151
  entropy of, 72, 98, 99, 102, 140, 150, 214
  equilibrium, 29, 51, 77, 158, 159, 161, 163, 164, 170, 191
  excess functions of, 60, 61, 66
  functions of, 61
  Gibbs energy of, 38, 56–59, 61–63, 66, 75, 76, 81, 87, 92, 94, 108, 111, 129
  isotherm, 38, 52, 55, 57, 61, 68, 69, 72, 74, 75, 77, 82, 86, 91, 92, 104, 109, 115, 142–144, 155
    BET, 92, 93, 129–131
    Fowler, 86, 87
    Gibbs adsorption, 52, 54, 144
    Guttmann, 81–83, 85, 86, 91
    Langmuir–McLean, 38, 52, 55, 61, 74, 75, 77–79, 86, 115, 142–144, 155, 196
    truncated BET, 129, 130
  macrosegregation, 3
  microsegregation, 3
  nanosegregation, 3
  non-equilibrium, 4, 26, 51, 153, 158–164
  prediction of, 38, 91, 125, 135, 146, 147, 149, 151, 170, 178
  radiation induced, 159, 162–164
  standard functions of, 66, 74
  stress induced, 159, 165, 184
  Suzuki, 38, 94, 120
  thermally induced, 159, 163, 164
  volume of, 65
Selenium, 86, 93, 173, 196
Silicon, 8, 21, 32, 34, 36, 38, 41–43, 64, 67, 68, 77, 83, 85, 114–118, 120–124, 127, 130, 137–140, 147, 173, 176, 177, 179, 186
Silver, 33, 99, 126, 142, 180, 186, 196
Sintering, 4
Site competition, 77, 84, 85, 106, 155, 178
Sodium, 32, 180
Special, 80
Spectroscopy
  Auger electron (AES), 27, 28
  electron-energy loss (EELS), 35
  extended X–ray absorption fine structure (EXAFS), 32
  fast atom bombardment – mass (FAB–MS), 32
  glow discharge mass (GDMS), 32
  glow discharge optical emission (GDOES), 32
  ion scattering (ISS, LEIS), 32
  Mössbauer, 34
  nuclear reaction analysis (NRA), 32
  proton-induced X–ray emission (PIXE), 32
  Rutherford backscattering (RBS), 32
  secondary ion mass (SIMS)
    static, 31
  secondary neutron mass (SNMS), 32
  X–ray photoelectron (XPS, ESCA), 32
Standard chemical potentials, 57, 58, 60, 63
Stress relief cracking, 184
Strontium, 114, 142, 143
Structural unit model, 10
Structure
  bcc, 10, 14, 17

fcc, 14
Sulphur, 29, 30, 33, 35, 36, 41, 43, 54, 77, 78, 83, 93, 109, 112, 114, 116, 118, 119, 131, 141, 142, 147, 155, 173, 176–179, 181, 182, 199
Surface
analysis techniques, 27, 69
System, 14, 25, 35, 64, 73, 76, 106, 146, 194
binary, 53, 62, 72, 74–76, 81, 83, 86, 90, 93, 94, 96, 99, 103, 104, 106, 116, 125, 130, 131, 144, 147, 148, 153
ideal, 14, 16, 17, 46, 48, 59, 60, 63, 64, 66, 73, 74
multicomponent, 82, 84
pseudobinary, 77, 106, 130
ternary, 105, 106, 108, 155

Tellurium, 86, 87, 173, 196
Temper, 79
Thermodynamics, 5–20, 22–25, 40, 44, 45, 47, 51–53, 55, 57–61, 63, 65–68, 70–74, 79, 90–92, 94–103, 107, 111, 112, 114, 121, 132, 136, 137, 141, 143, 148, 149, 175, 192, 195
Tin, 33, 36, 41, 48, 54, 76, 77, 79, 83, 88, 93, 109–112, 115–118, 130, 131, 147, 173, 177, 184
Titanium, 36, 38, 41, 85, 113, 114, 165, 179
Triple junctions, 36, 198, 199, 201
Tungsten, 38, 43, 91, 93, 112, 113, 125, 158, 177, 180

Vacancy–solute pair, 26
Vanadium, 85

$YBa_2Cu_3O_{7-\delta}$, 36, 114, 119
Yttria, 32
Yttrium, 32, 34, 36

Zinc, 173, 184
Zirconia, 32
Zirconium, 34, 38, 114, 165

Printed in the USA
CPSIA information can be obtained
at www.ICGtesting.com
CBHW071949260924
14890CB00017B/160

9 783642 125041